Springer Handbook of Auditory Research

Volume 67

AF823353

More information about this series at http://www.springer.com/series/2506

The ASA Press

The ASA Press imprint represents a collaboration between the Acoustical Society of America and Springer dedicated to encouraging the publication of important new books in acoustics. Published titles are intended to reflect the full range of research in acoustics. ASA Press books can include all types of books published by Springer and may appear in any appropriate Springer book series.

ASA
PRESS

Micheal L. Dent • Richard R. Fay
Arthur N. Popper
Editors

Rodent Bioacoustics

ASA PRESS

Springer

Editors
Micheal L. Dent
Department of Psychology
University at Buffalo, SUNY
Buffalo, NY, USA

Richard R. Fay
Loyola University Chicago
Chicago, IL, USA

Arthur N. Popper
Department of Biology
University of Maryland
College Park, MD, USA

ISSN 0947-2657 ISSN 2197-1897 (electronic)
Springer Handbook of Auditory Research
ISBN 978-3-030-06444-0 ISBN 978-3-319-92495-3 (eBook)
https://doi.org/10.1007/978-3-319-92495-3

This Springer imprint is published by the registered company Springer Nature Switzerland AG
The registered company address is: Gewerbestrasse 11, 6330 Cham, Switzerland

This volume is dedicated to Dr. Robert J. Dooling. While Bob's research has focused on birds, his scholarly approach and important findings have impacted all modern comparative research on vertebrate hearing. In addition to being a superlative scholar, Bob continues to be an exceptional mentor and a valued collaborator and friend to all of the editors of this volume. (Photo credit: John T. Consoli/University of Maryland).

Acoustical Society of America

The purpose of the Acoustical Society of America (ASA; www.acousticalsociety.org) is to generate, disseminate, and promote the knowledge of acoustics. The ASA is recognized as the world's premier international scientific society in acoustics, and counts among its more than 7000 members, professionals in the fields of bioacoustics, engineering, architecture, speech, music, oceanography, signal processing, sound and vibration, and noise control.

Since its first meeting in 1929, the ASA has enjoyed a healthy growth in membership and in stature. The present membership of approximately 7000 includes leaders in acoustics in the United States of America and around the world. The ASA has attracted members from various fields related to sound, including engineering, physics, oceanography, life sciences, noise and noise control, architectural acoustics; psychological and physiological acoustics; applied acoustics; music and musical instruments; speech communication; ultrasonics, radiation, and scattering; mechanical vibrations and shock; underwater sound; aeroacoustics; macrosonics; acoustical signal processing; bioacoustics; and many more topics.

To assure adequate attention to these separate fields and to new ones that may develop, the Society establishes technical committees and technical groups charged with keeping abreast of developments and needs of the membership in their specialized fields. This diversity and the opportunity it provides for interchange of knowledge and points of view has become one of the strengths of the Society.

The ASA's publishing program has historically included *The Journal of the Acoustical Society of America, JASA-Express Letters, Proceedings of Meetings on Acoustics*, the magazine *Acoustics Today*, and various books authored by its members across the many topical areas of acoustics. In addition, ASA members are involved in the development of acoustical standards concerned with terminology, measurement procedures, and criteria for determining the effects of noise and vibration.

Acoustical Society of America

Series Preface

Springer Handbook of Auditory Research

The following preface is the one that we published back in 1992. As anyone reading the original preface, or the many users of the series, will note, we have far exceeded our original expectation of eight volumes. Indeed, with books published to date and those in the pipeline, we are now set for over 65 volumes in SHAR, and we are still open to new and exciting ideas for additional books.

We are very proud that there seems to be consensus, at least among our friends and colleagues, that SHAR has become an important and influential part of the auditory literature. While we have worked hard to develop and maintain the quality and value of SHAR, the real value of the books is very much because of the numerous authors who have given their time to write outstanding chapters and to our many co-editors who have provided the intellectual leadership to the individual volumes. We have worked with a remarkable and wonderful group of people, many of whom have become great personal friends of both of us. We also continue to work with a spectacular group of editors at Springer. Indeed, several of our past editors have moved on in the publishing world to become senior executives. To our delight, this includes the current president of Springer US, Dr. William Curtis.

But the truth is that the series would and could not be possible without the support of our families, and we want to take this opportunity to dedicate all of the SHAR books, past and future, to them. Our wives, Catherine Fay and Helen Popper, and our children, Michelle Popper Levit, Melissa Popper Levinsohn, Christian Fay, and Amanda Fay Sierra, have been immensely patient as we developed and worked on this series. We thank them and state, without doubt, that this series could not have happened without them. We also dedicate the future of SHAR to our next generation of (potential) auditory researchers – our grandchildren – Ethan and Sophie Levinsohn, Emma Levit, and Nathaniel, Evan, and Stella Fay.

Series Preface

Springer Handbook of Auditory Research

Preface 1992

The Springer Handbook of Auditory Research presents a series of comprehensive and synthetic reviews of the fundamental topics in modern auditory research. The volumes are aimed at all individuals with interests in hearing research, including advanced graduate students, post-doctoral researchers, and clinical investigators. The volumes are intended to introduce new investigators to important aspects of hearing science and to help established investigators to better understand the fundamental theories and data in fields of hearing that they may not normally follow closely.

Each volume presents a particular topic comprehensively, and each serves as a synthetic overview and guide to the literature. As such, the chapters present neither exhaustive data reviews nor original research that has not yet appeared in peer-reviewed journals. The volumes focus on topics that have developed a solid data and conceptual foundation rather than on those for which a literature is only beginning to develop. New research areas will be covered on a timely basis in the series as they begin to mature.

Each volume in the series consists of a few substantial chapters on a particular topic. In some cases, the topics will be ones of traditional interest for which there is a substantial body of data and theory, such as auditory neuroanatomy (Vol. 1) and neurophysiology (Vol. 2). Other volumes in the series deal with topics that have begun to mature more recently, such as development, plasticity, and computational models of neural processing. In many cases, the series editors are joined by a co-editor having special expertise in the topic of the volume.

SHAR logo by Mark B. Weinberg, Potomac, Maryland, used with permission

Richard R. Fay, Chicago, IL, USA
Arthur N. Popper, College Park, MD, USA

Volume Preface

Rodents are one of the largest of all mammalian taxa, and a number of rodent species are among the most important model systems for biomedical research, including the study of hearing. Indeed, rodents have been featured "subjects" in many volumes in the Springer Handbook of Auditory Research (SHAR) series because so much of what we know about issues covered in the various volumes was gained through the use of rodents as model systems. Thus, since rodents are so important for what we know about hearing, it became clear that bringing together an overview of what is known (and not known) about rodent bioacoustics would be of considerable value to the auditory (and bioacoustic) research community. Most rodent research has focused on just a few species (mice, rats, and, to a lesser degree, chinchilla) among the 2000 rodent species. This volume, however, includes a wealth of bioacoustic data on less frequently used rodent species and provides the basis for a broader understanding of rodents. Chapters in this volume describe rodent bioacoustics from several different approaches.

Chapter 1 by Micheal L. Dent provides an overview of the wide range of rodent taxa as well as a summary of the contents of the book. In Chap. 2, Kazao Okanoya and Laurel A. Screven describe acoustic signals in air in both laboratory-raised and wild rodents in both the laboratory and the field, making this one of the most diverse chapters in terms of the numbers of rodent species studied. This is followed by a consideration of acoustic communication in subterranean rodents in Chap. 3 by Cristian Schleich and Gabriel Francescoli, who provide a more neuroethological perspective than other chapters in this volume. Chapter 4 by Micheal L. Dent, Laurel A. Screven, and Anastasiya Kobrina summarizes behavioral and physiological auditory acuity of all rodent species tested to date. An important part of hearing is binaural processing and sound localization, and these issues are considered in Chap. 5 by Amanda M. Lauer, James H. Engel, Jr., and Katrina Schrode.

The chapters on vocalizations and hearing are followed by a series of more mechanistic chapters. In Chap. 6, M. Fabiana Kubke and J. Martin Wild discuss the anatomy of vocalization and hearing. This is followed by a discussion of rodent models for genetic and age-related hearing issues in Chap. 7 by Kevin K. Ohlemiller.

Finally, the roles of internal state and context in vocal communication are discussed in both laboratory and wild rodents by Laura M. Hurley and Matina C. Kalcounis-Rueppell in Chap. 8.

Micheal L. Dent, Buffalo, NY, USA
Richard R. Fay, Chicago, IL, USA
Arthur N. Popper, College Park, MD, USA

Contents

Contributors

Micheal L. Dent Department of Psychology, University at Buffalo, SUNY, Buffalo, NY, USA

James H. Engel Department of Otolaryngology, Center for Hearing and Balance, and David M. Rubenstein Center for Hearing Research, Johns Hopkins University School of Medicine, Baltimore, MD, USA

Gabriel Francescoli Sección Etología, Facultad de Ciencias, Universidad de la República, Montevideo, Uruguay

Laura M. Hurley Department of Biology, Indiana University, Bloomington, IN, USA

Matina C. Kalcounis-Rueppell Biology Department, University of North Carolina at Greensboro, Greensboro, NC, USA

Anastasiya Kobrina Department of Psychology, University at Buffalo, SUNY, Buffalo, NY, USA

M. Fabiana Kubke Department of Anatomy and Medical Imaging, and Eisdell Moore Centre, University of Auckland, Auckland, New Zealand

Amanda M. Lauer Department of Otolaryngology, Center for Hearing and Balance, and David M. Rubenstein Center for Hearing Research, Johns Hopkins University School of Medicine, Baltimore, MD, USA

Kazuo Okanoya Department of Life Sciences, The University of Tokyo, Tokyo, Japan

Kevin K. Ohlemiller Fay and Carl Simons Center for Biology of Hearing and Deafness, Central Institute for the Deaf at Washington University, Saint Louis, MO, USA

Department of Otolaryngology, Washington University, School of Medicine, Saint Louis, MO, USA

Cristian Schleich IIMyC-Conicet, Universidad Nacional de Mar del Plata, Mar del Plata, Argentina

Katrina Schrode Department of Otolaryngology, Center for Hearing and Balance, and David M. Rubenstein Center for Hearing Research, Johns Hopkins University School of Medicine, Baltimore, MD, USA

Laurel A. Screven Department of Psychology, University at Buffalo, SUNY, Buffalo, NY, USA

J. Martin Wild Department of Anatomy and Medical Imaging, and Eisdell Moore Centre, University of Auckland, Auckland, New Zealand

Chapter 1
An Introduction to Rodent Bioacoustics

Micheal L. Dent

Abstract Rodents are a relatively diverse order of mammals that are found in abundance virtually all over the globe. The behavior of wild rodents is less well understood than that of laboratory rodents. Aboveground juvenile and adult rodents produce vocalizations that are used for communicating information about predators, mating readiness, hunger, and food availability. Subterranean rodents not only produce vocalizations but also drum their feet and bang their heads against burrows to communicate. The auditory system of rodents allows for detecting signals in quiet, discriminating between characteristics of communication signals, categorizing signals, and localizing sounds in space. Genetically manipulating laboratory rodents has elucidated much of what is known about auditory perception in mammals. Finally, the context and state of the rodent can have an influence on both the signal produced and the signal received. A common theme of the chapters in this volume is that a lot is known about bioacoustics in just a few species of rodents, while absolutely nothing is known about communication by most rodent species, presenting an opportunity for laboratory and field bioacousticians alike.

Keywords Acoustic communication · Animal communication · Chinchilla · Context · Discrimination · Hamster · Hearing · Mongolian gerbil · Mouse · Rat · Rodent anatomy · Sound localization · Subterranean communication · Ultrasonic vocalizations

1.1 Introduction

The fact that the Springer Handbook of Auditory Research (SHAR) has 65+ volumes without a rodent bioacoustics book is probably surprising to many followers of the series. However, even though rodents have not earned their own book until now, most, if not all, researchers in auditory laboratories are familiar with at least

M. L. Dent (✉)
Department of Psychology, University at Buffalo, SUNY, Buffalo, NY, USA
e-mail: mdent@buffalo.edu

M. L. Dent et al. (eds.), *Rodent Bioacoustics*, Springer Handbook of Auditory Research 67, https://doi.org/10.1007/978-3-319-92495-3_1

two species of rodents used as models of hearing or communication. Indeed, studies of rodent bioacoustics are cited in the majority of the books in the SHAR series.

A vast majority of the hearing and vocal communication studies in nonhumans has been conducted on rodents. As a consequence, this book ties together several forms of acoustical research on rodents, bringing together neuroethologists, animal behaviorists, biologists, and clinicians to improve understanding of this mammalian group that has become so important to the scientific community. It is the intent of the authors of this volume that investigators will learn more about acoustic communication in their rodent species and thus will be able to utilize them in better (and perhaps additional) ways.

Studies of bioacoustics in rodents have taken two separate trajectories over the years. In the first trajectory, researchers have studied the behavior of numerous species of wild rodents in their natural habitats. These studies are somewhat scarce compared to the second trajectory in which researchers have used a few species of laboratory rodents as models for human hearing and speech. This has led to a very narrow group of rodent species that have been utilized to study hearing and acoustic communication signals. Many of these laboratory rodents are inbred so as to maintain genetic similarity and decrease between-subject variability. An understanding of the diversity of communication systems in rodents is lost as a result, which is fine for scientists interested in factors such as the genes encoding hearing loss, but the loss of diversity is problematic for comparative researchers interested in factors such as the influence of an animal's environment on auditory processing.

Chapters in this volume describe rodent bioacoustics from several different approaches, spanning the two trajectories mentioned above. Acoustic signals in air have been measured in both laboratory and wild rodents in both the field and in the laboratory, making Chap. 2 by Okanoya and Screven one of the most diverse chapters in terms of the number of rodent species studied. Subterranean communication is studied both in the laboratory and in the wild, although the laboratory studies usually involve rodents that were caught in the wild. Since underground recordings are so difficult, fieldwork is often avoided for these underground rodents. Chapter 3 by Schleich and Francescoli on subterranean signaling, like the chapter on rodent signals in air, thus takes a more neuroethological perspective than some of the other chapters in this volume.

The hearing chapter (Dent, Screven, and Kobrina, Chap. 4) reports laboratory behavioral and physiological auditory acuity of all rodents tested to date. While simple measures of hearing have been conducted on laboratory rodents and many species of nonlaboratory rodents, detailed analyses of hearing have not been measured in nonlaboratory rodents. Sound localization and binaural hearing (Lauer, Engel, and Schrode, Chap. 5) in rodents has suffered a similar fate: simple studies of directional hearing have been conducted on numerous rodent species, but the underlying anatomy and physiology and the more complicated measures of spatial hearing have been limited to a few laboratory rodent models.

The chapters on the anatomy of vocal communication and hearing (Kubke and Wild, Chap. 6) and on rodent models for genetic and age-related hearing loss (Ohlemiller, Chap. 7) also rely largely on data from laboratory rodents. Finally, the roles of state and context in vocal communication are discussed in both laboratory and wild rodents by Hurley and Kalcounis-Rueppel (Chap. 8), with particular importance placed on the effects of the natural environment on acoustic communication in the rodent species included in this chapter.

1.1.1 The Evolution and Lifestyles of Rodents

Rodents are a diverse order and, as mentioned in Sect. 1.1, little is known about acoustic communication in most species. The evolutionary history is fairly well mapped out, however. The order Rodentia contains over 2000 species, which is more than 40% of all mammalian species. The 2000 rodent species are spread across five suborders (Fig. 1.1) and thirty-six families. The suborders are: (1) Sciuromorphs (squirrel-shaped), which include dormice, groundhogs, chipmunks, and squirrels; (2) Castorimorphs, which include beavers, pocket gophers, kangaroo rats, and kangaroo mice; (3) Anomaluromorphs, which include springhares; (4) Myomorphs (mouse-shaped), which include hamsters, rats, and mice; and (5) Hystricomorphs (porcupine-shaped), which include porcupines, mole rats, chinchillas, and degus (Feldhamer et al. 2015). This classification is by no means universally agreed upon (reviewed in Honeycutt et al. 2007).

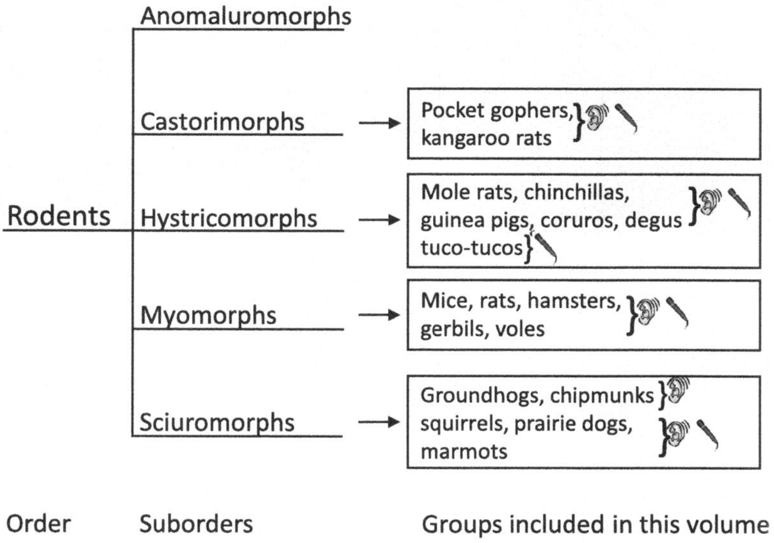

Fig. 1.1 The five suborders of the rodentia order and groups of rodents included in this volume. No groups of anomaluromorphs have been studied in the field of bioacoustics. Multiple species from the other orders have been included in the chapters of this volume. Auditory acuity measurements in a group are denoted with the ear icon and vocalization measurements in a group are denoted with the microphone icon

Fig. 1.2 Diversity of rodents. (**a**) The smallest mammal, the pygmy jerboa (*Allactaga tetradactyla*), and (**b**) the largest mammal, the capybara (*Hydrochoerus hydrochaeris*); (**c**) giant flying squirrel (*Petaurista sp.*); (**d**) beaver (*Castor sp.*). (**a** Digital image from upload.wikimedia.org/wikipedia/commons/3/3c/Allactaga_elater_Plzen_zoo_02.2011.jpg; **b** Digital image from www.flickr.com/photos/lorentey/5592629831; **c** Digital image from img.burrard-lucas.com/china/full/giant_flying_squirrel.jpg; **d** Digital image from https://c2.staticflickr.com/8/7153/6667988291_efec293851_b.jpg)

Rodents as a group exhibit vast amounts of diversity and adaptability. The smallest rodents, for example, the pygmy jerboa (*Allactaga sp.*; Myomorph suborder) (Fig. 1.2a), are under 5 g, while the largest rodents, for example, the capybara (*Hydrochoerus hydrochaeris*; Hystricomorph suborder), are over 50 kg (Fig. 1.2b) (e.g., Feldhamer et al. 2015). Rodents are found wild on all continents except Antarctica, and thus they have adapted to living in wet and dry, hot and cold, aboveground and subterranean, and light and dark environments, primarily solitary or in large social groups (Feldhamer et al. 2015). Rodents with special adaptations include giant flying squirrels (*Petaurista sp.*; Sciuromorph suborder), who have evolved folds of skin that allow them to glide from tree to tree to avoid predation (Fig. 1.2c), and beavers (*Castor sp.*; Castorimorph suborder), who have structural specializations allowing them to keep their mouths open while underwater so that they can carry branches and gnaw (Fig. 1.2d) (e.g., Vaughan 1985).

Convergent evolution of lifestyles is also common across rodents. For instance, subterranean rodents are found in Africa, Asia, and North and South America. Various morphological and physiological specializations for surviving underground, including shorter limbs, flat skulls, and tolerance of high carbon dioxide environments, have evolved numerous times (reviewed in Honeycutt et al. 2007). Overall, rodents are remarkably adaptive creatures who, as a group, have undergone fewer extinctions relative to other mammals (Honeycutt et al. 2007).

The defining rodent characteristic is having a single pair of upper and lower continuously growing incisors. Rodents have thus been referred to as "gnawing machines" (Druzinsky 2015). The incisors are used not only to eat but also to excavate tunnels and for defense. Most rodents are either herbivores or opportunistic omnivores. Some rodents, like the grasshopper mouse (genus *Onychomys*), are predatory carnivores, preying on insects and other mice. Rodents of all sizes are also food sources for a variety of predators and use the incisors to keep those predators away.

Rodents range from being relatively social animals (e.g., prairie voles, *Microtus ochrogaster*) to living solitary lifestyles (e.g., woodchucks, *Marmota monax*) (Lacey and Sherman 2007). Mating systems of rodents also vary widely, from monogamous California mice (*Peromyscus californicus*) to promiscuous Gunnison's prairie dogs (*Cynomys gunnisoni*) (Waterman 2007).

Social rodents communicate information about territories, sexual receptivity, predators, and food through multiple sensory channels, including chemosensory, visual, and auditory modalities. The sense of olfaction in most rodents is known to be excellent (e.g., Hurst et al. 2001). Olfaction is important for social, sexual, and antipredator behaviors in rodents (Arakawa et al. 2008). Visual acuity, on the other hand, varies widely among rodent species (Francescoli 2000). Auditory abilities fall somewhere in between vision and olfaction in terms of the range of acuities across species (Dent, Screven, and Kobrina, Chap. 4).

1.1.2 Rodent Auditory Behavior

The highly successful (at least from an evolutionary perspective) and diverse mammalian order of rodents is probably not thought of as an especially vocal group of animals for several reasons. First, as mentioned in Sect. 1.1.1, rodent communication is heavily olfactory. Chemical signals have advantages over acoustic signals in that they remain in the air even after signalers have left, allowing them to depart from a potentially dangerous situation but still enabling them to warn others (Bradbury and Vehrencamp 2011). Since rodents serve as a food source for many birds, larger mammals, and reptiles, effective anti-predatory communication is vital for survival, and olfactory signaling is just a part of a rodent's communicative repertoire. Rodents also use scent signals for marking territories (e.g., blind mole-rats, *Spalax ehrenbergi*) and for sexual selection (e.g., prairie voles, *Microtus ochrogaster*) (Roberts 2007).

A second reason rodents may have been overlooked as a vocal group is because many of their acoustic signals are ultrasonic (above 20 kHz) (Portfors 2007), infrasonic (below 20 Hz), or seismic (vibrating the ground) (Narins et al. 2016). Thus, without special recording equipment, many of the signals used by rodents for communication are not detectable by humans. Signals that are undetectable to humans tend to receive less attention relative to studies of signals that fall into the human auditory range. Nonetheless, it is known that acoustic communication in rodents is widespread. Rodents produce vocalizations during play, for mate attraction, as alarms during a dangerous situation, and as pups to urge their mother's return, to name just a few examples (e.g., Portfors 2007). Acoustic signals, like their olfactory counterparts, do not require daylight for communication to occur, allow the animals producing them to be somewhat cryptic when signaling, for example, narrowband signals which are often difficult to localize (e.g., Bradbury and Vehrencamp 2011). Acoustic signals also allow for great flexibility in communication because the animals can change the frequency range, duration, or other temporal characteristics to adjust the meaning of calls (e.g., Bradbury and Vehrencamp 2011). The auditory system of rodents allows them to be able to discriminate these changes for effective communication (Dent, Screven, and Kobrina, Chap. 4).

1.1.3 Laboratory Rodent Bioacoustics

The human–rodent connection is strong, and most of the habitats inhabited by humans are also inhabited by rodents (Feldhamer et al. 2015). Rodents such as mice have been benefiting from humans for thousands of years. Previously, farming was believed to have led to the influx of mice in human settlement locations. Fossil records suggest that mouse domestication began closer to 15,000 years ago, much earlier than the advent of agriculture, and was correlated with decreases in human mobility (Weissbrod et al. 2017). As human populations grew and became sedentary, so, too, did house mouse populations. House mice and other rodents (e.g., rats) have long been thought of negatively as pests who eat farmers' crops and as vectors for a number of infectious diseases (Feldhamer et al. 2015). However, some species can also have positive impacts on humans, for example, muskrats (*Ondatra zibethicus*) and beavers (*Castor canadensis*) that have been used for fur (Feldhamer et al. 2015), giant African pouched rats (*Cricetomys gambianus*) that are trained to detect landmines and tuberoculosis (Poling et al. 2010, 2011), and the various rodent species used in laboratories around the world for valuable medical research of all types. Thus, the human–rodent relationship has both positive and negative aspects.

Rodents, especially the Norway rat (*Rattus norvegus*), became popular for laboratory studies of social, feeding, reproductive, and emotional behaviors, of their sensory capabilities, and for studies of conditioning and learning in the early 1900s (Beach 1950). A number of characteristics makes rats (and mice) especially attractive for laboratory research, including their short lifespans, small size, large litter sizes, and ease of handling (Feldhamer et al. 2015). They are also relatively easy to train. Most scientists would say that rats and mice have become indispensable to research programs around the world. Thus, much of what is known about hearing and vocalizations in rodents comes from just those two species that have been tested in the laboratory.

The sequencing of the mouse genome in 2002 (Waterston et al. 2002), followed by the rat genome in 2004 (Gibbs et al. 2004), only increased the utility of these animals as research subjects since genetically engineered strains mimicking human diseases and disorders could be developed more easily. In the laboratory, rats and mice are used as models for human communication and hearing disorders and are involved in studies on hearing loss and prevention, hormones, and auditory plasticity, to name just a few topics. The importance of acoustic communication to rodents and the significance of these animals to biomedical research are summarized in the chapters in this book. When possible, field studies on acoustic communication in wild rodents are also included.

1.2 Rodent Vocalizations

Okanoya and Screven (Chap. 2) describe the diversity of vocalizations produced by rodents, both in terms of the contexts and environments in which the vocalizations are produced and in their acoustic characteristics. Physical factors in the environment, such as vegetation and high frequency attenuation, affect sound propagation and probably influence the vocalizations produced under different contexts. Specific events, such as isolation distress in pups, social and sexual coordination, nearby predators, anticipation of an aversive occurrence, aggressive encounters, play, and finding food, all give rise to vocalizations. This chapter summarizes the rodents' complex acoustic communication system for coordinating many types of behaviors and social situations.

1.3 Subterranean Communication

Schleich and Francescoli (Chap. 3) describe three decades of studies on acoustic communication in subterranean rodents. More than 250 of the 2000+ species of rodents spend most of their lives underground. These rodents range from strictly subterranean, which live completely underground, to fossorial, which live underground but also engage in other activities at the surface. The underground burrows inhabited by rodents are physically very different from the outside, leading to morphological specializations in subterranean rodents and leading to stark differences in communication in these species relative to aboveground rodents. This chapter discusses the vocalizations and seismic signals of subterranean rodents, correlating them to context, social group structure, and habitat.

1.4 Hearing

Dent, Screven, and Kobrina (Chap. 4) describe hearing in rodents. The ability to detect, discriminate, localize, and identify acoustic signals is important for all animals, and rodents are no exception. Rodents can be trained using operant conditioning to report detecting a sound, a change in a repeating background of sounds, or indicate to which category a sound belongs. The training process for some rodents (e.g., mice) is more time consuming than the training process for other rodents (e.g., rats). Thus, some researchers have also measured auditory acuity using much speedier reflex behavioral or physiological methods, although these methods have their limitations. Audiograms, frequency selectivity, frequency discrimination, intensity discrimination, temporal resolution, and the perception of complex sounds have all been measured in various species of rodents and are summarized in this chapter.

1.5 Sound Localization and Spatial Hearing

Lauer, Engel, and Schrode (Chap. 5) describe both the abilities of rodents to locate sounds in space and the underlying neural circuitry involved in the process. While small rodents must rely on interaural level differences at high frequencies to localize sounds, larger rodents may utilize interaural time differences. This chapter summarizes the sound localization acuity and underlying neural circuitry that support localization in rodents and compares it to other mammals.

1.6 Anatomy of Vocal Communication and Hearing

Kubke and Wild (Chap. 6) describe how rodents produce and detect acoustic signals. Even though some rodents are known to produce ultrasonic vocalizations, an unusual characteristic among mammals other than bats and some marine mammals, the muscles and nerves involved in the production of vocalizations are similar across rodents and other mammals. Some small rodents have unusual middle ears that are probably specializations for high frequency hearing, but most rodents have generally unspecialized auditory pathways. The auditory midbrain and cortex of rodents are responsive to natural vocalizations. This chapter summarizes what is known about other factors influencing the coding of vocalizations, including social experience.

1.7 Rodent Models for Genetic and Age-Related Hearing Loss

Ohlemiller (Chap. 7) describes genetic and age-related pathologies of hearing. As in the other chapters, the effects of genes and age on hearing have only been studied in a few species of rodents. There is quite a bit of overlap in the genes known to cause deafness in humans and rodents, making rodents useful models of hearing loss in most cases. Both inbred and outbred rodents are used in studies of hearing and hearing loss. While no models are perfect, different models are good at mimicking different diseases or pathologies. This chapter summarizes how these models have been invaluable for teasing apart the genetic, aging, sex, and anatomical contributions to hearing loss.

1.8 State and Context in Vocal Communication

Hurley and Kalcounis-Rueppell (Chap. 8) describe the influences of state and context in vocal communication of rodents. To fully understand acoustic communication in rodents, the characteristics of the sender and receiver must be studied, but the communication "scene" needs to be understood as well. The scene, or context, for each member involved in the communication process can include the environment where the animals live, the reproductive phase of the parties engaging in communication, and the developmental life history of the subjects. Understanding the context reduces the ambiguity that may be present in a vocalization, allowing the animal to take the appropriate action. The scene can be broken further into internal and external factors. Internal factors include self-identity and internal state (e.g., motivation levels). External factors include environmental conditions, such as the weather and the identity of a partner. This chapter summarizes the factors involved in communication in rodents and describes their impacts on both the sender and the receiver.

1.9 Summary

The chapters in this book highlight what is known about rodent bioacoustics. Given the great diversity in communication signals and auditory processing of some other vertebrates (e.g., birds), one might also expect to see such diversity in rodent acoustic communication. Unfortunately, not much is known about the production and reception of acoustic signals by most rodent species. The following questions remain unanswered for most rodents: Have temperature, humidity, and vegetation driven evolutionary changes in the properties of acoustic signals? Has anthropogenic noise altered vocalizations of rodents in similar ways as it has in other animals? Are all rodents specialized to discriminate between their own vocalizations?

Does the size of the rodent play a role in binaural acuity? Characterizing the natural behavior of wild rodents is important for the conservation of these mammals. It is hoped that this book will inspire scientists to heed the advice of Frank Beach (1950): "It would be much better if some of our well-trained experimentalists were encouraged to do a little pioneering."

Compliance with Ethics Requirements: Micheal Dent declares that she has no conflict of interest.

References

Arakawa, H., Blanchard, D. C., Arakawa, K., Dunlap, C., & Blanchard, R. J. (2008). Scent marking behavior as an odorant communication in mice. *Neuroscience and Biobehavioral Reviews*, 32(7), 1236–1248.

Beach, F. A. (1950). The snark was a boojum. *American Psychologist*, 5(4), 115–124.

Bradbury, J. W., & Vehrencamp, S. L. (2011). *Principles of animal communication*. Sunderland, MA: Sinauer Associates.

Druzinsky, R. E. (2015). The oral apparatus of rodents: Variations on the theme of a gnawing machine. In P. G. Cox & L. Hautier (Eds.), *Evolution of the rodents: Advances in phylogeny, functional morphology, and development* (pp. 323–349). Cambridge, UK: Cambridge University Press.

Feldhamer, G. A., Drickamer, L. C., Vessey, S. H., Merritt, J. F., & Krajewski, C. (2015). *Mammalogy: Adaptation, diversity, ecology*. Baltimore, MD: Johns Hopkins University Press.

Francescoli, G. (2000). Sensory capabilities and communication in subterranean rodents. In E. A. Lacey, J. L. Patton, & G. N. Cameron (Eds.), *Life underground: The biology of subterranean rodents* (pp. 111–144). Chicago: The University of Chicago Press.

Gibbs, R. A., Weinstock, G. M., Metzker, M. L., Muzny, D. M., et al. (2004). Genome sequencing of the brown Norway rat yields insights into mammalian evolution. *Nature*, 428, 493–521.

Honeycutt, R. L., Frabotta, L. J., and Rowe, D. L. (2007). Rodent evolution, phylogenetics, and biogeography. In J. O. Wolff & P. W. Sherman (Eds.), *Rodent societies: An ecological and evolutionary perspective* (pp. 8–27). Chicago: The University of Chicago Press.

Hurst, J. L., Payne, C. E., Nevison, C. M., Marie, A. D., et al. (2001). Individual recognition in mice mediated by major urinary proteins. *Nature*, 414, 631–634.

Lacey, E. A., & Sherman, P. W. (2007). The ecology of sociality in rodents. In J. O. Wolff & P. W. Sherman (Eds.), *Rodent societies: An ecological and evolutionary perspective* (pp. 243–254). Chicago: The University of Chicago Press.

Narins, P. M., Stoeger, A. S., & O'Connell-Rodwell, C. (2016). Infrasonic and seismic communication in the vertebrates with special emphasis on the Afrotheria: An update and future directions. In R. A. Suthers, W. T. Fitch, R. R. Fay, & A. N. Popper (Eds.), *Vertebrate sound production and acoustic communication* (pp. 191–227). New York: Springer International Publishing.

Poling, A., Weetjens, B. J., Cox, C., Beyene, N. W., & Sully, A. (2010). Using giant African pouched rats (*Cricetomys gambianus*) to detect landmines. *The Psychological Record*, 60, 715–728.

Poling, A., Weetjens, B. J., Cox, C., Beyene, N. W., et al. (2011). Tuberculosis detection by giant African pouched rats. *The Behavior Analyst*, 34(1), 47–54.

Portfors, C. V. (2007). Types and functions of ultrasonic vocalizations in laboratory rats and mice. *Journal of the American Association for Laboratory Animal Science*, 46(1), 28–34.

Roberts, S. C. (2007). Scent marking. In J. O. Wolff & P. W. Sherman (Eds.), *Rodent societies: An ecological and evolutionary perspective* (pp. 255–266). Chicago: The University of Chicago Press.

Vaughan, T. A. (1985). *Mammalogy*. Orlando, FL: Saunders College Publishing.
Waterman, J. (2007). Male mating strategies in rodents. In J. O. Wolff & P. W. Sherman (Eds.), *Rodent societies: An ecological and evolutionary perspective* (pp. 27–41). Chicago: The University of Chicago Press.
Waterston, R. H., Lindblad-Toh, K., Birney, E., Rogers, J., et al. (2002). Initial sequencing and comparative analysis of the mouse genome. *Nature*, 420, 520–562.
Weissbrod, L., Marshall, F. B., Vall, F. R., Khalaily, H. et al. (2017). Origins of house mice in ecological niches created by settled hunter-gatherers in the Levant 15,000 y ago. *Proceedings of the National Academy of Sciences of the United States of America*, 114(16), 4099–4104.

Chapter 2
Rodent Vocalizations: Adaptations to Physical, Social, and Sexual Factors

Kazuo Okanoya and Laurel A. Screven

Abstract This chapter introduces representative studies in acoustic communication in rodents. By using rodents as a model in which to study the evolution of vocal communication, researchers are able to utilize their diversity in physical habitats, social complexity, and sexual rituals. The widespread use of rodents as subjects of acoustic communication research is largely because many such species are the most successful mammalian group in terms of speciation. Much attention has been paid to isolation calls, alarm calls, and contact (or signature) calls in several species of rodents, with emphasis on the physical, social, and sexual variables involved in their production. Emergence of song-like vocalizations in both mother-infant contexts and male-female mating contexts are also discussed. Furthermore, the chapter focuses on the degree of plasticity in perception, production, and usage of these vocalizations in relation to the organization of neural structures related to hearing and vocalizations in rodents. Finally, these observations are integrated to suggest a general hypothesis on the evolution of vocal communication in rodents.

Keywords Acoustic communication · Acoustic environment · Animal communication · Degu · Ground squirrel · Mouse · Naked mole rat · Prairie dog · Rat · Ultrasonic vocalization · Vocal communication

K. Okanoya (✉)
Department of Life Sciences, The University of Tokyo, Tokyo, Japan
e-mail: okanoya@brain.riken.jp

L. A. Screven
Department of Psychology, University at Buffao, SUNY, Buffalo, NY, USA
e-mail: laurelsc@buffalo.edu

M. L. Dent et al. (eds.), *Rodent Bioacoustics*, Springer Handbook of Auditory Research 67, https://doi.org/10.1007/978-3-319-92495-3_2

2.1 Introduction

Communication is the transmission of information from a sender to a receiver (Di Paolo 1997). Although rodents use olfactory signals for communication, they also show high levels of vocal behavior (Bradbury and Vehrencamp 2012). Olfactory communication is powerful, but it mostly conveys information related to sexual reproduction, and its rate of transmission is slow. On the other hand, vocal signals are utilized in a variety of social contexts and have an instantaneous rate of transmission when animals are reasonably close to one another, making sound a more efficient mode of communication compared to olfaction.

Acoustic communication signals have a wide variety of functions in rodents, including eliciting approach behavior by conspecifics, alerting nearby animals of the presence of a predator, and attracting the attention of mothers by their pups. For rodents to transmit a wide array of information, acoustic communication has evolved to include diverse spectrotemporal characteristics in the signals. Nevertheless, the complexity of rodent vocalizations has long been overlooked, partially because many vocalizations that rodents produce are ultrasonic (>20 kHz) and are well outside the range of human hearing. Rodent vocalizations, however, have increasingly gathered the interest of scientists over the last few decades, largely due to their acoustical and functional varieties (Brudzynski 2014).

Rodents produce vocalizations ranging anywhere from 200 Hz to 100 kHz. Some rodents produce both low- and high-frequency vocalizations, depending on the context. Rodent vocalizations can be acoustically simple, such as rat (*Rattus rattus*) distress cries, which are pure tone-like calls that contain few frequency modulations (e.g., Brudzynski 2009); other calls can be very complex, such as mouse (*Mus musculus*) mating calls, which have rich harmonics and frequency modulations (e.g., Portfors 2007). Figure 2.1 provides an overview of common rodent vocalization types from multiple species.

The variation in rodent vocalizations across species may be due to several factors. The first is physical. Rodents live in a wide range of habitats: some, such as the prairie vole (*Microtus ochrogaster*), live on large, flat expanses such as prairies; others, such as the naked mole rat (*Heterocephalus glaber*), live in underground colonies. Sound propagation characteristics in each environment are quite diverse, favoring low frequencies in underground territories (Heth et al. 1986) compared to the wider range of frequencies used by rodents living in open spaces.

The second factor in the variation in vocal production across rodent species is social. In group-living rodents, it is crucial to develop signals for antipredator defense to coordinate escape behavior. As a result, some species of rodents have developed predator-specific alarm calls. For example, red squirrels (*Tamiasciurus hudsonicus*) emit acoustically distinct calls in response to aerial or terrestrial predators (Greene and Meagher 1998).

The last factor is sexual, as some rodents use vocalizations during sexual rituals. These vocalizations often form song-like structures and contain a variety of acoustical notes arranged in nonrandom orders. Song-like vocalizations can be sung in the

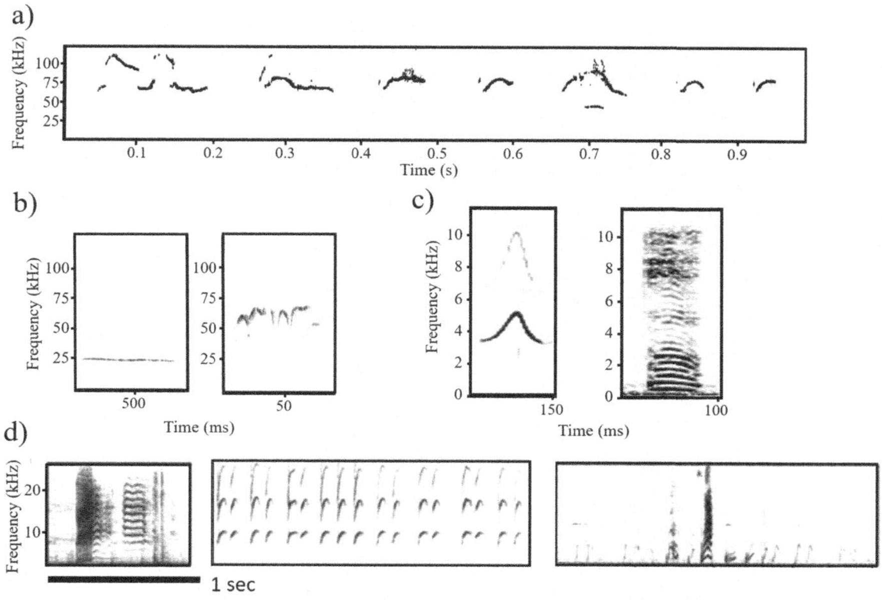

Fig. 2.1 Rodent vocalization types from various species. (**a**) Courtship ultrasonic vocalizations (USVs) from a male mouse. (**b**) Aversive 22 kHz call (*left*) and appetitive 50 kHz call (*right*) from a rat. (**c**) Contact call (*left*) and alarm call (*right*) from a naked mole rat. (**d**) Greeting call (*left*), pain calls (*center*), and grooming calls (*right*) from a degu

audible range, as in the naked mole rat (Pepper et al. 1991) and the degu (*Octodon degus*) (Long 2007), or in the ultrasonic range, as in mice (Holy and Guo 2005).

The physical, social, and sexual factors interact with each other, making both the evolution and usage of vocalizations even more complex. This chapter first addresses each factor separately to provide a general idea about the evolutionary adaptation of rodent vocalizations to each of these components. The chapter will also discuss how these factors influence each other. Studies of the diversity of rodent vocalizations can provide general insights into the evolution of communication systems of many animals beyond rodents.

2.2 Physical Factors

The fact that the physical environment affects sound propagation is well known. The medium, vegetation, temperature, and humidity can all influence transmission (e.g., Bradbury and Vehrencamp 2012). Because of the small interaural distances of most rodents, intra- and inter-species communication requires high frequency sounds for localizability using binaural cues (Ehret 2005; Lauer, Engle, and Schrode, Chap. 5) and that is likely to be a contributing factor to why many species have calls that are in the ultrasonic range. For example, peak modal frequency and range of

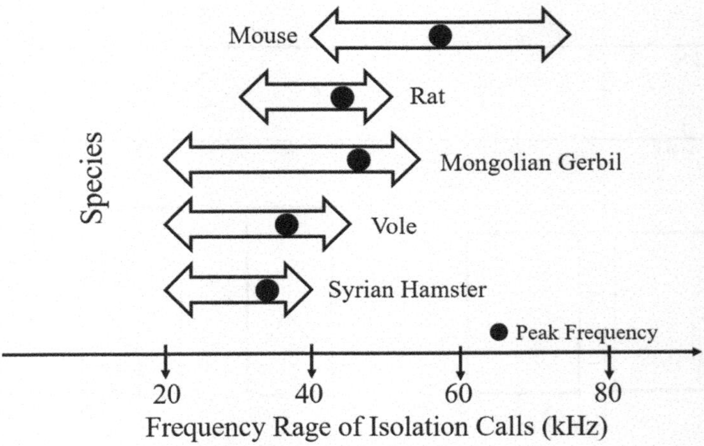

Fig. 2.2 Peak frequency (*black circles*) and frequency range (*arrows*) of isolation calls from five species of rodents. (Data replotted from Motomura et al. (2002))

isolation calls in five species of rodents are shown in Fig. 2.2 (Motomura et al. 2002). All species depicted produce vocalizations at frequencies higher than 20 kHz, reaching as high as 75 kHz in mice.

2.2.1 *Living in the Burrow*

Rodents living in burrows produce communication signals from 2–8 kHz (Heth et al. 1986; Schleich and Francescoli, Chap. 3). These animals have increased auditory sensitivity for sounds in this relatively low frequency range (Lange et al. 2007; Dent, Screven, and Kobrina, Chap. 4), which is likely to improve detection of their own vocalizations. Naked mole rats live in complex networks of underground rooms connected by tunnels. Because of the small width of the tunnels, signals higher than 20 kHz are not able to propagate long distances. Thus, selection pressures have favored the development of communication signals that are low in frequency compared to aboveground rodents. Frequencies higher than 8 kHz do not travel more than 5 m inside the mole rats' burrows (Okanoya et al. unpublished data), suggesting that it is adaptive for this species to produce low-frequency signals.

2.2.2 *Vegetation and Pitch of Vocalizations*

Gunnison's prairie dogs (*Cynomys gunnisoni*) emit alarm calls to warn conspecifics of the presence of predators. The acoustic structure of the calls differs across colony habitat types (Perla and Slobodchikoff 2002). The dominant frequency of the alarm call of a given prairie dog colony correlates with the percentage of vegetative cover in the habitat. The more vegetation, the higher the dominant frequency of the alarm call. Playback experiments confirm that the habitat vegetation correlates with the frequency ranges conveyed along greater distances. This adaptive modulation of alarm call acoustics could be due to usage learning, rather than production learning (see Sect. 2.5.1) (Kikusui et al. 2011; Arriaga et al. 2012). This issue is discussed further in Sect. 2.5.2.

2.3 Social Factors

Rodents seldom vocalize in the absence of other animals. The receivers of vocalizations are usually conspecifics, but sometimes they can be members of other species, including predators. The social environment of rodents appears to be one of the most important determining factors that leads to variability in rodent vocalizations.

2.3.1 Isolation Calls

In rodents, isolation calls are mostly produced by infants (Hashimoto et al. 2004). These calls function to help mothers find the location of infants that have moved away from the nest. Infants cannot maintain their body temperature and will experience hypothermia if they are out of their nests for an extended period. Hypothermia automatically triggers isolation calls in many species of rodents (Brunelli et al. 1994; Ehret 2005).

Isolation calls may not necessarily be produced purposefully. Some investigators have argued that these calls are the byproduct of a process called *laryngeal braking*, a method of respiration used to improve circulation of blood back to the heart (Blumburg and Sokoloff 2001). This process occurs when the pups are hypothermic and is part of the effort to increase internal body temperature. Whether or not production of these calls is purposeful or just a byproduct of laryngeal braking, they still function as a method of acoustic communication between pups and mothers. Motomura and colleagues (2002) compared isolation calls from five species of rodents, including Mongolian gerbils (*Meriones unguiculatus*), Syrian hamsters (*Mesocricetus auratus*), common voles (*Microtus arvalis*), rats, and mice. The isolation calls range from 20–70 kHz and are emitted from birth to about 2–3 weeks of age (Fig. 2.2). Once pups are able to thermoregulate more effectively, they stop producing isolation calls.

Mice emit isolation calls when separated from their mothers that elicit searching and retrieval behavior by the mother to return them back to the nest (e.g., Ehret and Haack 1981, 1984). Mouse isolation calls range from 70 kHz to 100 kHz and generally take the form of frequency-modulated sweeps emitted in a series (Liu et al. 2003). Rats also emit isolation calls between the ages of 4 and 16 days old (Sales and Pye 1974), but these calls are lower in frequency than the isolation calls of mice, with peak energy around 40 kHz. Rat pup calls are frequency-modulated sweeps between 30 kHz and 65 kHz, but rat pups can also make more complex calls that contain multiple components (Hashimoto et al. 2004).

Pine vole (*Microtus pinetorum*) and Syrian hamster pup isolation calls both have inverted-U shapes (also called chevrons), although each occupy different frequency ranges (Hashimoto et al. 2004). The pine vole emits 20 ms calls that range from 40 kHz to 60 kHz, although some calls have a bandwidth of only 40–50 kHz. The Syrian hamsters' ultrasonic chevron calls are from 30 kHz to 50 kHz, last for 20 ms, and contain more noise than the simple whistles observed in pine voles. Additionally, Syrian hamsters produce a broadband sonic call. Finally, the Mongolian gerbils emit frequency-modulated sweep calls that take various shapes and range from roughly 40 kHz to 55 kHz.

2.3.2 Contact or Signature Calls

Some species of rodents use contact calls for social coordination. Naked mole rats, when passing one another in tunnels, emit *soft chirps* (Pepper et al. 1991). These calls are individually specific, meaning that the variations of calls emitted by an individual are smaller than the variations in calls of different individuals (Yosida et al. 2007). Therefore, soft chirps may function not only as contact calls but also as signature calls, providing information about the caller's identity. The fundamental frequencies of these calls are correlated with body size, which is further associated with their caste (Yosida and Okanoya 2009), or social ranking, implying that the soft chirps are also used as caste signatures. In fact, naked mole rats are sensitive to the fundamental frequency of the soft chirp (e.g., Muller and Burda 1989). When a call from one naked mole rat is higher than another's call, it communicates that the higher frequency mole rat is of a lower rank than the other caller.

When two mole rats encounter one another in a tunnel, the frequency information in their calls affects their behavior—the higher-ranked mole rat passes above the lower-ranking mole rat (Yosida and Okanoya 2009). To do this, they exchange calls according to certain rules. The first mole rat to call emits a soft chirp and waits 300 ms for the other animal to respond. When a mole rat hears a soft chirp, it responds with its own call within 100–200 ms (Yosida et al. 2007). Examples of such antiphonal events are shown in Fig. 2.3. While antiphonal calling is reported in bats (*Diaemus youngi*) (Carter et al. 2008), gibbons (*Hylobates muelleri*) (Inoue et al. 2013), and crows (*Corvus macrorhynchos*) (Kondo et al. 2010) among others, the naked mole rat is the only rodent reported to produce antiphonal calls (Yosida et al. 2007). Whether the eusociality that has evolved in the naked mole rat can account for the emergence of individually specific, antiphonal contact calls must be further examined in other eusocial rodents. Evidence for this has been observed in the Damaraland mole rat (*Fukomys damarensis*), another eusocial burrow-living rodent species, as these animals might also have signature-like chirps (Burda 1995).

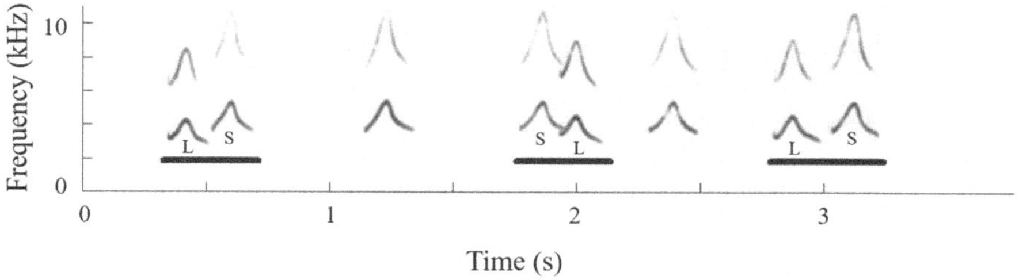

Fig. 2.3 Sonogram of antiphonal contact calls exchanged by two naked mole rats. Calls with a lower fundamental frequency are those from the larger animal (*L*); calls with a higher fundamental frequency were emitted by the smaller animal (*S*). Calls from both animals contain a higher frequency harmonic. *Lines* are placed below overlapping call exchange that occurred when two animals emitted calls at approximately the same time; calls without a line are a single call with fundamental and harmonic. (Data replotted from Yosida and Okanoya (2009))

As illustrated with naked mole rats, some vocalizations can provide a great deal of information to receivers about various aspects of the caller, including sex, body size, age, individual identity, kinship, and arousal level. Listeners can use this information to their advantage by engaging in certain behavioral responses only when callers fit specific criteria. Yellow-bellied marmot (*Marmota flaviventris*) calls significantly differ by both age and sex (Blumstein and Munos 2004). Individually specific characteristics of calls were perceived even when the receiver was at a great distance from the vocalizing animal and the calls were degraded. Blumstein and Armitage (1997) discovered that two spectrotemporal characteristics convey information about the caller: duration and frequency composition. Additionally, Blumstein and Daniel (2004) used playback paradigms to illustrate that yellow-bellied marmots can discriminate between individuals using alarm calls alone.

Sex, age, and individual differences in vocalizations are also observed in great gerbils (*Rhombomys opimus*) (Randall et al. 2005) in which female calls have significantly higher minimum and mean frequencies than male calls. These frequency differences provide reliable information that receivers could use to differentiate the sex of the caller. Further, adult great gerbil vocalizations also differ significantly in several frequency characteristics compared to those emitted by juveniles, including (but not limited to) peak frequency and mean frequency of calls. Receivers would be able to use this information to determine the age of the caller and to decide whether the calls were from adults, which are likely to produce more reliable calls than juveniles. Avoiding unnecessary behavioral responses is advantageous for listeners as they take time away from more profitable activities, such as foraging or pursuing mates.

Finally, mouse vocalizations were recorded to determine if they also have individualistic features that could help identify specific callers. Hoffmann and colleagues (2012) found that courtship calls emitted by males contained characteristics that could provide identity and relatedness information to listening females. In particular, the ultrasonic vocalizations (USVs) emitted by related males were more similar than calls emitted by unrelated males. These differences are advantageous to females as they can avoid mating with closely related males, which ultimately prevents inbreeding. Hoffman and colleagues (2012) also showed that male mice have individual signatures within their calls that could provide listeners with information about which specific individual is calling. It is likely that the advantages that these calls provide (e.g., an increase in either direct or inclusive fitness) have led to the evolution of these signals in some rodent species.

2.3.3 Alarm Calls

Rodents usually live in social colonies and are often highly susceptible to predation (e.g., Dent, Chap. 1). As a result, many species emit alarm calls to warn conspecifics about the presence of nearby predators. Alarm calling in the great gerbil is unique in rodents because it is accompanied by another acoustic signal called *footdrumming* (Randall et al. 2005). When a great gerbil spots a predator, it assumes a bipedal posture while footdrumming and simultaneously emits one of three distinct types of alarm vocalizations (Randall and Rogovin 2002). To produce the footdrumming signal, the great gerbils thump their hind feet on the ground while jumping. Footdrumming is produced by great gerbils of all ages and is believed to function primarily as a way to alert juveniles to the presence of danger. The accompanying vocalizations are chevron calls, intense calls, and whistle calls. Rhythmic chevron calls are produced in bouts with an intercall interval of 200 to 500 ms. If the threat persists, great gerbils will emit the intense call. Conspecifics show greater antipredator behavior when hearing intense calls compared to the rhythmic chevron calls. Finally, great gerbils produce whistle calls, which are short in duration and often precede the caller running into its own burrow to escape a predator. Listeners also display high levels of antipredator behavior when hearing these whistle calls.

Chinchillas (*Chinchilla chinchilla*) produce bark calls to communicate predation risk to conspecifics. These calls also have spectrotemporal elements that may provide individual identity information (Moreno-Gómez et al. 2015). Alarm calls in chinchillas are spectrotemporally complex, containing multiple components (Hunyady 2008). Additionally, these calls are produced at high intensity and likely travel considerable distances to warn conspecifics of the presence of danger.

Yellow-bellied marmots emit harmonic alarm calls with a relatively low fundamental frequency of about 1.5 kHz and durations around 20 ms (Blumstein and Recapet 2009). These calls were studied using playback experiments in naturally foraging animals to determine the effect of adding nonlinearity to their calls. Adding a short burst of noise in the middle of the call reduced the time that the animals spent foraging, while replacing the middle section with silence did not affect foraging time. These results illustrate that nonlinearities in these calls will increase their effectiveness at communicating fear or alarm, likely because naturally occurring nonlinearities in vocalizations could indicate a greater level of fear in the caller.

Because rodents are targeted by several types of predators, it would be useful if the rodents could produce specific calls to specific predators. Predator specificity is a phenomenon that has been observed most notably in vervet monkeys (*Chlorocebus pygerythrus*) (Seyfarth et al. 1980). However, the rodents' ability to produce nonlocalizable vocalizations is equally important in order to warn conspecifics without being targeted by the predator. Rodent alarm calls also were analyzed in terms of nepotism: alarm calls might be emitted if beneficial to relatives but will not be emitted when there are no relatives nearby (e.g., Sherman 1977).

Predator specificity in rodent alarm calls was first identified in Belding's ground squirrels (*Urocitellus beldingi*), a species that emits acoustically distinct alarm calls to aerial versus terrestrial predators (Sherman 1977). In Belding's ground squirrels, aerial alarm calls are characterized by single-note whistles, whereas terrestrial alarm calls are multiple-note trills (Sherman 1977). Sherman also discovered selective nepotism in this species, whereby Belding's ground squirrels emit more terrestrial alarm calls when relatives were around than when only nonrelatives are present. Aerial alarm calls, however, were produced regardless of the kinship status of nearby rodents.

Predator specificity also was shown in California ground squirrels (*Otospermophilus beecheyi*) (Owings and Virginia 1978) and arctic ground squirrels (*Spermophilus parryii*) (Melchior 1971). Similar to Belding's ground squirrels, California and arctic ground squirrels emit chatter alarm calls to terrestrial predators and whistle alarm calls to aerial predators. The chatter calls provide more information about the location of the caller to the predator than whistle calls. This might explain why Belding's ground squirrels emit aerial alarm calls regardless of kinships with other animals, as aerial alarm calls pose less risk to the caller. The acoustic characteristics of terrestrial chatter alarm calls and aerial whistle alarm calls by arctic ground squirrels are illustrated in Fig. 2.4.

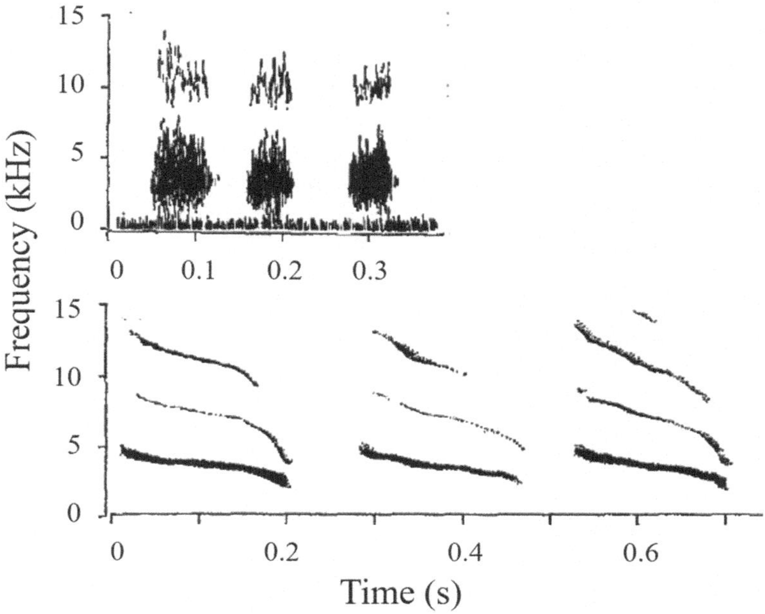

Fig. 2.4 Chatter (*upper*) and whistle (*lower*) calls in a species of ground squirrel. (Data replotted from Melchior (1971))

Table 2.1 Morton's motivation-structural rules (Morton 1977). An increase in fear raises the pitch of the call, while an increase in aggression increases the bandwidth (FM, frequency modulated; HF, high frequency; LF, low frequency; MF, middle frequency)

	Increasing aggression →			
		Low	Moderate	High
←Increasing fear	Low	LF Narrow bandwidth Up or down	MF Wide bandwidth Harsh	LF Wide bandwidth Harsh
	Moderate	MF Narrow bandwidth FM	MF Narrow bandwidth FM	MF Wide bandwidth Harsh
	High	HF Narrow bandwidth Harsh	HF Narrow bandwidth FM	HF Wide bandwidth Harsh

Richardson's ground squirrels (*Urocitellus richardsonii*) use 50 kHz alarm calls to warn conspecifics of an approaching predator (Wilson and Hare 2004). Ultrasonic alarm calls are not audible to most predators and are highly directional, so the Richardson's ground squirrels can warn nearby conspecifics without the threat of detection by the predators.

2.3.4 *Expression of Affect*

Animal calls often convey two kinds of information: the affective state of the caller and the purpose of the call (Marler et al. 1992). Stimulating areas of the brain that are associated with affect, such as the periaqueductal gray, leads to the production of several types of vocalizations in primates and rodents (Zhang et al. 1994), suggesting vocalizations are likely modulated by affective state.

Morton (1977) proposed that the acoustic structure of vocalizations contains affective state information by the *motivation-structural rules* (Table 2.1). In humans, specific characteristics of acoustic signals can be attributed to different emotional states. For example, if a person experiences fear, they will produce sounds with increased frequency and amplitude modulations. Alternatively, to communicate aggression, humans produce speech in low frequency ranges. It is possible that these rules may be applied to rodents, to a certain extent, though they might not be applied to ultrasonic vocalizations, as discussed in Sect. 2.8.

2.3.4.1 Aversive Calls

Aversive calls are emitted by rodents in situations that induce fear or anxiety (Wöhr and Schwarting 2013). These vocalizations likely communicate important, potentially situation-specific, information to listeners about possible danger. In laboratory rats, calls associated with negative affective state are usually around 22 kHz, narrowband, and long lasting (500–1000 ms; Litvin et al. 2007). These calls are emitted when rats anticipate the delivery of an aversive stimulus (e.g., Blanchard et al. 1991; van der Poel et al. 1989), such as an electric shock.

Rats produce these same 22 kHz vocalizations following defeat in an aggressive encounter, which likely functions to prevent further attacks from the victor (Sales and Pye 1974) as they are emitted in conjunction with submissive postures displayed by the caller. This call type is thought to warn conspecifics about the presence of a predator (Blanchard et al. 1991). Rats may have a predisposition to associate 22 kHz calls with negative affective states, as suggested by Endres and colleagues (2007), due in part to their ability to communicate fear among conspecifics (e.g., Wöhr and Schwarting 2008). Not all 22 kHz calls signal negative affect, however, as male rats emit 22 kHz calls during sexual encounters, a situation that would not be considered aversive (Barfield and Geyer 1972). Receivers might be able to distinguish the negative affect 22 kHz vocalizations from the vocalizations emitted following ejaculation by male rats because these post-copulation calls have increased frequency modulation (van der Poel et al. 1989).

Guinea pigs (*Cavia porcellus*) emit particular vocalizations in aversive situations. Following injury or social defeat, guinea pigs produce squeal calls or scream calls (Berryman 1976). Squeal calls are associated with less extreme negative events, such as slight injury, or in response to a bite from a conspecific; scream calls are usually emitted after an aggressive encounter with a dominant guinea pig. While producing the scream calls, the losing guinea pig assumes a defensive posture and increases the rate and frequency of the calls if the aggressor approaches. These calls likely function similarly to the submissive 22 kHz calls emitted by rats to prevent further injury by the dominant animal following an aggressive interaction.

Although aversive calls were not attributed to mice in negative situations historically (Portfors 2007), these types of vocalizations have been documented more recently. Chabout and colleagues (2012) first identified these aversive calls when mice experienced restraint-induced stress, although mice produce significantly more vocalizations during social interactions than during restraint. Chabout and colleagues found that mice emit more short and composite vocalizations during restraint than any other call variety, in stark contrast to the rich variation of call production during social encounters. In agreement, Grimsley and colleagues (2016) recorded vocalizations from mice under several types of restraint stress, including headpost and jacket restraint, and they found that mice emit mid-frequency vocalizations in these conditions, which are presumed to communicate increased stress or a negative state of arousal. These vocalizations (usually 9–15 kHz) are much lower than the usual ultrasonic vocalizations they produce under other circumstances (e.g., Portfors 2007). Context-specific vocalizations produced by mice during negative affective states are still not well understood.

2.3.4.2 Appetitive Calls

Rats emit short 50 kHz calls in situations assumed to elicit positive affect, including during mating, social play, and when anticipating the delivery of a positive stimulus (e.g., food). There are many variations of 50 kHz calls in rats, especially in the degree of frequency modulation. These calls have been divided into 14 different categories based on their spectrotemporal characteristics (Wright et al. 2010). Frequency modulation might be an expression of reward prediction error, as unexpected positive outcomes are associated with 50 kHz calls with frequency modulation (Yuki and Okanoya 2014).

Rats emit 50 kHz vocalizations during "heterospecific play," as suggested by Panksepp and Burgdorf (2000). Researchers "tickled" rats and recorded 50 kHz calls in response to the tickling, leading the researchers to believe that the 50 kHz calls could be analogous to laughter (Panksepp and Burgdorf 1999, 2003). Tickling was operationalized by Panksepp and Burgdorf as manual stimulation of the dorsal or ventral surface or the full body of the rats in 15 s intervals by the researcher. Rats were recorded during different types of manual tickling stimulations, and the 50 kHz vocalizations differed significantly across stimulation locations. "Laughter" occurred the most when the rats were tickled on their full bodies compared to when tickling was localized to a specific part of the body.

2.3.4.3 Agonistic Calls

Although many rodent species live in social colonies, some rodent species hold their own territories. Several species of tree squirrels (*Tamiasciurus hondsonicus* and *T. douglasii*) defend their own territories and use vocal displays to alert trespassing conspecifics that the territory owner has seen them (Smith 1978). Territory owners emit rattle and screech calls to trespassers; the receivers typically respond by fleeing from the territory (Smith 1968). Additionally, tree squirrels will produce growl calls when they are chasing or fighting with a trespassing squirrel. These calls are produced at very low amplitudes, probably to decrease the likelihood of being overheard by a predator while attempting to fight off the trespassing animal. Rats produce USVs during aggressive interactions with conspecifics. During agonistic interactions, rats produce 50 kHz calls (Takeuchi and Kawashima 1986), despite observations that these calls are typically associated with positive affect. It is likely that specific characteristics of the 50 kHz calls produced during agonistic interactions differ from those emitted during positive encounters. Burgdorf and colleagues (2008) investigated the conditions under which rats would emit 50 kHz calls and compared their spectrotemporal parameters. Compared to calls produced during appetitive behaviors, such as rough-and-tumble play or mating, the 50 kHz calls emitted during aggressive interactions lack frequency modulation, leading the authors to speculate that the addition of frequency modulation serves as a contextual cue.

2.3.4.4 Affective Contagion

Playback studies revealed rodent calls evoke specific behaviors in listeners that mirror the affective state of the caller (Burgdorf et al. 2008). For example, playback of 22 kHz calls to rats causes a decrease in approach to the sound source by the listener, whereas playback of 50 kHz calls increases such activity (Wöhr and Schwarting 2007). This could be a demonstration of affective contagion in rats: the affective state conveyed by the caller induced a similar state in the receiver. These calls, however, could simply be functioning as key stimuli to release fixed action patterns, as in the paradigm in classical ethology called the "the innate releasing mechanism" (Tinbergen and Perdeck 1951).

One procedure that resolves this dilemma is the use of a *cognitive bias task*. Cognitive bias tasks utilize the animal's tendency to perceive the same stimulus as positive *or* negative depending on their affective state (Mendl et al. 2009). Saito and colleagues (2016) found rats were more likely to perceive a stimulus as positive after exposure to the 50 kHz calls and more likely to perceive a stimulus as negative following exposure to the 22 kHz calls. This indicates the affective state of rats influences their perception of ambiguous stimuli, providing evidence for the idea of affective contagion in rats.

2.3.5 Song-Like Vocalizations in Social Contexts

When the term song is used to describe vocal behavior in animals, it is generally considered analogous to the complex songs of other species, such as birds or humpback whales. When applied to animal behavior, song is defined by its physical characteristics as vocalizations that consist of several acoustical units, ordered in a systematic way. Song could also be defined by its social characteristics: songs are mostly used in mating contexts, but they might also be used in other social settings for social affiliation. Based on these definitions, some rodent vocalizations do fall under the umbrella of song-like acoustic communication.

Holy and Guo (2005) were the first to use the term "song" for mouse vocalizations. Mouse USVs were elicited from animals using female olfactory signals, and the authors showed that the male mice produced sequences of USV syllables in nonrandom orders. From these results, the authors posit that the vocal production fits the definition of song commonly used to classify vocalizations of birds, amphibians, and insects. Additionally, Mongolian gerbils produce songs in some situations. Kobayasi and Riquimaroux (2012) found Mongolian gerbils produced calls in particular sequences with particular syntactic rules. The concept of song production in rodents is still largely unexplored, but it does not appear to be directly analogous to bird songs, which are necessary for reproduction (e.g., Hammerschmidt et al. 2009).

2.4 Sexual Factors

In some rodent species, vocalizations are used to calm a sexual partner before mating. Vocalizations are emitted by many species of rodents during courtship, throughout mating, and following ejaculation. The vocalizations rodents emit during courtship and mating have likely been sexually selected for, since females in some species of mice prefer male mice that vocalize compared to nonvocalizing males (e.g., Pomerantz et al. 1983).

2.4.1 *Vocalizations Involved in Mating*

Both male and female mice produce USVs in mating situations (Holy and Guo 2005) and in same-sex social encounters (Gourbal et al. 2004). The acoustic morphology of the USVs used in each of these situations differs (Matsumoto and Okanoya 2016; Matsumoto 2017). In general, short calls are used in the exploratory phase and acoustically more complex notes are emitted in the mating and copulatory phases (Fig. 2.5) (Matsumoto and Okanoya 2016). Ablation of the amygdala results in a reduction of complex song components in mice (Matsumoto et al. 2012). One interpretation is that lesions in the amygdala reduce sexual motivation. As a result, production of complex calls, which are controlled by sexual motivation, decreases although the motor programs that generate complex calls are not in the amygdala.

Mice produce a wide variety of calls, most of which are believed to be involved in mating (Portfors 2007). Researchers have attempted to parse these calls into various categories based on their spectrotemporal parameters, with the number of calls ranging between 7 and 14, depending on the study (e.g., Portfors 2007; Hoier

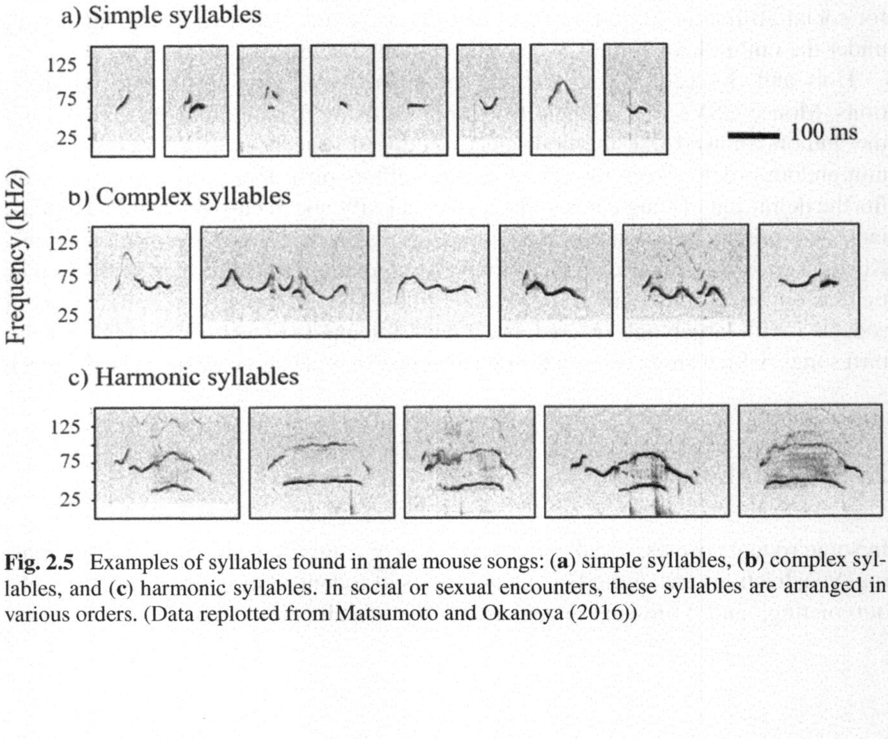

Fig. 2.5 Examples of syllables found in male mouse songs: (**a**) simple syllables, (**b**) complex syllables, and (**c**) harmonic syllables. In social or sexual encounters, these syllables are arranged in various orders. (Data replotted from Matsumoto and Okanoya (2016))

et al. 2016). Various categories of USVs are produced by both male and female mice and are often inferred to communicate information to conspecifics that increases the likelihood of mating. It is unclear if mice divide their vocalizations into categories, however. These categories have been created by researchers and, therefore, may not be an accurate representation of how the mice perceive their calls. Neilans and colleagues (2014) tested whether mice were able to discriminate vocalizations and found that mice could not discriminate between calls that were spectrotemporally similar compared to calls that were less similar. This supports the idea that calls that are more similar are parsed into one category, whereas calls that are dissimilar would be considered a different category. Discrimination of vocalizations is discussed in more detail by Dent, Screven, and Kobrina (Chap. 4). Further evidence for call categories comes from Matsumoto and Okanoya (2016). Mice produce specific calls during discrete phases of mating, indicating that these calls could be involved in eliciting different behaviors throughout courtship and mating sequences.

Rats also emit vocalizations during specific phases of mating, although the USVs that rats produce are not as complex as those from mice. Rat vocalizations contain little frequency modulation and have a limited repertoire of syllables compared to those of mice. Barfield and colleagues (1979) described USV emission during courtship and mating in rats and noted that during both the initial phases of courting and mounting, male and female rats emit 50 kHz USVs. Following ejaculation, there is a transition from 50 kHz to 22 kHz vocalizations. Although 22 kHz USVs are typically associated with negative situations, these post-ejaculatory calls are thought to maintain contact with the female during the male's refractory period. Vocalizations emitted during specific phases of courtship and mating have also been found in Mongolian gerbils (Holman 1980).

Berryman (1976) found both male and female guinea pigs emit purr calls during behaviors associated with mating, including circling and swaying, which is a behavioral sequence called the *rumba*. Purr calls have a fundamental frequency ranging between 200 kHz and 750 Hz and often include up to seven harmonics. They are typically 25–60 ms long and often occur in bouts. Guinea pigs produce purr calls throughout the courtship and mating phases, including during intromission and ejaculation.

Historically, male rodents have been assumed to be the producers of the majority of vocalizations during the mating sequence (e.g., Whitney et al. 1973). More recently, however, researchers have shown that females also call during these interactions. For example, in mice, females produce USVs when interacting with males. Neunuebel and colleagues (2015) recorded USVs in male–female pairs to determine which mouse in the dyad was calling. Males and females both contributed to USV displays. Their results are supported by Hammerschmidt and colleagues (2012a), who showed that calls of male and female mice do not differ significantly. This means that when researchers record male–female dyads, it would not be possible to differentiate USVs emitted by the male from those emitted by the female.

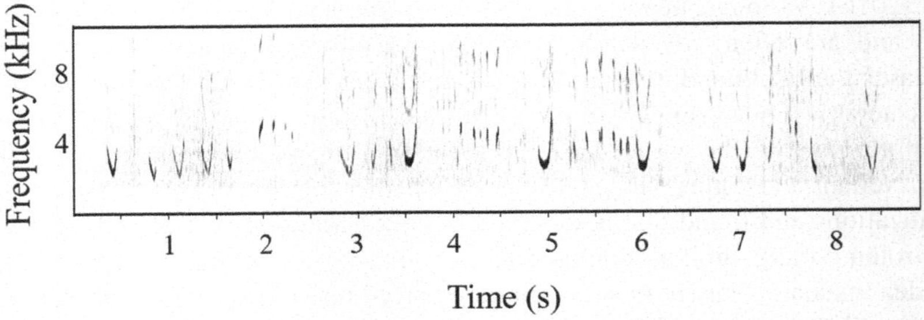

Fig. 2.6 A song bout of a naked mole rat queen. The song consisted mainly of V-trills (Pepper et al. 1991), but pitch and tempo of the train of syllables had high variability. (Okanoya, unpublished data)

2.4.2 *Mating Song and Infant Babbles in Degus and Naked Mole Rats*

Baby degus produce trains of short calls (infant babbles) when interacting with parents. Adult males also produce similar sounds continuously in mating rituals while following females. Baby calls are louder and higher pitched than adult male calls (Okanoya, unpublished observation). A similar train of sounds was produced by a naked mole rat queen when she was mating with a male (Fig. 2.6) (Yosida and Okanoya, unpublished observation). These calls resemble the V-trills described by Pepper and colleagues (1991), but variations in pitch and note morphology are much greater.

2.4.3 *Song and Sexual Selection*

Mouse courtship songs appear to have three main syllable categories: simple, complex, and harmonic. These categories can be further subdivided into the rich variety of USVs shown in the literature (Sect. 2.4.1) (Portfors 2007). Complex syllables produced by male mice may play a role in eliciting approach behavior in females, and more spectrotemporally simple USVs function to maintain contact once a female is nearby (Chabout et al. 2015). The USVs produced by male mice are very high in frequency and require high rates of repetition to attract females, a behavior likely to be energetically costly to the male. Thus, complex and harmonic syllables could be a sexually selected trait in the mouse, as they elicit approach behavior in females in preference experiments (e.g., Pomerantz et al. 1983). Complex and harmonic syllables should require more motor coordination and thus more neural resources compared to simple syllables.

Both male and female laboratory mice emit ultrasonic songs in various social encounters, and the acoustical and syntactical organizations are not different between the sexes (Hammerschmidt et al. 2012a). Wild mouse strains may be singing even more complex songs than laboratory mice, however, even in same-sex interactions (Von Merten et al. 2014). These findings argue against the possibility that courtship songs are under sexual selection pressure. More in-depth analyses that compare strains of mice, communication context, acoustic complexity, and the cost of production of each syllable are necessary before a conclusion can be drawn.

2.5 Plasticity of Rodent Vocalizations

Some rodent species have large repertoires of vocalizations. For example, degus have 15 categories of calls (Long 2007) and naked mole rats and giant mole rats both have 14 types (Pepper et al. 1991; Bednářová et al. 2013). For these species, particular calls are emitted in particular situations. The variety of vocal signals in rodents is large even when compared with primates (Winter et al. 1966) or birds (Ficken et al. 1978). Due to the wide range of vocal production in rodents some degree of plasticity associated with acoustic communication is possible.

2.5.1 Production Learning

Mouse vocalizations may be under the control of *production learning mechanisms*, meaning that young mice would require the experience of listening to vocalizations from adults to produce proper calls as they mature. There is little evidence that mice learn their songs as juveniles, although a cross-fostering paradigm demonstrated that experience did have an effect on the frequency of calls produced in one instance (Arriaga et al. 2012). It is possible, therefore, that mice might experience some degree of vocal learning. Arriaga and colleagues related this finding to the sparse, but identified, corticobulbar projection from the motor cortex to the nucleus retroambiguus of the medulla oblongata. This projection has been found in other vocal learning species, including humans (Jurgens 1979, 2002) and birds (Deacon 1998).

The evidence for vocal learning in mice, however, has not been extensive. Genetically engineered deaf mice have normal, strain-specific USVs, suggesting that auditory feedback is not required for the production of calls (Hammerschmidt et al. 2012b; Mahrt et al. 2013), contrary to the findings of Arriaga and colleagues (2012). Additionally, removal of the cerebral cortex did not have a negative influence on call production (Hammerschmidt et al. 2015). Finally, Kikusui and colleagues (2011) investigated the calls of fostered juveniles, also using a cross-fostering paradigm. Cross-fostered mice did not change their vocalization patterns as a result of the acoustic experience they had as pups, arguing against vocal learning in these animals. Given the evidence against production learning, there is currently a consensus that mouse songs are innate, although there is the possibility of minor plasticity (Arriaga and Jarvis 2013).

2.5.2 Usage and Perceptual Learning

Although production learning is limited in rodents, there is some evidence supporting *perceptual learning*, which is learning when or how to use their vocalizations for communication. In other words, production of these calls is likely innately programmed, but proper usage may require experience. For example, two types of alarm calls have been identified in degus: flat and frequency modulated (FM). In playback experiments, juveniles only emit flat (not FM) alarm calls and did not display the appropriate vigilance behavior in response to the FM calls typically produced by adults (Fig. 2.7) (Nakano et al. 2013). As degus mature, they begin producing and responding to the FM alarm calls, suggesting that behavioral responses to different alarm calls might be learned in this species.

It is also possible that rats learn to associate their 22 kHz calls with aversive situations. When adult rats were trained to associate several signals with electric shock, the 22 kHz signal was most effective in learning this association (Endres et al. 2007). These findings suggest that rats might be predisposed to learn to associate 22 kHz vocalizations and fear-related situations.

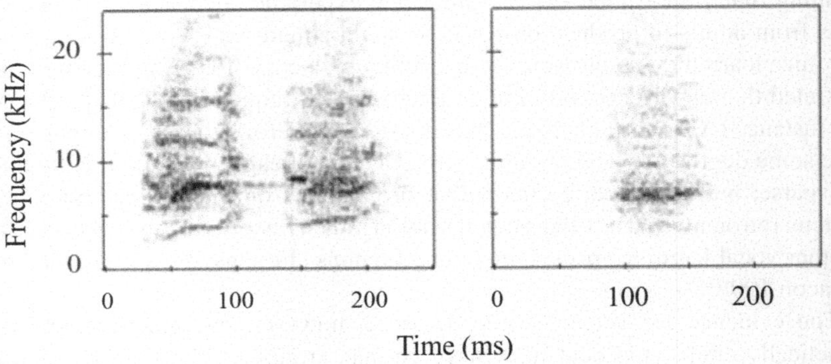

Fig. 2.7 Alarm calls of adult (*left*) and juvenile (*right*) degus. Adult alarm calls are frequency modulated; those of juveniles are not. Frequency modulated alarm calls evoke a freezing response from adult animals, but the same call has no effect for juveniles. (Data replotted from Nakano and colleagues (2013))

2.6 Interactions of Physical, Social, and Sexual Factors

The physical, social, and sexual factors involved in acoustic communication in rodents, as described in Sects. 2.2, 2.3, and 2.4, can also interact and have a profound effect on rodent communication. For example, the use of alarm calls in ground squirrels depends on kinship with listeners—squirrels will produce easily locatable calls only when relatives are nearby (Sherman and Holmes 1985). This pattern of alarm calling in squirrels is likely due to an increase in vulnerability to predation, a risk they are only willing to take when the survival of their kin is at stake. Similarly, complex and harmonic USV syllables in mice are easy to localize, and mice often use these calls during courtship and mating. Sexual selection favors traits that make producers more susceptible to predation (Zahavi 1975), because surviving with such traits is a strong indicator of fitness of the individual. This is an example of the interaction between physical (localizability) and sexual (attractiveness) factors.

The courtship song of the naked mole rat queen might represent the combination of all three of these factors. Because naked mole rats are burrow dwellers, their opportunity to find genetically distant mates is limited. As a result, genetic relatedness is high in this species (Reeve et al. 1990), resulting in eusociality. In naked mole rats, only the colony queen is allowed to reproduce, and only a few males are able to copulate with the queen. Naked mole rat queens produce several kinds of vocalizations (e.g., V-trills and song) that are believed to be involved in soliciting males to mate (Pepper et al. 1991). These calls are influenced by the interaction of physical (spectrotemporal parameters of calls), social (experience with courtship calls), and sexual (attractiveness of the queen and competition between males) factors.

2.7 Neural Mechanisms Underlying Call Production and Perception

One way researchers have attempted to understand the underlying mechanisms controlling acoustic communication in animals has been to investigate neural control of vocalization production. By examining which areas of the brain are controlling the emission of calls, it is possible to draw conclusions about what factors are influencing them, such as fear or aggression.

2.7.1 Neural Mechanisms Underlying Mouse Song

As mentioned in Sect. 2.4.1, partial lesions of the lateral amygdala eliminated most of the complex and harmonic calls produced by male mice during courtship and mating and reduced the male's motivation to mount (Matsumoto et al. 2012). It is unclear if eliminating the males' sexual motivation by ablating the amygdala led to the cessation of the production of complex and harmonic calls of courtship songs, but it is clearly a consequence of this manipulation.

Although the amygdala controls the motivation for mating, other areas, including the motor cortex (Arriaga and Jarvis 2013) and the periaqueductal grey (Behbehani 1995), are likely to be involved in controlling motor production of courtship song. Song syllable morphology was affected by ablation of the aural area of the motor cortex in male mice (Arriaga et al. 2012). Further, gene expression studies show that the motor cortex and periaqueductal gray are activated by singing. The periaqueductal gray is considered to be involved in innate vocal control. The involvement of the motor cortex suggests that there might be learning-related plasticity in mouse song, but this point is still under contention (Arriaga and Jarvis 2013), as discussed in Sect. 2.5.1.

2.7.2 Neural Mechanisms Underlying Rat Calls

Neural mechanisms underlying the 22 kHz and 50 kHz affective calls in rats (see Sect. 2.3.4) have been the subject of hundreds of studies. To examine general activation patterns of the brain by both calls, Sadananda and colleagues (2008) used immediate early gene expression as a measure of activity in different areas of the brain following stimulation by either 22 kHz or 50 kHz calls. The 50 kHz calls induced sparse Fos-like immunoreactivity throughout much of the brain, including areas assumed to be associated with activities that often accompany production of 50 kHz appetitive calls. The 22 kHz calls, in contrast, elicited c-Fos expression in brain areas associated with submission and fear conditioning, such as the ventrolateral periaqueductal gray and various parts of the amygdala, respectively. These results are in agreement with known neuroanatomical substrates for processing negative emotions (Sadananda et al. 2008).

Neural control of affective calls also was examined by recording neural activity to call playbacks to rats. Hearing 22 kHz calls elicited different responses than hearing 50 kHz calls. Responses to 22 kHz calls were characterized by tonic increases in firing rate, while responses to 50 kHz calls were characterized by a tonic decrease in firing rate (Parsana et al. 2012). These results are consistent with behavioral effects of these calls: hearing 22 kHz calls leads to a decrease in locomotor activity in rats, while hearing 50 kHz calls leads to an increase in locomotor activity (e.g., Wöhr and Schwarting 2013). Further, Kagawa and colleagues (2017) investigated the brain's responses to natural versus artificial stimuli. In the amygdala, only negative contexts show an overlap in activity to ethological and unnatural stimuli.

2.7.3 Neural Mechanisms Underlying Guinea Pig Calls

Guinea pigs produce a wide variety of call types (Berryman 1976). Neural responses to species-specific vocalizations have been investigated throughout the central auditory system, including the inferior colliculus, medial geniculate body (Syka et al. 1997), and auditory cortex (Šuta et al. 2013). In the inferior colliculus, guinea pig calls were encoded by their spectral and temporal characteristics. Additionally, neurons in the auditory cortex responded to spectrotemporal features of calls, illustrating that the different calls that these animals emit can be differentiated within the brain.

2.8 The Evolution of Rodent Vocalizations

Most rodent vocalizations are produced by the larynx and they can be related to respiratory action (Riede 2013). Respiration refers to the flow of air through the throat on the way to or out of the lungs. As an animal becomes more aroused, the respiration rate becomes faster, while the respiration rate is slower in conditions of low arousal. Noise associated with respiration, therefore, could be used as a signal conveying the caller's arousal level. If the trachea and airway produced respiration-associated noise, the resonance frequency of the calls would be higher in smaller animals and lower in larger animals. As a result, the pitch of respiration-associated vocalizations could be used as an honest indicator of the size of the animal producing the sound.

2.8.1 Respiration Sounds as Signals

Vocalizations may have evolved as indicators of affect or physical traits such as body size. Once the communicative co-option of vocalizations began, vocal signals could have increased in complexity according to the physical, social, and sexual factors they experienced. This is partially because the physiological cost of sound production is rather low, while the effective signal space of sound is rather high (Catchpole and Slater 2003). Mice produce USVs by forcing air through the larynx, similar to a whistle. The whistle-like quality of USVs from mice was demonstrated in vitro by dissecting out the larynges of mice and blowing air through them to create sounds that were very similar to calls produced in vivo (Mahrt et al. 2016). Those sounds can be modulated by the resonating properties of the supralaryngeal cavity or by changing respiratory pressure or tracheal tension.

2.8.2 Use of the Ultrasonic Range

Evolutionarily, animals next started to employ their vocalizations in new and innovative ways. For example, calls evolved to warn conspecifics about nearby predators and to locate pups that had wandered from the nest. Producing these calls in the ultrasonic range was adaptive for several reasons. First, most predators of rodents have poor auditory sensitivity at high frequencies. Using ultrasonic calls, therefore, enables rodents to use alarm signals without the risk of attracting the attention of predators (Wilson and Hare 2004). Second, high frequency sounds enable rodents to use binaural phase and intensity cues for sound localization. Additionally, the high directionality of ultrasonic sounds provides another cue to aid in localization by rodents. Although how the mechanisms for ultrasonic sound production evolved are not fully understood, the benefits of using ultrasonic vocalizations during communication are clear.

2.8.3 *Origin of Rodent Isolation Vocalizations*

Isolation calls from infants span the audible and ultrasonic frequency ranges, depending on the size and hearing abilities of the species. Since isolation calls are so widespread among rodents, it is possible that isolation calls were one of the first rodent vocalizations to evolve. Following the development of pup isolation calls, rodents likely took advantage of the physiological mechanisms to produce and perceive isolation calls, leading to the expansion of vocal signals to include adult vocalizations to coordinate social and sexual interactions. In mice, for example, sequences of short calls are used in mating rituals (Matsumoto and Okanoya 2016) in addition to other social interactions between mice of the same sex (Chabout et al. 2012, 2015).

2.8.4 *Origin of Rodent Alarm Vocalizations*

A second origin of the expansion of rodent vocal communication may have been alarm calls. Like isolation calls, alarm calls are ubiquitous in rodents and extend from the audible to ultrasonic frequency ranges. Production of alarm calls is likely critical to the survival of rodents, as they are subject to high levels of predation. The heightened state of arousal in rodents when a predator is nearby likely increases respiration, allowing rodents to take advantage of initially unintentional sounds created as a byproduct of arousal. Similar to laryngeal braking in rodent pups, it is possible that, despite animals not purposefully emitting these sounds originally, nearby listeners could still have used this information as a signal of an approaching predator and responded with evasive behavior.

2.8.5 *Evolution of Vocalizations*

Taking these potential origins of vocal behavior together puts forth a possible mechanism of rodent vocal evolution (Fig. 2.8). The evolution of vocalizations potentially occurred through three processes: ritualization, exploitation, and sexual

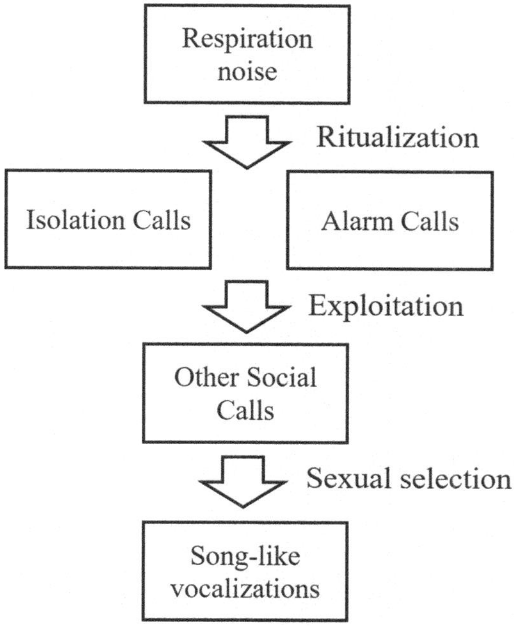

Fig. 2.8 A potential mechanism of rodent vocal evolution. Vocal signals are believed to originate from noises associated with respiration. These sounds were ritualized as isolation calls and alarm calls under specific selection pressures. By exploiting the neural substrates developed for isolate or alarm calls, other classes of calls, such as contact calls, emerged. These calls were used in succession by juveniles as affiliative signals. Adult males in a courtship context also utilized singing in succession

selection. Vocalizations may have originated as a byproduct of respiration. Sounds that were associated with hypothermia in isolated juveniles were high-pitched, reaching well into the ultrasonic frequency range. These respiration-induced sounds ultimately became ritualized as isolation calls, eliciting retrieval behavior from mothers. Additionally, sounds that were associated with predation-enhanced arousal were likely ritualized into alarm calls. Neural substrates that evolved that were associated with these ritualized vocalizations were exploited to develop more complex social calls, such as contact or signature calls. This likely occurred for affiliation calls in degus as well as for courtship songs in mice. Throughout this process, vocalizations were likely subjected to sexual selection pressures that increased the complexity of calls, both acoustically and sequentially.

2.9 Summary

The evolution of rodent vocalizations from the interaction of physical, social, and sexual factors has been complex, and there is much remaining to be discovered. From this interaction, rodents developed rich auditory communication systems, enabling them to coordinate social, sexual, and agonistic situations. Researchers believe several rodent species have evolved context-dependent and syntactically organized call sequences that may be similar to song production in birds, amphibians, and insects. Their vocal communication systems adapted to multiple habitats and situations by altering the spectrotemporal parameters of their calls, most notably frequency, and took advantage of ultrasonic frequency ranges, which avoided detection by predators and improved localizability. Because of the wide range of adaptive radiations in rodents, rodent vocalizations have come forward as a useful model to study the evolutionary forces and underlying neural mechanisms that have shaped the diversity of behavior.

Acknowledgments This work was supported by MEXT/JSPS KAKENHI Grant Number #4903, JP17H06380 to K.O. We thank Dr. Yui Matsumoto for drawing Fig. 2.5.

Compliance with Ethics Requirements Kazou Okanoya declares he has no conflict of interest.
Laurel A. Screven declares she has no conflict of interest.

References

Arriaga, G., & Jarvis, E. D. (2013). Mouse vocal communication system: Are ultrasounds learned or innate? *Brain and Language*, 124(1), 96–116.

Arriaga, G., Zhou, E. P., & Jarvis, E. D. (2012). Of mice, birds, and men: The mouse ultrasonic song system has some features similar to humans and song-learning birds. *PLoS One*, *7*(10), e46610.

Barfield, R. J., Auerbach, P, Geyer, L. A., & McIntosh, T. K. (1979). Ultrasonic vocalizations in rat sexual behavior. *American Zoologist*, 19(2), 469–480.

Barfield, R. J., & Geyer, L. A. (1972). Sexual behavior: Ultrasonic postejaculatory song of the male rat. *Science*, 176(4041), 1349–1350.

Bednářová, R., Hrouzková-Knotková, E., Burda, H., Sedláček, F., & Šumbera, R. (2013). Vocalizations of the giant mole-rat (*Fukomys mechowii*), a subterranean rodent with the richest vocal repertoire. *Bioacoustics*, 22(2), 87–107.

Behbehani, M. M. (1995). Functional characteristics of the midbrain periaqueductal gray. *Progress in Neurobiology*, 46(6), 575–605.

Berryman, J. C. (1976). Guinea pig vocalizations: Their structure, causation and function. *Ethology*, 41(1), 80–106.

Blanchard, R. J., Blanchard, D. C., Agullana, R., & Weiss, S. M. (1991). Twenty-two kHz alarm cries to presentation of a predator by laboratory rats living in visible burrow systems. *Physiology & Behavior*, 50(5), 967–972.

Blumburg, M. S., & Sokoloff, G. (2001). Do infant rats cry? *Psychological Review*, 108(1), 83–95.

Blumstein, D. T., & Armitage, K. B. (1997). Alarm calling in yellow-bellied marmots. I. The meaning of situationally specific calls. *Animal Behaviour*, 53(1), 143–171.

Blumstein, D. T., & Daniel, J. C. (2004). Yellow-bellied marmots discriminate among the alarm calls of individuals and are more responsive to the calls of juveniles. *Animal Behaviour*, 68(1), 1257–1265.

Blumstein, D. T., & Munos, O. (2004). Individual, age and sex-specific information is contained in yellow-bellied marmot alarm calls. *Animal Behaviour*, 69(1), 353–361.

Blumstein, D. T., & Recapet, C. (2009). The sound of arousal: The addition of novel nonlinearities increases responsiveness in marmot alarm calls. *Ethology*, 115(11), 1074–1081.

Bradbury, J. W., & Vehrencamp, S. L. (2012). *Principles of animal communication*. Sunderland, MA: Sinauer Associates, Inc.

Brudzynski, S. M. (2009). Communication of adult rats by ultrasonic vocalization: Biological, sociobiological, and neuroscience approaches. *Institute for Laboratory Animal Research Journal*, 50(1), 43–50.

Brudzynski, S. M. (2014). Social origin of vocal communication in rodents. In G. Witzany (Ed.), *Biocommunication of animals* (pp. 63–79). New York: Springer.

Brunelli, S. A., Shair, H. N., & Hofer, M. A. (1994). Hypothermic vocalizations of rat pups (*Rattus norvegicus*) elicit and direct maternal search behavior. *Journal of Comparative Psychology*, 108(3), 298–303.

Burda, H. (1995). Individual recognition and incest avoidance in eusocial common mole-rats rather than reproductive suppression by parents. *Cellular and Molecular Life Sciences*, 51(4), 411–413.

Burgdorf, J., Kroes, R. A., Moskal, J. R., Pfaus, J. G., et al. (2008). Ultrasonic vocalizations of rats (*Rattus norvegicus*) during mating, play, and aggression: Behavioral concomitants, relationship to reward, and self-administration playback. *Journal of Comparative Psychology*, 122(4), 357–367.

Carter, G. G., Skowronski, M. D., Faure, P. A., & Fenton, B. (2008). Antiphonal calling allows individual discrimination in white-winged vampire bats. *Animal Behaviour*, 76(4), 1343–1355.

Catchpole, C., & Slater, P. (2003). *Bird song: Biological themes and variations*. Cambridge, UK: Cambridge University Press.

Chabout, J., Serreau, P., Ey, E., Bellier, L., et al. (2012). Adult male mice emit context-specific ultrasonic vocalizations that are modulated by prior isolation or group rearing environment. *PLoS One*, 7(1), e29401.

Chabout, J., Sarkar, A., Dunson, D. B., & Jarvis, E. D. (2015). Male mice song syntax depends on social contexts and influences female preferences. *Frontiers in Behavioral Neuroscience*, 9(1), 76.

Deacon, T. W. (1998). *The symbolic species: The co-evolution of language and the brain*. New York: WW Norton & Company.

Di Paolo, E. A. (1997). An investigation into the evolution of communication. *Adaptive Behavior*, 6(2), 285–324.

Ehret, G. (2005). Infant rodent ultrasounds—a gate to the understanding of sound communication. *Behavior Genetics*, 35(1), 19–29.

Ehret, G., & Haack, B. (1981). Categorical perception of mouse pup ultrasound by lactating females. *Naturwissenschaften*, 68(4), 208–209.

Ehret, G., & Haack, B. (1984). Motivation and arousal influence sound-induced maternal pup-retrieving behavior in lactating house mouse. *Ethology*, 65(1), 25–39.

Endres, T., Widmann, K., & Fendt, M. (2007). Are rats predisposed to learn 22kHz calls as danger-predicting signals? *Behavioural Brain Research*, 185(2), 69–75.

Ficken, M. S., Ficken, R. W., & Witkin, S. R. (1978). Vocal repertoire of the black-capped chickadee. *The Auk*, 95(1), 34–48.

Gourbal, B. F., Barthelemy, M., Petit, G., & Gabrion, C. (2004). Spectrographic analysis of the ultrasonic vocalisations of adult male and female BALB/c mice. *Naturwissenschaften*, 91(8), 381–385.

Greene, E., & Meagher, T. (1998). Red squirrels, *Tamiasciurus hudsonicus*, produce predator-class specific alarm calls. *Animal Behaviour*, 55(3), 511–518.

Grimsley, J. M. S., Sheth, S. Vallabh, N., & Grimsley, C. A. (2016). Contextual modulation of vocal behavior in mouse: Newly identified "mid-frequency" vocalization emitted during restraint. *Frontiers in Behavioral Neuroscience*, 10(1), 38.

Hammerschmidt, K., Radyushkin, K., Ehrenreich, H., & Fischer, J. (2009). Female mice respond to male ultrasonic 'songs' with approach behavior. *Biology Letters*, 5(5), 589–592.

Hammerschmidt, K., Radyushkin, K., Ehrenreich, H., & Fischer, J. (2012a). The structure and usage of female and male mouse ultrasonic vocalizations reveal only minor differences. *PLoS One*, 7(7), e41133.

Hammerschmidt, K., Reisinger, E., Westekemper, K., Ehrenreich, L., et al. (2012b). Mice do not require auditory input for the normal development of their ultrasonic vocalizations. *BMC Neuroscience*, 13(1), 40. https://doi.org/10.1186/1471-2202-13-40

Hammerschmidt, K., Whelan, G., Eichele, G., & Fischer, J. (2015). Mice lacking the cerebral cortex develop normal song: Insights into the foundations of vocal learning. *Scientific Reports*, 5(1), 8808.

Hashimoto, H., Moritani, N., Aoki-Komori, S., Takana, M. & Saito, T. R. (2004). Comparison of ultrasonic vocalizations emitted by rodent pups. *Experimental Animals*, 53(5), 409–416.

Heth, G., Frankenberg, E., & Nevo, E. (1986). Adaptive optimal sound for vocal communication in tunnels of a subterranean mammal (*Spalax ehrenbergi*). *Experientia*, 42(11-12), 1287–1289.

Hoffmann, F., Mulsolf, K., & Penn, D. J. (2012). Spectrographic analyses reveal signals of individuality and kinship in the ultrasonic courtship vocalizations of wild house mice. *Physiology & Behavior*, 105(3), 766–771.

Hoier, S., Pfeifle, C., von Merten, S., & Linnenbrink, M. (2016). Communication at the garden fence: Context dependent vocalization in female house mice. *PLoS One*, 11(3), e0152255.

Holman, S. D. (1980). Sexually dimorphic, ultrasonic vocalizations of Mongolian gerbils. *Behavioral and Neural Biology*, 28(2), 183–192.

Holy, T. E., & Guo, Z. (2005). Ultrasonic songs of male mice. *PLoS Biology*, 3(12), e386.

Hunyady, H. (2008). *Vocal sounds of the chinchilla*. MS Thesis, Bowling Green State Univeristy, OH. https://etd.ohiolink.edu/!etd.send_file%3Faccession%3Dbgsu1206318183%26disposition%3Dinline (Accessed Feb. 2018)

Inoue, Y., Sinun, W., Yosida, S., & Okanoya, K. (2013). Intergroup and intragroup antiphonal songs in wild male Mueller's gibbons (*Hylobates muelleri*). *Interaction Studies*, 14(1), 24–43.

Jurgens, U. (1979). Neural control of vocalization in nonhuman primates. In H. D. Steklis & J. Raleigh (Eds.), *Neurobiology of social communication in primates: An evolutionary perspective* (pp. 11–44). London: Academic Press, Inc.

Jurgens, U. (2002). Neural pathways underlying vocal control. *Neuroscience & Biobehavioral Reviews*, 26(2), 235–258.

Kagawa, H., Seki, Y., & Okanoya, K. (2017). Affective valence of neurons in the vicinity of the rat amygdala: Single unit activity in response to a conditioned behavior and vocal sound playback. *Behavioural Brain Research*, 324(1), 109–114.

Kikusui, T., Nakanishi, K., Nakagawa, R., Nagasawa, M., et al. (2011). Cross fostering experiments suggest that mice songs are innate. *PLoS One*, *6*(3), e17721.

Kobayasi, K. I., & Riquimaroux, H. (2012). Classification of vocalizations in the Mongolian gerbil, *Meriones unguiculatus*. *The Journal of the Acoustical Society of America*, 131(2), 1622–1631.

Kondo, N., Watanabe, S., & Izawa, E-I. (2010). A temporal rule in vocal exchange among large-billed crows *Corvus macrorhynchos* in Japan. *Ornithological Science*, 9(1), 83–91.

Lange, S., Burda, H., Wegner, R. E., Dammann, P., et al. (2007). Living in a "stethoscope": Burrow-acoustics promote auditory specializations in subterranean rodents. *Naturwissenschaften*, 94(2), 134–138.

Litvin, Y., Blanchard, D. C., & Blanchard, R. J. (2007). Rat 22 kHz ultrasonic vocalizations as alarm cries. *Behavioural Brain Research*, 182(2), 166–172.

Liu, R. C., Miller, K. D., Merzenich, M. M., & Schreiner, C. E. (2003). Acoustic variability and distinguishability among mouse ultrasound vocalizations. *The Journal of the Acoustical Society of America*, 114(6), 3412–3422.

Long, C. (2007). Vocalisations of the degu *Octodon degus*, a social caviomorph rodent. *Bioacoustics*, 16(3), 223–244.

Mahrt, E. J., Agarwal, A., Perkel, D., Portfors, C., Elemans, C. P. H. (2016). Mice produce ultrasonic vocalizations by intra-laryngeal planar impinging jets. *Cell Biology*, 26(1), R880–881.

Mahrt, E. J., Perkel, D. J., Tong, L., Rubel, E. W., & Portfors, C. V. (2013). Engineered deafness reveals that mouse courtship vocalizations do not require auditory experience. *The Journal of Neuroscience*, 33(13), 5573–5583.

Marler, P., Evans, C. S., & Hauser, M. D. (1992). Animal signals: Motivational, referential, or both. In: H. Papoušek, U. Jürgens, & M. Papoušek (Eds.), *Nonverbal vocal communication: Comparative and developmental approaches* (pp. 66–86). Cambridge, UK: Cambridge University Press

Matsumoto, Y. K., (2017). *Types and functions of ultrasonic vocalizations in adult mice*. Ph. D. dissertation. The University of Tokyo, Tokyo, Japan.
Matsumoto, Y. K., & Okanoya, K. (2016). Phase-specific vocalizations of male mice at the initial encounter during the courtship sequence. *PLoS One*, 11(2), e0147102.
Matsumoto, Y. K., Okanoya, K., & Seki, Y. (2012). Effects of amygdala lesions on male mouse ultrasonic vocalizations and copulatory behaviour. *Neuroreport*, 23(11), 676–680.
Melchior, H. R. (1971). Characteristics of arctic ground squirrel alarm calls. *Oecologia*, 7(2), 184–190.
Mendl, M., Burman, O. H., Parker, R. M., & Paul, E. S. (2009). Cognitive bias as an indicator of animal emotion and welfare: Emerging evidence and underlying mechanisms. *Applied Animal Behaviour Science*, 118(3), 161–181.
Moreno-Gómez, F. N., Leon, A., Velásquez, N., & Delano, P. (2015). Individual and sex distinctiveness in bark calls of domestic chinchillas elicited in a distress context. *The Journal of the Acoustical Society of America*, 138(3), 1614–1622.
Morton, E. S. (1977). On the occurrence and significance of motivation-structural rules in some bird and mammal sounds. *The American Naturalist*, 111(981), 855–869.
Motomura, N., Shimizu, K., Shimizu, M., Aoki–Komori, S., et al. (2002). A comparative study of isolation-induced ultrasonic vocalization in rodent pups. *Experimental Animals*, 51(2), 187–190.
Muller, C., & Burda, H. (1989). Restricted hearing range in subterranean rodent, *Cryptomys hottentotus*. *Naturwissenschaften*, 76(3), 134–135.
Nakano, R., Nakagawa, R., Tokimoto, N., & Okanoya, K. (2013). Alarm call discrimination in a social rodent: Adult but not juvenile degu calls induce high vigilance. *Journal of Ethology*, 31(2), 115–121.
Neilans, E. G., Holfot, D. P., Radziwon, K. E., & Dent, M. L. (2014). Discrimination of ultrasonic vocalizations by CBA/CaJ mice (*Mus musculus*) is related to spectrotemporal dissimilarity of vocalizations. *PLoS One*, 9(1), e85405.
Neunuebel, J. P., Taylor, A. L., Arthur, B. J., & Egnor, S. E. R. (2015). Female mice ultrasonically interact with males during courtship displays. *eLife*, 4, e06203.
Owings, D. H., & Virginia, R. A. (1978). Alarm calls of California ground squirrels (*Spermophilus beecheyi*). *Ethology*, 46(1), 58–70.
Panksepp, J., & Burgdorf, J. (1999). Laughing rats? Playful tickling arouses high frequency ultrasonic chirping in young rodents. In S. R. Hameroff, A. W. Kaszniak, & D. J. Chalmers (Eds.), *Toward a science of consciousness III* (pp. 231–244). Cambridge: The MIT Press.
Panksepp, J. & Burgdorf, J. (2000). 50-kHz chirping (laughter?) in response to conditioned and unconditioned tickle-induced reward in rats: Effects of social housing and genetic variables. *Behavioural Brain Research*, 115(1), 25–38.
Panksepp, J., & Burgdorf, J. (2003). "Laughing" rats and the evolutionary antecedents of human joy? *Physiology & Behavior*, 79(3), 533–547.
Parsana, A. J., Li, N., & Brown, T. H. (2012). Positive and negative ultrasonic social signals elicit opposing firing patterns in rat amygdala. *Behavioural Brain Research*, 226(1), 77–86.
Pepper, J. W., Braude, S. H., Lacey, E. A., & Sherman, P. W. (1991). Vocalizations of the naked mole-rat. In P. W. Sherman, J. U. M. Jarvis, & R. D. Alexander (Eds.), *The biology of the naked mole-rat* (pp. 243–274). Princeton: Princeton University Press.
Perla, B. S., & Slobodchikoff, C. (2002). Habitat structure and alarm call dialects in Gunnison's prairie dog (*Cynomys gunnisoni*). *Behavioral Ecology*, 13(6), 844–850.
Pomerantz, S. M., Nunez, A. A., & Bean, N. J. (1983). Female behavior is affected by male ultrasonic vocalizations in house mice. *Physiology & Behavior*, 31(1), 91–96.
Portfors, C. V. (2007). Types and functions of ultrasonic vocalizations in laboratory rats and mice. *Journal of the American Association for Laboratory Animal Science*, 46(1), 28–34.
Randall, J. A., & Rogovin, K. A. (2002). Variation in and meaning of alarm calls in a social desert rodent. *Ethology*, 108(6), 513–527.

Randall, J. A., McCowan, B., Collins, K. C., Hooper, S. L., & Rogovin, K. (2005). Alarm signals of the great gerbil: Acoustic variation by predator context, sex, age, individual, and family group. *The Journal of the Acoustical Society of America*, 118(4), 2706–2714.

Reeve, H. K., Westneat, D. F., Noon, W. A., Sherman, P. W., & Aquadro, C. F. (1990). DNA "fingerprinting" reveals high levels of inbreeding in colonies of the eusocial naked mole-rat. *Proceedings of the National Academy of Sciences of the United States of America*, 87(7), 2496–2500.

Riede, T. (2013). Stereotypic laryngeal and respiratory motor patterns generate different call types in rat ultrasound vocalization. *Journal of Experimental Zoology Part A: Ecological Genetics and Physiology*, 319(4), 213–224.

Sadananda, M., Wöhr, M., & Schwarting, R. K. (2008). Playback of 22-kHz and 50-kHz ultrasonic vocalizations induces differental c-fos expression in rat brain. *Neuroscience Letters*, 435(1), 17–23.

Saito, Y., Yuki, S., Seki, Y., Kagawa, H., & Okanoya, K. (2016). Cognitive bias in rats evoked by ultrasonic vocalizations suggests emotional contagion. *Behavioural Processes*, 132(1), 5–11.

Sales, G. & Pye, D. (1974). *Ultrasonic communication by animals*. Netherlands: Springer-Verlag.

Seyfarth, R. M., Cheney, D. L., & Marler, P. (1980). Vervet monkey alarm calls: Semantic communication in a free-ranging primate. *Animal Behaviour*, 28(4), 1070–1094.

Sherman, P. W. (1977). Nepotism and the evolution of alarm calls. *Science*, 197(4310), 1246–1253.

Sherman, P. W., & Holmes, W. G. (1985). Kin recognition: Issues and evidence. In B. Holldobler & M. Lindauer (Eds.), *Experimental behavioral ecology and sociobiology* (pp. 437–460). Cambridge, UK: Cambridge University Press.

Smith, C. C. (1968). The adaptive nature of social organization in the genus of three squirrels *Tamiasciurus*. *Ecological Monographs*, 38(1), 31–64.

Smith, C. C. (1978). Structure and function of the vocalizations of tree squirrels (*Tamiasciurus*). *Journal of Mammalogy*, 59(4), 793–808.

Šuta, D., Popelář, J., Burianová, J., & Syka, J. (2013). Cortical representation of species-specific vocalizations in Guinea pig. *PLoS One*, 8(6), e65432.

Syka, J., Popelář, J., Kvašňák, E., Šuta, J., & Jilek, M. (1997). Processing of species-specific vocalizations in the inferior colliculus and medial geniculate body of the guinea pig. In J. Syka (Ed.), *Acoustical signal processing in the central auditory system* (pp. 431–441), Boston: Springer-Verlag.

Takeuchi, H., & Kawashima, S. (1986). Ultrasonic vocalizations and aggressive behavior in male rats. *Physiology & Behavior*, 38(4), 545–550.

Tinbergen, N., & Perdeck, A. C. (1951). On the stimulus situation releasing the begging response in the newly hatched herring gull chick (*Larus argentatus argentatus* Pont.). *Behaviour*, 3(1), 1–39.

Van der Poel, A. M., Noach, E. J. K., & Miczek, K. A. (1989). Temporal patterning of ultrasonic distress calls in the adult rat: Effects of morphine and benzodiazapines. *Psychopharmacology*, 97(2), 147–148.

von Merten, S., Hoier, S., Pfeifle, C., & Tautz, D. (2014). A role for ultrasonic vocalisation in social communication and divergence of natural populations of the house mouse (*Mus musculus domesticus*). *PLoS One*, 9(5), e97244.

Whitney, G., Cable, J. R., Stockton, M. D., & Tilson, E. F. (1973). Ultrasonic emissions: Do they facilitate courthip of mice? *Journal of Comparative and Physiological Psychology*, 84(3), 445–452.

Wilson, D. R., & Hare, J. F. (2004). Animal communication: Ground squirrel uses ultrasonic alarms. *Nature*, 430(6999), 523–523.

Winter, P., Ploog, D., & Latta, J. (1966). Vocal repertoire of the squirrel monkey (*Saimiri sciureus*), its analysis and significance. *Experimental Brain Research*, 1(4), 359–384.

Wright, J. M., Gourdon, J. C., & Clarke, P. B. S. (2010). Identification of multiple call categores within the rich repertoire of adult rat 50-kHz ultrasonic vocalizatons: Effects of amphetamine and social context. *Psychopharmacology*, 211(1), 1–13.

Wöhr, M., & Schwarting, R. K. (2007). Ultrasonic communication in rats: Can playback of 50-kHz calls induce approach behavior? *PLoS One*, 2(12), e1365. https://doi.org/10.1371/journal.pone.0001365

Wöhr, M., & Schwarting, R. K. (2008). Maternal care, isolation-induced infant ultrasonic calling, and their relations to adult anxiety-related behavior in the rat. *Behavioral Neuroscience*, 122(2), 310–330.

Wöhr, M., & Schwarting, R. K. (2013). Affective communication in rodents: Ultrasonic vocalizations as a tool for research on emotion. *Cell and Tissue Research*, 354(1), 81–97.

Yosida, S., Kobayasi, K. I., Ikebuchi, M., Ozaki, R., & Okanoya, K. (2007). Antiphonal vocalization of a subterranean rodent, the naked mole-rat (*Heterocephalus glaber*). *Ethology*, 113(7), 703–710.

Yosida, S., & Okanoya, K. (2009). Naked mole-rat is sensitive to social hierarchy encoded in antiphonal vocalization. *Ethology*, 115(9), 823–831.

Yuki, S., & Okanoya, K. (2014). Behavioral correlates of 50-kHz ultrasonic vocalizations in rats: Progressive operant discrimination learning reduces frequency modulation and increases overall amplitude. *Animal Behavior and Cognition*, 1(4), 452–463.

Zahavi, A. (1975). Mate selection—a selection for a handicap. *Journal of Theoretical Biology*, 53(1), 205–214.

Zhang, S., Davis, P. J., Bandler, R., & Carrive, P. (1994). Brain stem integration of vocalization: Role of the midbrain periaqueductal gray. *Journal of Neurophysiology*, 72(3), 1337–1356.

Chapter 3
Three Decades of Subterranean Acoustic Communication Studies

Cristian Schleich and Gabriel Francescoli

Abstract The subterranean environment has strongly influenced the evolution of the sensory biology of subterranean rodents. While dark and monotonous tunnels have led to reductions in the visual capabilities of some species, other senses appear to be highly developed in contrast. Among them, the emission of acoustic and seismic signals plays a major role in communication and alertness of subterranean mammals. In this chapter, the ecological and evolutionary conditions that influence the characteristics of vibrational communication in subterranean rodents are reviewed. First, the characteristics of rodents' burrows and how they dictate the methods used to study subterranean communication are discussed. Second, the properties and roles of vocalizations and seismic signals in subterranean species are examined, including the main hypotheses about the evolution of these signals. Third, what is understood about social and vocal complexity in subterranean rodents is summarized and the similarities and differences in the vocal repertoires of social and solitary groups are analyzed. The chapter ends with a short summary and a consideration of future challenges in the field of subterranean acoustic communication in rodents.

Keywords Bathyergids · Burrow acoustics · Ctenomyids · Geomyids · Seismic signals · Sociality · Sound transmission · Spalacines · Rodent vocalizations · Vocal complexity

C. Schleich (✉)
IIMyC-Conicet, Universidad Nacional de Mar del Plata, Mar del Plata, Argentina
e-mail: cschleic@mdp.edu.ar

G. Francescoli
Sección Etología, Facultad de Ciencias, Universidad de la República, Montevideo, Uruguay
e-mail: gabo@fcien.edu.uy

M. L. Dent et al. (eds.), *Rodent Bioacoustics*, Springer Handbook of Auditory Research 67, https://doi.org/10.1007/978-3-319-92495-3_3

3.1 Introduction

The vital role of acoustic communication in the life of solitary and social mammals has been addressed and enlightened by many behavioral, ecological, and evolutionary studies (e.g., Bradbury and Vehrencamp 1998; Owings and Morton 1998). Numerous situations, like mate choice, intrasexual competition, parental care, territorial defense, and predation avoidance, all rely upon information transfer in acoustic signals transmitted between senders and receivers. As a consequence, acoustic communication, which integrates a diverse and extensive range of disciplines from neurobiology and biophysics to evolution and psychology, has become a central focus of mammalogical researchers.

Distributed on almost all continents, more than 250 rodent species spend a considerable amount of their lives underground (Begall et al. 2007a). These species are classified as fossorial when they inhabit underground tunnels but display varied surface activity, or they are classified as strictly subterranean when most events (e.g., foraging, mating, and breeding) take place underground (Nevo 1999; Lacey et al. 2000). All of these species have been strongly influenced by the highly specialized and challenging underground environment (Burda et al. 2007). The physical conditions of the burrows are characterized by high humidity and low gas ventilation with high levels of CO_2 and low O_2. The animals are deprived of most sensory cues available above ground (Burda et al. 2007). This has led to morphological and physiological modifications in the sensory systems of the group, particularly in those species that carry out most of their activities inside their tunnels. As a consequence, the sensory biology of subterranean rodents became the focus of a growing number of studies that addressed diverse aspects of behavior, including olfaction, audition, vision, magnetoreception, and somatosensation (Begall et al. 2007a). Some of those studies became captivating readings for both scientific communities and the general public, for example, the ones concerning the regression of the visual systems of the blind (*Spalax ehrenbergi*) and naked mole rats (*Heterocephalus glaber*) (Cooper et al. 1993; Nemec et al. 2004) and those that demonstrated the exploitation of odorous substances released from plants in order to localize soil in which plants have been growing and to identify the palatability of particular plants (Heth et al. 2000, 2002; Lange et al. 2005, Schleich and Zenuto 2007). Also, the use of seismic or magnetic signals by the blind mole rat and Ansell's mole rat to orient in space (Burda et al. 1990a, b, 1991) added subterranean rodents to the fascinating groups of vertebrates that display clear seismic communication (e.g., elephants) and magnetoreception (e.g., turtles, migratory and homing birds) (Wiltschko and Wiltschko 1995).

The study of acoustic communication in subterranean rodents started with the seminal work on the vocal repertoire of the blind mole rat (Capranica et al. 1974). The field experienced a big expansion in the last three decades when several researchers from diverse places became immersed in the fascinating underground world. What were the reasons for that increase in acoustics research that continues today? In addition to the obvious particularities of the subterranean environment and the importance of studying acoustic communication as a useful tool to understand and elucidate general evolutionary principles, two main motives can be found behind most communication studies in this group of mammals. First, subterranean rodents are widespread and diverse in their life history characteristics, occupying habitats of dissimilar soil composition, quality, and productivity and exhibiting different commitments to their tunnels, ranging from strictly subterranean to species displaying some aboveground activity. For example, while Bathyergids (African mole rats) and blind mole rats inhabit self-constructed burrow systems completely isolated from the surface and feed almost exclusively on underground storage organs of plants (Nevo 1999), Ctenomyids (South American tuco-tucos) display daily aboveground activity when they emerge from burrow openings to feed on grasses and perennial forbs (Busch et al. 2000; del Valle et al. 2001).

In addition to their differential commitment to the underground environment, subterranean rodents occupy a wide range of habitats. In general, they occupy non-forest biomes such as grasslands, savannas, steppes, and deserts (Busch et al. 2000). These environments differ in soil hardness, humidity, ambient temperature, and food resources, resulting in ecological and demographic variation among subterranean species. This ecological diversity served as the starting point of many communication studies from an evolutionary perspective, particularly in four subterranean families: Spalacines (Eurasian blind mole rats), Geomyids (American pocket gophers), Bathyergids, and Ctenomyids.

A second reason for the increase in research on these rodents is because the social structure of these subterranean mammals represented an excellent opportunity for testing the *social complexity hypothesis* for vocal communication—the concept that increased vocal complexity can be found in groups with more complex social behavior (Blumstein and Armitage 1997). Subterranean rodents comprise solitary species like *Ctenomys talarum* (individuals occupy their own burrow, the female cares for her offspring alone, and young disperse after weaning) (Malizia et al. 1995) to eusocial species like the naked mole rat (juveniles never disperse and are lifelong nonbreeding helpers) (Brett 1991). Consequently, studying vocal and seismic communication in subterranean rodents has become a valuable tool for understanding the adaptive responses of these species to the limitations imposed by life in underground burrows.

This chapter is divided into three main sections. The first one (Sect. 3.2) describes the main physical and acoustical characteristics of subterranean burrows, analyzing how they influenced the methodologies used to record subterranean rodents. The second part (Sect. 3.3) addresses the physical characteristics of vocalizations and the behavioral contexts associated with the emission of them in solitary and social species of subterranean rodents. Also, the use of mechanical vibrations to transfer information between individuals is discussed in this section. The third part (Sect. 3.4) starts with a concise examination of what is understood about social and vocal complexity and that topic is followed by a comparison of the vocal repertoires in social and solitary species of subterranean rodents. Finally, the conclusion summarizes the main findings and future challenges in the field of subterranean acoustic communication. Overall, this chapter offers a broad overview of the major findings and theories in the field of subterranean acoustic communication to motivate new students and researchers to engage in the amazing underground world of these often ignored rodents.

3.2 Communication and Burrow Acoustics Studies

Since the first studies of acoustic communication in subterranean rodents (Capranica et al. 1974; Pepper et al. 1991), the rapid evolution of bioacoustics technologies offered researchers a broad spectrum of affordable and high quality recording equipment and signal processing and analysis tools (Blumstein et al. 2011). Despite this huge advance in acoustic recording, the complexity and extension of the burrow systems (Hickman 1990; Šumbera et al. 2012) has remained an insurmountable obstacle for obtaining complete vocal repertoires directly from the tunnels. As a consequence, most work on acoustic communication in subterranean rodents has depended on laboratory assays, and comparatively few studies have been able to record acoustic signals in the field.

Taking a brief look at the solitary species of subterranean rodents, in most of the cases, animals were directly recorded from terrariums where individuals were allowed to interact, and acoustic signals were obtained concomitantly with their associated behaviors (Francescoli 1999; Knotková et al. 2009). In other cases, and in an attempt to imitate the structure and social complexity of the species in the wild, semi-natural enclosures with various artificial burrows connected to a common space were used to obtain acoustic recordings (Schleich and Busch 2002a; Amaya et al. 2016). Captive situations that permit an approximation of the natural social structure of the target species are likely to minimize stress among individuals (Gannon and Sikes 2007). Therefore, these last approaches, although requiring longer recording periods to obtain a complete vocal profile of the studied species, permit the proper acclimatization of the animals to the experimental device and identification of the artificial burrow as their own territory. Both are relevant to obtain a more accurate behavioral and acoustic repertoire of the subterranean species.

Of the social species of subterranean rodents, several in the family Bathyergidae and one South American member of the Octodontidae family have been studied in detail. In most of the cases, the different colonies were housed in tunnel systems, provided with open plastic boxes, and were recorded using ad libitum observations of the colony or after the incursion of a predator or a foreign conspecific to the colony (Pepper et al. 1991; Dvořáková et al. 2016). Some of the colonies used in those studies had been established several years before the study (e.g., Bednářová et al. 2013). Taking into account the various effects of long-lasting housing on individual motivation, exploratory activity, and even the development of stereotypic behaviors (Gannon and Sikes 2007; Calisi and Bentley 2009), it is likely that individuals of old colonies tend to present behavioral profiles that do not match those of their wild counterparts, for example, narrower vocal repertoires and less frequent vocalizations. Therefore, it would be interesting to compare the vocal repertoires and nonvocal behaviors in colonies of social species at different times after transporting them to the biotheriums. This certainly would improve the understanding of the influence of captivity on the establishment and development of social interactions in general and on the vocal repertoires of subterranean rodents in particular.

Comparatively, laboratory studies outnumber the studies from researchers who have attempted to study some acoustic aspect of subterranean rodents in the field. Among field studies, it is possible to differentiate those studies that recorded rodents' vocalizations in the proximity of their burrows (Francescoli 2002; Amaya et al. 2016) and those that addressed the propagation of airborne sounds in the tunnels of subterranean species (Heth et al. 1986). As mentioned before, burrow systems of subterranean rodents display great variation in their architecture and composition at both the intra- and inter-specific levels, which could affect the propagation of airborne sounds through their tunnels.

Several factors, such as soil type, compaction, tunnel diameter, and number of tunnel branches, may have a strong influence on the physical characteristics of sounds that carry information inside burrows. However, and despite this huge diversity in their structure and composition, few studies have examined the propagation of airborne sounds in burrows of subterranean rodents. For example, Heth and colleagues (1986) analyzed the transmission of synthetic calls of various frequencies through natural tunnels of the blind mole rat to determine if the ones that propagated farther matched with the frequencies of that species' vocalizations. Courtship calls of blind mole rats range from 500 Hz to 4500 Hz, with main frequencies around 500 Hz. Researchers found that sounds of 440 Hz propagated better than lower or higher frequency sounds in burrows of this subterranean species, although transmission was effective only over short distances (not more than 5 m).

A similar methodology was used by Lange and colleagues (2007) to test the propagation of sounds of diverse frequencies (at one intensity) in natural burrows of two species of Zambian mole rats (*Fukomys mechowii* and *F. kafuensis*) whose burrows have different diameters. Interestingly, they found not only that low-frequency sounds emitted inside the burrows were attenuated less than high frequency ones, but they also found that their amplitude was naturally increased—a "stethoscope effect" (Lange et al. 2007). Also, sound propagation was quite similar among burrows of different diameters, depths, and soil characteristics.

Another study related emission efficiency to the structural characteristics of the burrows' entrances, which tuco-tucos (*Ctenomys talarum*) utilize when they vocalize. Schleich and Antenucci (2009) analyzed the effects of the burrow entrance's structure on the sound emission efficiency by producing different sounds (similar to the territorial vocalization of this species) in a tunnel with a comparable configuration to the natural burrows. Similar to the study described previously, low-frequency components of the artificial signal were not only less attenuated but also amplified when played back inside the tunnel. Low frequency components of the signal (400 Hz) were recorded 15 m away from the burrow where the signal had been broadcast. Therefore, despite the many factors that could affect the transmission of sounds inside subterranean rodents' burrows, the acoustic characteristics of the tunnels are similar among diverse subterranean species, opening the possibility of convergent adaptation in the acoustic biology of subterranean rodents.

Despite the severe limitations imposed by the subterranean habitat, it is clear that useful and valuable information can be obtained from field studies. Therefore, experiments in both lab and field must be encouraged to fully disentangle and understand the complexity of acoustic communication in solitary and social subterranean rodents.

3.3 Acoustic Signals

Subterranean rodents produce diverse types of acoustic signals for a variety of purposes, with vocalizations present in all species studied thus far (Schleich et al. 2007; Dvořáková et al. 2016). Some species also utilize the seismic channel for communication underground, a form of communication that occurs in various taxa ranging from spiders and insects to mammals (Randall 2014). This section describes and discusses the biological significance of both vocal and seismic signals in subterranean rodents. In particular, the types and roles of vocalizations in four different categories will be discussed: (1) agonistic vocalizations, (2) mating and reproductive vocalizations, (3) distress and contact calls, and (4) parent-offspring interactions. The characteristics of signal production and modalities of reception of seismic signals in subterranean rodents also will be examined.

3.3.1 Vocalizations

3.3.1.1 Agonistic Calls

Solitary tuco-tucos of the species *C. talarum*, *C. pearsoni*, and *Ctenomys sp.* (Anillaco tuco-tuco, presumably solitary but still to be confirmed) frequently emit a distinctive vocalization that gave the common name to the animal (tuco-tucos). This vocalization is characterized by high intensity and long duration, and most of the notes' energy is concentrated in the low-frequency range. The tuco-tuco vocalization is associated with aggressive spacing behaviors, suggesting its possible role as a territorial signal (Fig. 3.1; Table 3.1) (Schleich and Busch 2002a; Amaya et al. 2016). Males of *C. pearsoni* and *Ctenomys* sp. emit vocalizations more frequently than females, but females of *C. pearsoni* vocalize more in the reproductive season, suggesting that this call, in addition to informing the receiver about the emitter's location and sex, may also indicate reproductive condition (Francescoli 2011). Another agonistic vocalization present in species of this genus is a grunt, which is a harsh low-frequency sound that is emitted by mature individuals of both sexes during aggressive encounters (Francescoli 1999; Schleich and Busch 2002a).

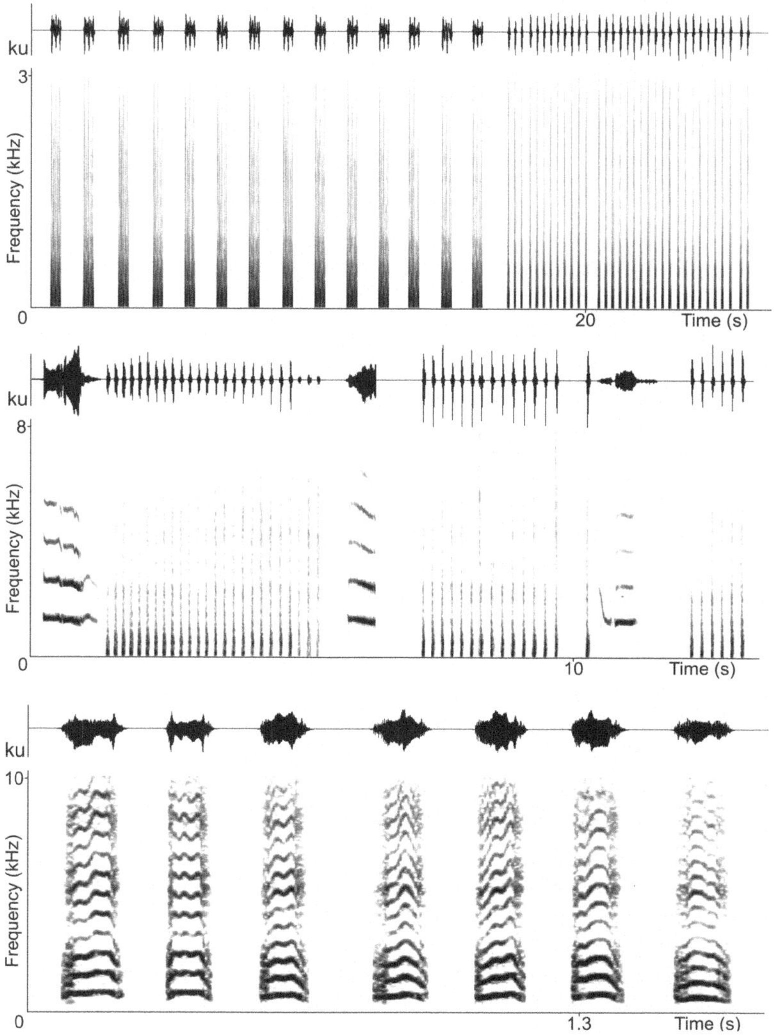

Fig. 3.1 Examples of envelope curves (*upper display of each pair*) and sonograms (*lower display of each pair*) used to analyze vocalizations. In this case, territorial (*top pair*), mating (*middle pair*), and juvenile vocalizations (*bottom pair*) of the Anillaco tuco-tuco (*Ctenomys* sp.) are shown. Note different time scales for each pair. *ku*, relative amplitude. (All sonograms provided by J. Amaya)

Agonistic calls are also found in the solitary silvery mole rat (*H. argenteocinereus*). Males emit low clucks during encounters with conspecifics, and individuals of both sexes emit hisses during aggressive encounters (Knotková et al. 2009). These vocalizations are also characterized by main frequencies in the low-frequency range. Interestingly, no agonistic calls are present in the subterranean Baird's pocket

Table 3.1 Vocal repertoire of solitary species of subterranean rodents. The range of the dominant frequency (kHz) of each category is indicated in parentheses

	Solitary species				
	Ctenomys talarum[1]	*Ctenomys pearsoni*[2]	*Ctenomys sp.*[3]	*Heliophobius argenteocinereus*[4]	*Geomys breviceps*[5]
Agonistic vocalizations (kHz)	Tuc-tuc (0.2–0.4) Grunts (0.2–0.4)	S-signal (0.2–0.3) G-signal (0.5–0.6)	Tuc-tuc (0.1–0.3)	Low-cluck (0.5–1.8) Hiss (0.3–1.1)	
Mating and reproductive vocalizations (kHz)	Male call (0.2–0.4) Female call (0.4–1)	C-signal (1–1.5)	Male call I (0.2–0.9) Male call II (0.3–2.6)	Female call (1.1–2.5) High cluck (0.7–2.1) Gabbling (0.9–3.2)	Purrs (1.1–2.5)
Distress vocalizations (kHz)				Squeals (1.5–3.4) Squeaks (1.9–2.9) Scream (1.5–3.3)	Squeaks (1.3–1.9)

[1]Schleich and Busch (2002a)
[2]Francescoli (1999)
[3]Amaya et al. (2016)
[4]Knotková et al. (2009)
[5]Devries and Sikes (2008)

gopher (*Geomys breviceps*) (Devries and Sikes 2008), even though this kind of vocalization is common in the repertoire of most subterranean rodents.

Much more is known about vocalizations in social than in solitary subterranean rodents (Table 3.2). Pepper and colleagues (1991) recorded low- and middle-frequency hisses in the naked mole rat during colony defense against predators or unfamiliar conspecifics. Moreover, during colony defense against predators, individuals of this species emit abrupt grunts. Aggressive encounters with conspecifics are also accompanied by upsweep trills. Finally, loud chirps, harsh calls, and noisy calls, are all produced during mild conflicts between colony mates.

Another social bathyergid, *Fukomys anselli*, produces six different agonistic vocalizations (Credner et al. 1997). These calls, mainly emitted by the dominant attacking animal, were classified as whistles, trills, hisses, and grunts. In the giant mole rat, *Fukomys mechowii*, three different vocalizations (high trill, swing trill, and scream) were produced during agonistic interactions, usually by the dominant animal (Bednářová et al. 2013). While the high trill was emitted at the start of contact by the dominant individual, the swing trill and screams appeared later, only if the interaction escalated.

While the Micklem's mole rat (*Fukomys micklemi*) produces only one agonistic vocalization, the grunt (Vanden Hole et al. 2014), the Mashona mole rat (*Fukomys*

Table 3.2 Vocal repertoire of social species of subterranean rodents. The range of the dominant frequency (kHz) of each category is indicated in parentheses

	Social species					
	Heterocephalusglaber[1]	*Fukomys anselli*[2]	*Fukomys mechowi*[3]	*Fukomys micklemi*[4]	*Fukomys darlingi*[5]	*Spalacopus cyanus*[6]
Agonistic vocalizations (kHz)	Hiss (0.4–1.3) Grunt Upsweep trill (1–9) Loud chirp (8–14)	Whistle (0.5–0.6) Trill (1.6–2.5) Trill II (2–3) Hiss (8–14) Grunt (1–2; 3) Grunt II (0.5–0.8; 2; 6)	High trill (0.4–1.2) Swing trill (0.6–1) Scream (1–3.6)	Grunt (0.3–0.4; 0.7–0.8)	Whistle (3.6–4.8) Squeak (3.2–5.6)	Cluck I (0.56–1.03) Cluck II (0.34–0.52)
Mating and reproductive vocalizations (kHz)	V-trill (2–4)	Cluck (1.6–2.5) Shriek (1.6–2.5) Cry[7] (0.6–0.9; 1.6–2.5; 4–8; 12)	Cluck (0.4–0.5) Shriek (0.4–0.9) Harsh (0.5–2.5)	Female cluck (1.5–2) Female shriek (1.5–2.5)	Cluck (0.6–1.9) Shriek (0.5–1.3)	Creaking (0.17–0.47) Scream (1.55)
Distress vocalizations (kHz)		Loud calls (1.25–16)	Cry (3–6)	Loud call (2–4.5) High call (5–7) Loud squeak (5–7) Chevron (1–1.5) Squeal (2.5–4) Adult chirp (1.5–3)	Squeal (2–5.3) Harsh call (2–4.9) Cry (3.7–5.4)	Cluck III (0.34–0.69)

(continued)

Table 3.2 (continued)

	Social species					
	Heterocephalusglaber[1]	*Fukomys anselli*[2]	*Fukomys mechowi*[3]	*Fukomys micklemi*[4]	*Fukomys darlingi*[5]	*Spalacopus cyanus*[6]
Contact vocalizations (kHz)	Soft chirp (2–4)	Grunt (5) Twitter (0.5–3.0)	Twitter (3.0–4.2) Twitter like (0.9–3.0) Long twitter (0.4–2.6) Gabbling (0.5–1.3) Squeak (0.3–2.0) Grunt (0.3–1.0)	Loud call (2.0–4.5) Soft call (2.0–4.5) High call (5–7) Soft squeak (~ 6–7)	Cheep I (3.5–5.0) Cheep II (1.9–4.6) Twitter (1.5–3.6)	Cooing (0.34–0.52) Twitter (1.03–4.31) Twitter II (1.55) Squeak (0.34) (0.34)

[1]Pepper et al. (1991)
[2]Credner et al. (1997)
[3]Bednářová et al. (2013)
[4]Vanden Hole et al. (2014)
[5]Dvořáková et al. (2016)
[6]Veitl et al. (2000)
[7]This cry has three equally intense frequency ranges

darlingi) exhibits two aggressive calls, the whistle and the squeak (Dvořáková et al. 2016). Both tonal vocalizations are emitted by the aggressive individual.

Finally, coruros (*Spalacopus cyanus*) emit aggressive vocalizations during and after encounters between unfamiliar individuals. Fights are accompanied by the emission of the cluck I, a brief sound with various harmonics and wide frequency range. Before and after fights, the individuals produced the cluck II (Veitl et al. 2000).

3.3.1.2 Mating and Reproductive Vocalizations

Both sexes of *C. talarum* emit vocalizations during courtship and mating (Schleich and Busch 2002a). Males produce a low-intensity sound during the first phases of courtship and copulation. The other mating vocalization is emitted by the female and consists of a soft sound similar to a cry or groan. Three different types of female mating vocalizations can be identified in this species. The vocalizations vary in the number of multiple bands and in the amount of frequency modulation, although no clear pattern of emission could be identified (Schleich and Busch 2002a). Females of the other solitary tuco-tuco studied, *C. pearsoni*, also exhibit vocalizations during sexual encounters. In this species, five different mating signals were recorded, differing in their spectral characteristics. Females emit the different types of C-signal notes during courtship, and there is no successful mount and penetration without their emission (Francescoli 1999). No male calls are present in this species during courtship and copulation (Francescoli 1999). However, males of the Anillaco tuco-tuco emit two different mating vocalizations (Amaya 2017).

The solitary blind mole rat also produces a courtship call (Nevo et al. 1987). It is mainly emitted by males and is characterized by low frequencies and a wide frequency range. Interestingly, genetically different populations of this species exhibit variations in the physical structure of their courtship call, which may act as a reproductive isolation mechanism (Nevo et al. 1987).

Individual Baird's pocket gophers produce two different sounds during mating, purrs and squeaks, although there is no clear or distinctive behavioral context associated with their emissions. The purrs, characterized by a fundamental frequency with a saw-toothed pattern, are emitted by male pocket gophers during copulation attempts (DeVries and Sikes 2008). Also, females produce squeaks during close-contact interactions between sexes. Although both vocalizations are emitted in sexual encounters, they are also recorded in other behavioral contexts (DeVries and Sikes 2008).

Silvery mole rat males produce high-frequency clucks during the first phases of courtship and mating (Knotková et al. 2009). Also, if mating is prolonged, males emit short sounds named "gabbling". Females of this species emit chirps, tonal calls, and highly modulated sounds when they are ready for mating and clucks at the end of courtship (Table 3.1) (Knotková et al. 2009).

Mating calls are understandably also present in social species. Receptive breeding naked mole rat females produce a distinctive sound (the V-trill) when soliciting copulations (Pepper et al. 1991). Courtship and mating are frequently complemented by intense vocalizations in *F. anselli* (Credner et al. 1997). During the initial phases, males emit a soft sound called the cluck. Meanwhile, females usually produce a call (shriek) with a wide frequency range. Finally, the cry, apparently emitted by females, accompanies the end of copulation (Credner et al. 1997).

Three different mating vocalizations are recorded from the giant mole rat: cluck, shriek, and the harsh (Bednářová et al. 2013). The cluck is a soft sound emitted by a female when a male sniffs at her anogenital region. This behavioral context is similar for the shriek, an intense call with longer duration. Finally, the harsh, a short and loud sound, accompanies contact behaviors and copulation (Bednářová et al. 2013). Females of the social Micklem's mole rat produce two mating sounds during pre-mating behavior, the female cluck (a short, low intensity sound) and the shriek (Vanden Hole et al. 2014). The Mashona mole rat generates two mating vocalizations: the cluck and the shriek (Dvořáková et al. 2016). The cluck by this species is a short, low-frequency sound, emitted by the female during courtship while the animals are sniffing each other's anogenital area. The shriek is produced in the last part of the courtship, when the male tries to jump on the female. Finally, coruro males produce creaking sounds before, during, and after copulation (Table 3.2) (Veitl et al. 2000). Also, screams are emitted at the end of the copulation, although the identity of the sender (the male or the female) producing these screams is unknown.

3.3.1.3 Distress and Contact Calls

In solitary tuco-tucos, grunts are also emitted during stressful situations, and therefore can also be considered as distress calls, apart from their classification as agonistic calls (Francescoli 1999; Schleich and Busch 2002a). Silvery mole rats produce cheeps at the beginning of an encounter or during fights by the submissive individual. These brief sounds are characterized by a broad frequency range (Knotková et al. 2009). This solitary mole rat also emits harsh and longer sounds (squeals) during close contacts. Females produce squeaks during encounters that frequently end in a fight. Finally, individuals under attack produce screams when fights intensify (Knotková et al. 2009).

Due to their way of life, contact calls generally are not described in solitary species of subterranean rodents. Only one exception can be mentioned here, the purr recorded in the pocket gopher, which, in addition to its role during courtship, may also function as a contact call (Devries and Sikes 2008).

Fukomys anselli produce loud calls under stressful situations. This sound, also interpreted as an appeasement call, contains a broad frequency range (Credner et al. 1997). Loud and brief cries are emitted by adult giant mole rats during competition for food and when one animal restricts the movement of the other (Bednářová et al. 2013). Several distress calls have been recorded in the Micklem's mole rat: loud call, high call, loud squeak, chevron, squeal, and the adult chirp (Vanden Hole et al. 2014). The loud and high calls are mostly heard during contact between unfamiliar animals. The loud squeak, a high intense broadband sound, is used in combination with loud calls during vocalized hopping. The chevron call is recorded in situations of disturbance. The squeal is produced during unfamiliar encounters with other individuals, often while sniffing. Finally, the adult chirp, recorded only a few times, is produced during conflicts involving food when encountering unknown conspecifics (Vanden Hole et al. 2014).

Three different distress vocalizations are present in the Mashona mole rat. Defending animals emit squeals or harsh atonal calls during encounters with unfamiliar individuals. Lastly, animals emit very loud and high tonal sounds, called cries, as a reaction to biting or when restricted in their movement (Dvořáková et al. 2016). Coruros also produce distress calls, described as clucks, when confronted with unfamiliar conspecifics. Potentially dangerous situations are accompanied by loud squeals (Veitl et al. 2000).

Contact vocalizations are extremely frequent and numerous in social species. The most frequent call in the naked mole rat is the soft chirp, which is emitted upon physical contact with colony mates (Pepper et al. 1991). Further studies on this vocalization showed that naked mole rats can identify and recognize the social status of other group members by means of this call (Yosida et al. 2007; Yosida and Okanoya 2009). Two different contact calls are found during friendly encounters of *F. anselli*: the grunt and the twitter. The former is also recorded from the breeding pair during greeting, and the twitter is produced during meetings of an adult animal with an unfamiliar juvenile (Credner et al. 1997). Several contact calls are described in the giant mole rat: the twitter, twitter-like call, long twitter, gabbling, squeak, and the grunt. The first three, classified as cheeping sounds, are emitted by a female lying in the nest when another animal enters. Gabbling, produced during welcoming rituals, is characterized by a short duration and the presence of various harmonics. It is generally followed by the squeak, which has relatively similar structure. Finally, the grunt, a loud contact sound, is usually followed by squeaks or gabbling (Bednářová et al. 2013).

The Micklem's mole rat produces four contact vocalizations: the loud call, soft call, high call, and soft squeak (Vanden Hole et al. 2014). The most frequent vocalizations are the loud and soft calls. The loud call is common during contact between unfamiliar animals, while the soft call, a very brief and soft sound, is often recorded during familiar contacts, such as sniffing, passing over, and allogrooming. High calls are usually produced during contacts with unfamiliar individuals. Encounters between both familiar and unfamiliar animals are accompanied by soft squeaks (Vanden Hole et al. 2014).

Two different cheeps were classified as contact calls in the Mashona mole rat (Dvořáková et al. 2016). Both are generated by females when approached by a male. The last contact call of this social species is the twitter, which was recorded when mole rats passed each other in tunnels (Dvořáková et al. 2016).

Coruros also display a variety of contact vocalizations (Veitl et al. 2000). Cooing calls are emitted when individuals engage in naso-nasal contact. During close naso-anogenital contact between two adults, a brief and broadband sound, the twitter, is produced. A similar sound (twitter II) is produced during encounters between individuals from two colonies. Squeaks, characterized by a longer duration, are emitted by two familiar male coruros of the same colony when meeting each other after a period of separation (Veitl et al. 2000).

As described in this section, contact or friendly vocalizations constitute a major component of the vocal repertoires of social species (Table 3.2). These calls, not observed in solitary species, may function to enhance group cohesion, although their differential development in social species with diverse familiar complexity (communal, social, or eusocial species) is still under debate (Sect. 3.4).

3.3.1.4 Juvenile Calls

Although probably existent in most species of subterranean rodents, juvenile vocalizations have only been described in the solitary species *C. pearsoni* (Francescoli 2001) and *C. talarum* (Schleich and Busch 2002b). They have also been described in the social naked mole rats (Pepper et al. 1991), *F. anselli* (Credner et al. 1997), coruros (Veitl et al. 2000), and Micklem's mole rats (Vanden Hole et al. 2014). Even if in some cases the context associated with the emission of these calls was dissimilar, the vocalizations were produced mostly during solicitation behaviors when the parents responded by providing resources to their offspring (Schleich et al. 2007). Schleich and Busch (2002b, 2004) studied diverse aspects of these vocalizations in *C. talarum* to obtain deeper knowledge about the functional significance of these juvenile vocal signals to their parents. To be considered as honest signals of offspring condition (that is, truly reflecting the offspring's need), begging calls must meet the predictions proposed by theoretical models (Godfray 1995; Kilner and Johnstone 1997). The first prediction states that the intensity of begging must reflect the individual's true need, and begging intensity is negatively correlated with offspring condition. In the case of tuco-tucos, the time that the pups spent begging during isolation decreased concomitantly with the development of thermoregulatory capacity and independent walking and eating, suggesting that begging intensity in C. *talarum* pups may reflect physical condition (Schleich and Busch 2002b).

The second prediction states that producing the signal (begging call) entails some cost, a situation not observed in this species, for which no significant energetic costs were associated with the emission of begging calls (Schleich and Busch 2004). These contradictory results reveal the complexity of the parent-offspring relationship and the difficulty of establishing a clear function for these juvenile vocalizations in subterranean rodents. More theoretical and experimental studies in other species of subterranean rodents will provide valuable knowledge to elucidate the function of these begging calls.

3.3.1.5 Vocalizations Summary

Two points become evident from these brief looks at subterranean vocalizations. First is the association between social organization and the size and diversity of the vocal repertoire, which is discussed further in Sect. 3.4. Second are the spectral characteristics of subterranean rodents' calls. Most of the vocalizations in subterranean species studied to date are shifted toward the mid- to low-frequency range relative to vocalizations of other rodents (Okanoya and Screven, Chap. 2). This feature leads to the elaboration of the hypothesis that the frequency range of subterranean rodent calls may be the result of convergent adaptation to the underground environment, where only low frequency sounds are able to propagate over large distances (Sect. 3.2). The acoustic characteristics of subterranean rodents' vocalizations were also found to be relatively coincident with the hearing capabilities of species studied thus far. Their hearing is characterized by a low-frequency peak and an overall low sensitivity (Begall et al. 2007b; Dent, Screven, and Kobrina, Chap. 4). Also, the absence of high-frequency vocalizations and ultrasonic calls in subterranean species, which do occur in surface-dwelling rodents, has been proposed as another feature supporting the convergent evolution of the vocal and auditory systems in subterranean rodents (Begall et al. 2007b). This last feature also deserves more attention in future studies, since to date no research on the vocal apparatus of subterranean rodents has been done. It is known that the fundamental frequency of vocal sounds is determined primarily by the anatomical length of the vocal folds, but modifying the air pressure in the trachea, as well as the length, stiffness, and tension of the vocal folds also influences sound production (Taylor et al. 2016; Kubke and Wild, Chap. 6). Therefore, understanding the full range and complexity of the mechanisms of vocal production in these species will serve as another test of the hypothesis of acoustic convergence in this group.

In addition, it is evident that a more extensive comparative approach, including more subterranean species and their phylogenetically close nonsubterranean relatives, is needed to fully support this evolutionary hypothesis and to obtain additional knowledge of the convergent and divergent characteristics of the acoustic and auditory systems of this underground group.

3.3.2 Seismic Communication

Producing vibrations by hitting the ground with feet, arms, legs, head, and even the chest is a common behavior in mammals (Randall 2010, 2014). Subterranean rodents are not the exception, and the use of seismic signals has been fully described in three species, the blind mole rat (Heth et al. 1987, 1991), the Cape mole rat, *Georychus capensis* (Narins et al. 1992), and *Tachyoryctes* sp., another solitary spalacid (Hrouzková et al. 2013). The production of substrate vibrations has also been observed in the social giant mole rat (Bednářová et al. 2013). Initial hypotheses suggested that the use of seismic signals to communicate was linked to the social system, with solitary species utilizing seismic waves for long distance intraspecific communication (Burda et al. 1990a; Narins et al. 1992). However, the recent description of vibration signals in a social species shows that the evolution of seismic communication is still a complex and unresolved subject.

The first evidence of the use of seismic signals to communicate in subterranean rodents came from the blind mole rat (Heth et al. 1987; Rado et al. 1987). This species generates vibration signals by knocking the anterodorsal surface of its head against the tunnel ceiling in a series of temporally patterned short bursts, a behavior supposedly derived from digging activities. This knocking produces a low frequency signal (100–250 Hz) that propagates through the substrate further than airborne signals, thereby providing an effective long distance communication modality between the solitary and territorial blind mole rats (Nevo et al. 1991). In the laboratory, seismic signals were associated with aggressive encounters and defense of artificial territories, providing support for their role as territorial signals. In addition to their function in interburrow communication, these seismic signals were also considered as an important premating isolating mechanism between populations (Heth et al. 1987) because of the structural variations (number of thumps in a pulse, length of pulses, and thumping frequency) among genetically distinct populations of the blind mole rat.

Although not involved in intraspecific communication, Kimchi and colleagues (2005) found that blind mole rats use reflected vibrations produced by knocking the head to detect objects in a type of "echolocation". Initial studies demonstrated that blind mole rats have the ability to avoid obstacles by digging accurate and energy-conserving bypass tunnels (Kimchi and Terkel 2003), which stimulated an investigation into the production of seismic signals during these behaviors. Interestingly, blind mole rats produce significantly more seismic waves while burrowing a bypass tunnel around an obstacle than while digging a straight tunnel, supporting the role of these signals in estimating the type and position of underground obstacles. In this case, low-frequency seismic signals produced during excavation differed from the ones used in intraspecific communication in the emission rate and frequency range (250–300 Hz). The perception of these seismic waves would be through the mole's paws, which contain 15–20 lamellate corpuscle mechanoreceptors per foot, without clear differences between the fore and hind feet. This subterranean echolocation system may serve to determine the most energy-conserving strategy for tunnel construction by the blind mole rat.

While the use of seismic signals for intraspecific communication in the blind mole rat was extensively studied and confirmed, the detection mechanism for these signals provoked much debate. Initially, Rado and colleagues (1989) suggested that this species uses its auditory system to detect seismic signals through a form of bone conduction by contact of the mandible with the tympanic bulla. This pathway was refuted by Nevo and colleagues (1991), who demonstrated that the mechanism of seismic reception was somatosensory and independent of the auditory mechanism. This conclusion was based on laboratory recordings, which showed that masking airborne auditory cues (produced by tapping) with white noise did not affect the evoked potentials recorded from the brain of a blind mole rat. Moreover, tapping on a tube separated from the one that the animal was in generated smaller evoked potentials. Later, Rado and colleagues (1998) demonstrated that bilateral deafening of the mole rat or high-intensity masking noise almost completely eliminated these brain stem and middle latency responses. Therefore, they concluded that blind mole rats received seismic vibrations by pressing their lower jaw against the burrows' wall, and they processed the information through the auditory system (jaw listening). Although the relative participation of both somatosensory and auditory routes for seismic sensitivity in *Spalax* is still uncertain (Mason and Narins 2010), the frequent jaw-listening behavior observed in this subterranean rodent may provide more support for the auditory processing of seismic signals.

The solitary cape mole rat creates sexually dimorphic patterns of drumming using their hind feet (Narins et al. 1992). Although similar in both sexes, foot drumming varies during the breeding season when males increase the delivery pattern compared to females (Bennett and Jarvis 1988). In addition to demonstrating that female cape mole rats responded to playback of mechanically generated thumps, Narins and colleagues (1992) showed that the airborne component of the generated artificial signal was severely attenuated by its passage through the soil, while the seismic component suffered less degradation, demonstrating the inadequacy of airborne sound as a medium for interburrow information transmission. Taken together, these results supported the existence of seismic communication between individuals of the cape mole rat. Regarding the mechanisms of seismic sensitivity, little is known about the existence of vibration-sensitive somatosensory receptors in this or other bathyergids (Mason and Narins 2010). Only some anecdotal comments suggested that these subterranean rodents potentially use their incisors for acoustical vibration reception (Poduschka 1978; Mason and Narins 2010).

Recently, the presence of substrate-borne vibrations was described in a subterranean mole rat of the genus *Tachyoryctes* (Hrouzková et al. 2013). This solitary member of the family Spalacidae produces substrate-borne vibrations by striking the head against the tunnel roof in a similar manner to the blind mole rat. Two different seismic signals that varied in emission rates were recorded in this species. The slow signal was observed in situations when the individuals were disturbed, suggesting that it may serve to decrease conflict between neighbors. The fast signal was produced during contact with conspecifics, suggesting a role in intraspecific communication. Also, the fast signals may encode information about the identity of the sender, making those signals suitable for spacing behavior and territorial advertisement (Hrouzková et al. 2013).

Mason and colleagues (2010), using a combination of dissection and computed tomography, analyzed the ear region of *Tachyoryctes* and two other spalacids (*Spalax*, *Eospalax*) and found three different, but not mutually exclusive, alternative routes for the transmission of bone-conducted vibrations to the inner ear in these species. One involved a direct transference of vibrations from the mandible to the bulla and then to the incus and malleus, resulting in the movement of the stapes within the oval window. The second implicated the transmission of vibrations directly from the mandibular condyle to the bony external auditory meatus, resulting in the reception of sound by the tympanic membrane and transmission to the inner ear. The last route, which is more likely to be present in *Spalax* and *Eospalax* than in *Tachyoryctes*, is transmission of vibrations through a fluid pathway not involving the middle ear. Although yet to be proven, these routes may be responsible for the observed seismic sensitivity in these species.

Although the production of vibratory signals in social subterranean rodents was suggested in *Fukomys damarensis* for the coordination of mating behavior (Jarvis and Bennett 1991), the use of this kind of signal during aggressive encounters has been described fully only in the social giant mole rat (Bednářová et al. 2013). Males of this species hit their chests against the bottom of their tunnels during violent encounters with unfamiliar males. Females also produced this chest beating during feeding but less frequently than males. Although its role in intraspecific communication has yet to be proven, the behavior that accompanied the production of these seismic signals supports its role in the transfer of information among individuals of this species.

One notable outcome from these works is that the ways of producing seismic signals vary among different species, suggesting that seismic communication has evolved independently in phylogenetically unrelated lineages. Nevertheless, the path of the evolution of vibrational communication in this group of mammals is far from understood. The diversity and complexity of vibrational communication in subterranean rodents is remarkable and is clearly one of the areas of subterranean communication that deserves more attention due to its importance for this and other groups of animals.

Knowing more about the vocal and seismic behavior of subterranean rodents also will improve our efforts to protect these species, some of which, like *Ctenomys*, are considered vulnerable to extinction (Mapelli et al. 2012). Conservation of small mammals requires both eco-ethological and genetic information to understand the vulnerability of some populations to extinction (Fernandes et al. 2007). Therefore, obtaining exhaustive data about life history characteristics of subterranean rodents and the probable impacts of human activities on their biology will contribute to the development of preservation initiatives.

3.4 Linking Social Organization and Vocal Repertoires

3.4.1 Defining Social and Vocal Complexity

Communication is central to sociality because most cooperative actions performed by animals rely on messages of some form and information exchanges, usually vital to coordinating actions among two or more individuals. Many studies show that complex communication systems play an important role in coordinating group activities and maintaining group cohesion (Blumstein and Armitage 1997; Freeberg et al. 2012). Indeed, many published reports show that there is a direct relationship between social complexity and communication complexity, usually expressed as an increase in repertoire complexity when social groups become more complex. In this sense, after a thorough analysis of the evidence gathered, Freeberg and coauthors (2012, p. 1797) state "... (i) that social complexity may play an important role in driving communicative complexity in animal species and (ii) that further tests of this possible role will be of great interest—and are much needed—to advance our understanding of communicative evolution."

However, there are some studies that argue that sociality is not a prerequisite for the evolution of communicative complexity (Ord and García-Porta 2012). Working with compiled data on variations in signal complexity among closely related species (e.g., in ants, frogs, and birds), Ord and García-Porta (2012) used phylogenetic methods to evaluate possible factors involved in the evolution of communication and found that social factors were presumably involved in the evolution of complex visual signals in lizards, while the evolution of complexity in frog and bird vocalizations was most likely explained by differences in ecology and allometry. Thus, as a conclusion from their work, the authors stated that social complexity could lead to the development of complex communication signals, but it is not the only factor at play.

This level of generality in the conclusions may have arisen because there are many ways to quantify communicative complexity: as the number of different signals (vocalizations, in this case); as different types of content emitted or different contexts of vocalizations; as the quantity of information encoded in each vocalization; as different categories of reactions obtained from the receiver; or as differences in the active space of signals, for example. On the other hand, there are also many ways of quantifying social complexity: as the number of individuals; as the number of different roles that could be occupied in the population; or as social matrixes, expressing the level of the relationship between different types of individuals.

In rodents, previous studies have tried to find a correlation between social complexity and communicative complexity using vocal communication (Blumstein and Armitage 1997; Pollard and Blumstein 2012). Some researchers tried to determine if the *acoustic adaptation hypothesis* as a form of environmental influence (Morton 1975) could explain an increase in communicative complexity in opposition to social complexity. Others attempted to find a way of measuring social complexity using characteristics other than colony size (Blumstein and Armitage 1997). The previously mentioned papers considered mostly fossorial rodents and alarm calls to investigate this subject and found evidence leading to the conclusion that acoustic adaptation is not related to communicative complexity, or at least acoustic adaptation could not explain the variation in communicative complexity in full (Blumstein 2003).

For subterranean rodents, Lacey (2000) proposed that sociality varies along a continuum from solitary to eusocial species. It is usually assumed that sociality evolved from solitary individuals becoming tolerant enough to allow grouping and social living. However, Burda and colleagues (2000) suggested sociality as the initial condition, a hypothesis challenged by studies in Bathyergidae (Walton et al. 2000; Faulkes and Bennett 2013). Sobrero and colleagues (2014) presented this challenge in a different way, stating that in Caviomorpha and Phiomorpha the social condition could be considered as ancestral.

The diversity of social situations occurring in social groups is in some way linked to the evolution of communication repertoires because of the appearance of new functional scenarios not existing for solitary species. These new functional scenarios may call for new signal categories to allow expression of new messages and for the probable increment in signal number in old categories not yet present in solitary species (Blumstein 2003; Freeberg et al. 2012).

3.4.2 Social Species Compared to Solitary Species

Subterranean rodents occur in every major continental land mass, but the distribution of the studies about their biological characteristics is not nearly as wide. As explained in the introduction, only some families or groups have been studied extensively. Thus, when considering the continuum from solitary to social/eusocial species, the three families with species that have been well studied will be summarized: the family Bathyergidae (southern Africa), the family Ctenomyidae (southern South America), and the subterranean member of the family Octodontidae (central Chile).

In the family Bathyergidae, the majority of species are catalogued as subterranean, with many social species, one eusocial, and some solitary ones. Both *Fukomys* species and the naked mole rat live in colonies: *F. anselli* has reported colony sizes of 13–29 individuals (Scharff et al. 2001; Šklíba et al. 2012), the giant mole rat has colony sizes of 10–20 (Bednarová et al. 2013), and Mashona's mole rat has 5–9

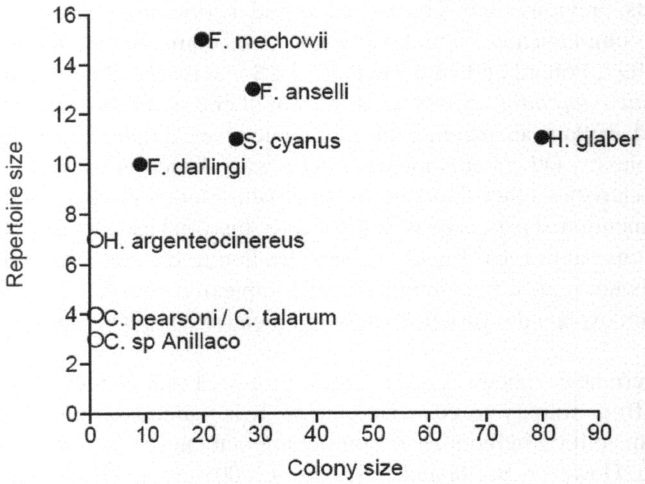

Fig. 3.2 Relationship between adult repertoire size and colony size (as a measure of social complexity) in subterranean rodents. *Open circles* indicate solitary species and *filled circles* indicate groups of various sizes. Colony size data for: *F. mechowii* (Sichilima et al. 2008), *F. anselli* (Šklíba et al. 2012), *F. darlingi* (Dvořáková et al. 2016), *S. cyanus* (Veitl et al. 2000), *H. glaber* (Pepper et al. 1991), *H. argenteocinereus* (Knotková et al. 2009), *C. pearsoni* (Francescoli 1999 and unpublished data), *C. talarum* (Schleich and Busch 2002a), and *Ctenomys sp.* "Anillaco" (Amaya 2017). Repertoire size data taken from references in Tables 3.1 and 3.2

animals per colony (Bennett et al. 1994). The naked mole rat, on the other hand, varies between 60 and 80 individuals in a colony (Pepper et al. 1991).

Looking at some of the species best known from the point of view of vocalizations and social structure (see Tables 3.1, 3.2; Fig. 3.2), it can be said that *F. anselli*, the giant mole rat, and the Mashona's mole rat, all social species, exhibit a base repertoire of 13, 15, and 10 adult signals, respectively (Dvořáková et al. 2016). The authors reported four juvenile vocalizations for *F. anselli*. The eusocial naked mole rat (the only eusocial rodent known) presents a very rich repertoire of vocal signals, with 11 adult vocalizations and 5 juvenile ones. The only solitary representative of the family that has been studied in some depth in relation to vocalizations is the silvery mole rat, which has a repertoire of seven different adult vocalizations (Knotková et al. 2009). This is the only solitary Bathyergidae studied from this point of view and the only one with which to compare the repertoires of the most studied social members of the family.

The social coruro has a colony size varying between 15 and 26 individuals, with males and females living together in the same burrow system (Begall and Gallardo 2000). Its vocal repertoire (see Table 3.1) consists of 11 adult signals (Veitl et al. 2000), while juveniles produce 4 vocalizations that are similar to those of adults, plus one that is a mix of two adult vocalizations, and another one unique to juveniles.

Ctenomyidae is a family with 50–60 species of subterranean rodents, mainly solitary, but some species are being recognized as social or as exhibiting multiple occupants of burrows (Tammone et al. 2012). None of the social species has been thoroughly studied from the point of view of their vocalizations, but a few of the solitary ones have. From these solitary species, three have been studied most intensively for their vocal system: *C. pearsoni*, *C. talarum*, and *Ctenomys sp*. For these species, up to four adult vocal signals, serving mostly mating and agonistic purposes, have been reported (see Table 3.1).

If, regardless of their phylogenetic relationships, one looks at summaries of the vocalization repertoires (see Table 3.1), the repertoires of social species are more extensive than the repertoires of the solitary species. This is in line with the idea that communication complexity (at least evaluated as number of signals) is higher in species that are more complex from the point of view of sociality (evaluated as number of individuals living together). Obviously, some of the signal categories employed in the classification are probably exclusive to social animals (like "contact"), but the general tendency detected seems to be in line with the idea that increased social complexity is correlated with increased communication complexity. These data suggest that social life is more demanding on communication systems than solitary life, a fact that has been presumed for a long time (Freeberg et al. 2012). These data also demonstrate that signals with mating and/or agonistic purposes are the basic types of signals any animal—regardless of whether they are solitary or social—should have in their repertoires, at least for what concerns subterranean rodents.

3.4.3 Similarities and Differences in the Vocal Repertoires of Social Groups

Social subterranean rodents have some similarities from the point of view of their vocal signal repertoires, which can be seen when scanning Table 3.1. One of the main conclusions that can be taken from the data is that social and solitary species seem to have the same number of mating signals in their repertoires, regardless of their phylogenetic relationships. Thus, it seems that the need for vocal signals used with the goal of reproducing is not different between social and solitary species. This is in line with the idea that reproduction is (almost) the only social activity performed by solitary subterranean rodents. In apparent opposition to reproduction, agonistic behavior seems to be more demanding on vocal signals for the social species, probably responding to the increase in the number of individuals that are part of social groups.

In conclusion, Table 3.2 shows that social species have similar repertoire sizes, at least when examining the two main categories referred to previously. Other categories are differentially developed or expanded in the examined species, with the category "contact" almost lacking for the naked mole rat (the eusocial species), and the "distress" category apparently restricted to the *Fukomys* genus. The "alarm" category is present in two social species examined (the giant mole rat and the naked mole rat) and absent in *F. anselli*, the Micklem's mole rat and the Mashona mole rat. This difference is difficult to explain based on current knowledge, and both evolutionary and environmental conditions may have shaped the differential development of this signal in bathyergids.

Another interesting note here is that the total number of vocal signals detected for the species surveyed is similar among the social species, and the total number is also similar when comparing the solitary species. However, the number of vocal signals differs when comparing social and solitary species, leading to the conclusion that social species have expanded repertoires and that, in this line of reasoning, increased social complexity is correlated with increased communication complexity. This pattern falls in line with what has been shown in other animals (Krams et al. 2012).

In future studies about subterranean rodents it will be extremely important to increase the number of species surveyed for vocalizations together with the analysis of the relationship between social complexity and communication complexity to comprehend all the variability in the parameters defining both categories. Only with this kind of analysis about an extensive number of species and families will it be possible to draw a coherent picture of the true relationships between social structures and communication variability in subterranean rodents.

3.5 Concluding Remarks

In this chapter, the acoustic signals used by subterranean rodents to communicate were described and the main hypotheses that have been developed to explain how natural selection has acted on the design of these animal signals were discussed. Despite the huge amount of information accumulated through the previous decades, it is evident that there is still much work to be done. Most of the knowledge about communication among subterranean rodents comes from relatively few species belonging to four families. Although this was useful for establishing some general assumptions about the evolution and functional meaning of the different acoustic signals employed by subterranean rodents to communicate, it is evident that is not enough to fully comprehend all the factors that have shaped the evolution of the acoustic behavior in this group of fascinating mammals.

Although positive and negative aspects can be found in both lab and field experiments, obtaining and understanding a complete profile of the vocal behavior of the different species requires a combination of both settings. Increasing the number of field studies and creating more natural environmental settings are two required perspectives to increase our knowledge of this particular group of rodents.

Finally, the mechanisms of vocal production are another gap in our understanding of acoustic communication in subterranean rodents. Knowing the physics and morphology behind the production of vocalizations constitutes a vital step toward the construction of theories concerning the evolution of communication signals. Although some terrestrial mammals display morphological specializations in their vocal tracts that allow them to produce unexpectedly wide-ranging calls (Taylor et al. 2016), as a general rule, larger animals produce lower frequency calls and smaller individuals emit higher frequency ones. This trend does not fully apply to subterranean rodents, suggesting that the underground environment has also influenced the evolution of the vocal anatomy in this group and emphasizing the need for these kinds of studies in underground mammals. In addition to presenting a complete and current revision of the acoustic behavior of subterranean rodents, this chapter should serve as an important source of inspiration for scientists interested in the evolution of the communication diversity seen among subterranean mammals.

Compliance with Ethics Requirements Cristian E. Schleich declares that he has no conflict of interest.

Gabriel Francescoli declares that he has no conflict of interest.

References

Amaya, J. P. (2017). Interacciones acústicas de una especie de *Ctenomys* de la provincia de La Rioja. Ph. D. dissertation. Facultad Ciencias Naturales y Museo. La Plata, Argentina.

Amaya, J. P., Areta, J. I., Valentinuzzi, V. S., & Zufiaurre, E. (2016). Form and function of long-range vocalizations in a Neotropical fossorial rodent: The Anillaco tuco-tuco (*Ctenomys sp.*). *PeerJ*, 4, e2559. https://doi.org/10.7717/peerj.2559

Bednářová, R., Hrouzková-Knotková, E., Burda, H., Sedláček, F., & Sumbera, R. (2013). Vocalizations of the giant mole rat (*Fukomys mechowii*), a subterranean rodent with the richest vocal repertoire. *Bioacoustics*, 22, 87–107.

Begall, S., & Gallardo, M. H. (2000). *Spalacopus cyanus* (Rodentia: Octodontidae): An extremist in tunnel constructing and food storing among subterranean mammals. *Journal of Zoology London*, 251, 53–60.

Begall, S., Burda, H., & Schleich, C. (2007a). *Subterranean rodents: News from underground*. Berlin: Springer Science+Business Media.

Begall, S., Lange, S., Schleich, C., & Burda, H. (2007b). Acoustics, audition and auditory system. In S. Begall, H. Burda, & C. Schleich (Eds.) *Subterranean rodents: News from underground* (pp. 97–111). Berlin: Springer Science+Business Media.

Bennett, N. C., & Jarvis, J. U. M. (1988). The social structure and reproductive biology of colonies of the mole rat *Cryptomys damarensis* (Rodentia, Bathyergidae). *Journal of Mammalogy*, 69, 293–302.

Bennett, N. C., Jarvis, J. U. M., & Cotterill, F. P. D. (1994). The colony structure and reproductive biology of the afrotropical Mashona mole rat, *Cryptomys darlingi*. *Journal of Zoology*, 234, 477–487.

Blumstein, D. T. (2003). Social complexity but not the acoustic environment is responsible for the evolution of complex alarm communication. In R. Ramousse, D. Allainé, & M. Le Berre (Eds.), *Adaptive strategies and diversity in marmots* (pp. 31–38). Montreux, Switzerland: International Network on Marmots.

Blumstein, D. T., & Armitage, K. B. (1997). Does sociality drive the evolution of communicative complexity? A comparative test with ground-dwelling sciurid alarm calls. *American Naturalist*, 150, 179–200.

Blumstein, D. T, Mennill, D. J., Clemins, P., Girod, L., et al. (2011). Acoustic monitoring in terrestrial environments using microphone arrays: Applications, technological considerations and prospectus. *Journal of Applied Ecology*, 48, 758–767.

Bradbury, J. W., & Vehrencamp, S. L. (1998). *Principles of animal communication*. Sunderland, MA: Sinauer.

Brett, R. A. (1991). The population structure of naked mole rat colonies. In P. W. Sherman, J. U. M. Jarvis, & R. D. Alexander (Eds.), *The biology of the naked mole rat* (pp. 97–136). Princeton: Princeton University Press.

Burda, H., Bruns, V., & Müller, M. (1990a). Sensory adaptations in subterranean mammals. In E. Nevo & O. A. Reig (Eds.), *Evolution of subterranean mammals at the organismal and molecular levels* (pp. 269–293). New York: Wiley-Liss.

Burda, H., Marhold, S., Westenberger, T., Wiltschko, R., & Wiltschko, W. (1990b). Magnetic compass orientation in the subterranean rodent *Cryptomys hottentotus*, Bathyergidae. *Experientia*, 46, 528–530.

Burda, H., Beiles, A., Marhold, S., Simson, S., et al. (1991). Magnetic orientation in subterranean mole rats of the superspecies *Spalax ehrenbergi*: Experiments, patterns and memory. *Israel Journal of Zoology*, 37, 182–183.

Burda, H., Honeycutt, R. L., Begall, S., Locker-Grütjen, O., & Scharff, A. (2000). Are naked and common mole rats eusocial and if so, why? *Behavioral Ecology and Sociobiology*, 47, 293–303.

Burda, H., Šumbera, R., & Begall, S. (2007). Microclimate in burrows of subterranean rodents. In S. Begall, H. Burda, & C. Schleich (Eds.), *Subterranean rodents: News from underground* (pp. 21–34). Berlin: Springer Science+Business Media.

Busch, C., Antinuchi, C. D., delValle, J. C., Kittlein, M. J., et al. (2000). Population ecology of subterranean rodents. In E. A. Lacey, J. L. Patton, & G. N. Cameron (Eds.), *Life underground* (pp. 183–226). Chicago: University of Chicago Press.

Calisi, R. M., & Bentley, G. E. (2009). Lab and field experiments: Are they the same animal? *Hormones and Behavior*, 56, 1–10.

Capranica, R. R., Moffat, J., & Nevo, E. (1974). Vocal repertoire of a subterranean rodent (*Spalax*). *The Journal of the Acoustical Society of America*, 55(2), 481. (Abstract)

Cooper, H. M., Herbin, M., & Nevo, E. (1993). Ocular regression conceals adaptive progression of the visual system in a blind subterranean mammal. *Nature*, 361, 156–159.

Credner, S., Burda, H., & Ludescher, F. (1997). Acoustic communication underground: Vocalization characteristics in subterranean social mole rats (*Cryptomys sp.*, Bathyergidae). *Journal of Comparative Physiology A: Sensory, Neural, and Behavioral Physiology*, 180, 245–255.

del Valle, J. C., Lohfelt, M. I., Comparatore, V. M., Cid, M. S., & Busch, C. (2001). Feeding selectivity and food preference of *Ctenomys talarum* (tuco-tuco). *Mammalian Biology*, 66, 165–173.

Devries, M. S., & Sikes, R. S. (2008). Vocalisation of a North American subterranean rodent *Geomys breviceps*. *Bioacoustics*, 18(1), 1–15.

Dvořáková, V., Hrouzková, E., & Šumbera, R. (2016). Vocal repertoire of the Mashona mole rat (*Fukomys darlingi*) and how it compares with other mole rats. *Bioacoustics*, 25, 253–266.

Faulkes, C. G., & Bennett, N. C. (2013). Plasticity and constraints on social evolution in African mole-rats: Ultimate and proximate factors. *Philosophical Transactions of the Royal Society B: Biological Sciences*, 368, 20120347.

Fernandes, F. A., Fernández-Stolz, G. P., Lopes, C. M., & Freitas, T. R. O. (2007). The conservation status of the tuco-tucos, genus *Ctenomys* (Rodentia: Ctenomyidae) in southern Brazil. *Brazilian Journal of Biology*, 67(4), 839–847.

Francescoli, G. (1999). A preliminary report on the acoustic communication in Uruguayan *Ctenomys* (Rodentia, Octodontidae): Basic sound types. *Bioacoustics*, 10, 203–218.

Francescoli, G. (2001). Vocal signals from *Ctenomys pearsoni* pups. *Acta Theriologica*, 46, 327–330.

Francescoli, G. (2002). Geographic variation in vocal signals of *Ctenomys pearsoni*. *Acta Theriologica*, 47, 35–44.

Francescoli, G. (2011). Tuco-tucos' vocalization output varies seasonally (*Ctenomys pearsoni*; Rodentia, Ctenomyidae): Implications for reproductive signaling. *Acta Ethologica*, 14(1), 1–6.
Freeberg, T. M., Dunbar, R. I. M., & Ord, T. J. (2012). Social complexity as a proximate and ultimate factor in communicative complexity. *Philosophical Transactions of the Royal Society B: Biological Sciences*, 367, 1785–1801.
Gannon, W. L., & Sikes, R. S. (2007). Guidelines of the american society of mammalogists for the use of wild mammals in research. *Journal of Mammalogy*, 88(3), 809–823.
Godfray, H. C. J. (1995). Evolutionary theory of parent-offspring conflict. *Nature*, 376, 133–138.
Heth, G., Frankenberg, E., & Nevo, E. (1986). Adaptive optimal sound for vocal communication in tunnels of a subterranean mammal (*Spalax ehrenbergi*). *Experientia*, 42(11), 1287–1289.
Heth, G., Frankenberg, E., Raz, A., & Nevo, E. (1987). Vibrational communication in subterranean mole rats (*Spalax ehrenberghi*). *Behavioral Ecology and Sociobiology*, 21, 31–33.
Heth, G., Frankenberg, E., Pratt, H., & Nevo, E. (1991). Seismic communication in the blind subterranean mole rat: Patterns of head thumping and their detection in the *Spalax ehrenbergi* superspecies in Israel. *Journal of Zoology, London*, 224, 633–638.
Heth, G., Todrank, J., & Nevo, E. (2000). Do *Spalax ehrenbergi* blind mole rats use food odours in searching for and selecting food? *Ecology Ethology & Evolution*, 12, 75–82.
Heth, G., Todrank, J., Begall, S., Zilbiger, Y., et al. (2002). Odours underground: Subterranean rodents may not forage "blindly". *Behavioral Ecology and Sociobiology*, 52, 53–58.
Hickman, G. C. (1990). Adaptiveness of tunnel system features in subterranean mammal burrows. In E. Nevo, & O. A. Reig (Eds.), *Evolution of subterranean mammals at the organismal and molecular levels* (pp 185–210). New York: Alan R. Liss, Inc.
Hrouzková, E., Dvoráková, V., Jedlicka, P., & Sumbera, R. (2013). Seismic communication in demon African mole rat *Tachyoryctes daemon* from Tanzania. *Journal of Ethology*, 31(3), 255–259.
Jarvis, J. U. M., & Bennett, N. C. (1991). Ecology and behavior of the family Bathyergidae. In P. W. Sherman, J. U. M. Jarvis, & R. D. Alexander (Eds.), *The biology of the naked mole rat* (pp. 66–96). Princeton: Princeton University Press.
Kilner, R., & Johnstone, R. A. (1997). Begging the question: Are offspring solicitation behaviours signals of need? *Trends in Ecology and Evolution*, 12, 11–15.
Kimchi, T., & Terkel, J. (2003). Mole rats (*Spalax ehrenbergi*) select bypass burrowing strategies in accordance with obstacle size. *Naturwissenschaften*, 90, 36–39.
Kimchi, T., Reshef, M., & Terkel, J. (2005). Evidence for the use of reflected self-generated seismic waves for spatial orientation in a blind subterranean mammal. *The Journal of Experimental Biology*, 208, 647–659.
Knotková, E., Veitlm S., Šumbera, R., Sedláček, F., & Burda, H. (2009). Vocalisations of the silvery mole rat: Comparison of vocal repertoires in subterranean rodents with different social systems. *Bioacoustics*, 18, 241–257.
Krams, I., Krama, T., Freeberg, T. M., Kullberg, C., & Lucas, J. R. (2012). Linking social complexity and vocal complexity: A parid perspective. *Philosophical Transactions of the Royal Society B: Biological Sciences*, 367, 1879–1891.
Lacey, E. A. (2000). Spatial and social systems of subterranean rodents. In E. A. Lacey, J. L. Patton, & G. N. Cameron (Eds.), *Life underground: The biology of subterranean rodents* (pp. 257–296). Chicago: The University of Chicago Press.
Lacey, E. A., Patton, J. L., & Cameron, G. N., Eds. (2000). *Life underground: The biology of subterranean rodents*. Chicago: The University of Chicago Press.
Lange, S., Neumann, B., Hagemeyer, P., & Burda, H. (2005). Kairomone-guided food location in subterranean Zambian mole rats (*Cryptomys spp*., Bathyergidae). *Folia Zoologica*, 54, 263–268.
Lange, S., Burda, H., Wegner, R. E., Dammann, P., et al. (2007). Living in a "stethoscope": Burrow acoustics promote auditory specializations in subterranean rodents. *Naturwissenschaften*, 94(2), 134–138.

Malizia, A. I., Zenuto, R. R., & Busch, C. (1995). Demographic and reproductive attributes of dispersers in two populations of the subterranean rodent *Ctenomys talarum* (tuco-tuco). *Canadian Journal of Zoology*, 73, 732–738.

Mapelli, M. J., Mora, M. S., Mirol, P. M., & Kittlein, M. J. (2012). Population structure and landscape genetics in the endangered subterranean rodent *Ctenomys porteousi*. *Conservation Genetics*, 13, 165–181.

Mason, M. J., & Narins, P. M. (2010). Seismic sensitivity and communication in subterranean mammals. In C. E. O'Connell-Rodwell (Ed.), *The use of vibrations in communication: Properties, mechanisms and function across taxa* (pp. 121–139). Kerala: Transworld Research Network.

Mason, M. J., Lai, F. W. S., Li, J. G., & Nevo, E. (2010). Middle ear structure and bone conduction in *Spalax*, *Eospalax* and *Tachyoryctes* mole rats (Rodentia: Spalacidae). *Journal of Morphology*, 271, 462–472.

Morton, E. S. (1975). Ecological sources of selection on avian sounds. *The American Naturalist*, 109(965), 17–34.

Narins, P. M., Reichman, O. J., Jarvis, J. U. M., & Lewis, E. R. (1992). Seismic signal transmission between burrows of the cape mole rat, *Georychus capensis*. *Journal of Comparative Physiology A: Sensory, Neural, and Behavioral Physiology*, 170, 13–21.

Nemec, P., Burda, H., & Peichl, L. (2004). Subcortical visual system of the African mole rat *Cryptomys anselli*: To see or not to see? *European Journal of Neuroscience*, 20, 757–768.

Nevo, E. (1999). *Mosaic evolution of subterranean mammals: Regression, progression and global convergence*. New York: Oxford Science Publications.

Nevo, E., Heth, G., Beiles, A., & Frankenberg, E. (1987). Geographic dialects in blind mole rats: Role of vocal communication in active speciation. *Proceedings of the National Academy of Sciences of the United States of America*, 84, 3312–3315.

Nevo, E., Heth, G., & Pratt, H. (1991). Seismic communication in a blind subterranean mammal: A major somatosensory mechanism in adaptive evolution underground. *Proceedings of the National Academy of Sciences of the United States of America*, 88, 1256–1260.

Ord, T. J., & García-Porta, J. (2012). Is sociality required for the evolution of communicative complexity? Evidence weighed against alternative hypotheses in diverse taxonomic groups. *Philosophical Transactions of the Royal Society B: Biological Sciences*, 367, 1811–1828.

Owings, D. H. & Morton, E. S. (1998). *Animal vocal communication: A new approach*. Cambridge, UK: Cambridge University Press.

Pepper, J. W., Braude, S. H., Lacey, E. A., & Sherman, P. W. (1991). Vocalizations of the naked mole rat. In P. W. Sherman, J. U. M. Jarvis, & R. D. Alexander (Eds.), *The biology of the naked mole rat* (pp. 243–274). Princeton: Princeton University Press.

Poduschka, W. (1978) Abwehrreaktion der Mullratte, *Cryptomys hottentotus* (Lesson, 1826). *Säugetierkunde Mitt*, 26, 260–268.

Pollard, K. A., & Blumstein, D. T. (2012). Evolving communicative complexity: Insights from rodents and beyond. *Philosophical Transactions of the Royal Society B: Biological Sciences*, 367, 1869–1878.

Rado, R., Levi, N., Hauser, H., Witcher, J., et al. (1987). Seismic signalling as a means of communication in a subterranean mammal. *Animal Behaviour*, 35, 1249–1251.

Rado, R., Himelfarb, M., Arensburg, B., Terkel, J., & Wollberg, Z. (1989). Are seismic communication signals transmitted in the Blind mole-rat by bone conduction? *Hearing Research*, 41, 23–30.

Rado, R., Terkel, J. & Wollberg, Z. (1998). Seismic communication signals in the blind mole rat (*Spalax ehrenbergi*): Electrophysiological and behavioral evidence for their processing by the auditory system. *Journal of Comparative Physiology: Sensory, Neural, and Behavioral Physiology*, 183, 503–511.

Randall, J. A. (2010). Drummers and stompers: Vibrational communication in mammals. In C. E. O'Connel-Rodwell (Ed.), *The use of vibrations in communication: Properties, mechanisms and functions across taxa* (pp 99–120). Trivandrum, India: Transworld Research Network.

Randall, J. A. (2014). Vibrational communication: Spiders to kangaroo rats. In G. Witzany (Ed.), *Biocommunication of animals* (pp. 103–133). Dordrecht, Netherlands: Springer Science+Business Media.

Scharff, A., Locker-Grütjen, O., Kawalika, M., & Burda, H. (2001). Natural history of the giant mole rat, *Cryptomys mechowi* (Rodentia: Bathyergidae), from Zambia. *Journal of Mammalogy*, 82, 1003–1015.

Schleich, C. E., & Busch, C. (2002a). Acoustic signals of a solitary subterranean rodent *Ctenomys talarum* (Rodentia: Ctenomyidae): Physical characteristics and behavioural correlates. *Journal of Ethology*, 20, 123–131.

Schleich, C. E., & Busch, C. (2002b). Juvenile vocalizations of *Ctenomys talarum* (Rodentia: Octodontidae). *Acta Theriologica*, 47, 25–33.

Schleich, C. E., & Busch, C. (2004). Energetic expenditure during vocalization in pups of the subterranean rodent *Ctenomys talarum*. *Naturwissenshaften*, 91, 548–551.

Schleich, C. E., & Zenuto, R. R. (2007). Use of vegetation chemical signals for digging orientation in the subterranean rodent *Ctenomys talarum* (Rodentia: Ctenomyidae). *Ethology*, 113, 573–578.

Schleich, C., Veitl, S., Knotkova´, E., & Begall, S. (2007). Acoustic communication in subterranean rodents. In S. Begall, H. Burda, & C. Schleich (Eds.), *Subterranean rodents: News from underground* (pp 113–127). Berlin: Springer Science+Business Media.

Schleich, C. E., & Antenucci, D. C. (2009). Sound transmission and burrow characteristics of the subterranean rodent *Ctenomys talarum* (Rodentia: Ctenomyidae). *Acta Theriologica*, 54(2), 165–170.

Sichilima, A. M., Faulkes, C. G., & Bennett, N. C. (2008). Field evidence for a seasonality of reproduction and colony size in the Afrotropical giant mole rat *Fukomys mechowii* (Rodentia: Bathyergidae). *African Ecology*, 43, 144–149.

Šklíba, J., Mazoch, V., Patzenhauerovám, H., Hrouzková, E., et al. (2012). A maze-lover's dream: Burrow architecture, natural history and habitat characteristics of Ansell's mole rat (*Fukomys anselli*). *Mammalian Biology*, 77, 420–427.

Sobrero, R., Inostroza-Michael, O., Hernández, C. E., & Ebensperger, L. A. (2014). Phylogeny modulates the effects of ecological conditions on group living across hystricognath rodents. *Animal Behaviour*, 94, 27–34.

Šumbera, R., Mazoch, V., Patzenhauerová, H., Matěj Lövy, M., et al. (2012). Burrow architecture, family composition and habitat characteristics of the largest social African mole rat: The giant mole rat constructs really giant burrow systems. *Acta Theriologica*, 57(2), 121–130.

Tammone, M. N., Lacey, E. A., & Relva, M. A. (2012). Habitat use by colonial tuco-tucos (*Ctenomys sociabilis*): Specialization, variation, and sociality. *Journal of Mammalogy*, 93(6), 1409–1419.

Taylor, A. M., Charlton, B. D., & Reby, D. (2016). Vocal production by terrestrial mammals: Source, filter, and function. In R. A. Suthers, W. T. Fitch, R. R. Fay, & A. N. Popper (Eds.), *Vertebrate sound production and acoustic communication* (pp. 229–260). Switzerland: Springer Science+Business Media.

Vanden Hole, C., Van Daele, P. A. A. G., Desmet, N., Devos, P., & Adriaens, D. (2014). Does sociality imply a complex vocal communication system? A case study for *Fukomys micklemi* (Bathyergidae, Rodentia). *Bioacoustics*, 23, 143–160.

Veitl, S., Begall, S., & Burda, H. (2000). Ecological determinants of vocalisation parameters: The case of the coruro *Spalacopus cyanus* (Octodontidae), a fossorial social rodent. *Bioacoustics*, 11, 129–148.

Walton, A. H., Nedbal, M. A., & Honeycutt, R. L. (2000). Evidence from Intron 1 of the nuclear transthyretin (prealbumin) gene for the phylogeny of African mole-rats (Bathyergidae). *Molecular Phylogenetics and Evolution*, 16, 467–474.

Wiltschko, R., & Wiltschko, W. (1995). *Magnetic orientation in animals*. New York: Springer-Verlag.

Yosida, S., Kobayas, K. I., Ikebuchi, M., Ozaki, R., & Okanoya, K. (2007). Antiphonal vocalization of a subterranean rodent, the naked mole rat (*Heterocephalus glaber*). *Ethology*, 113, 703–710.

Yosida, S., & Okanoya, K. (2009). Naked mole rat is sensitive to social hierarchy encoded in antiphonal vocalization. *Ethology*, 115, 823–831.

Chapter 4
Hearing in Rodents

Micheal L. Dent, Laurel A. Screven, and Anastasiya Kobrina

Abstract Hearing in rodents has been measured using both behavioral and physiological methods. Features of hearing that have been measured in rodents include auditory acuity in quiet and in noise, frequency selectivity and sensitivity, intensity resolution, temporal resolution, and complex sound perception. Generally, and especially for simple tone detection, behavioral thresholds are lower than physiological thresholds. Within behavioral studies, operant experiments using awake, behaving rodents produce lower thresholds than simple reflexive measures. Rodents generally have broader frequency filters than other mammals. Frequency and intensity resolution are similar but slightly elevated relative to other mammals. The few measures of complex sound perception performed to date show that at least some rodents have the capacity to distinguish between spectrotemporal characteristics of acoustic signals for communication. Most studies have typically employed domesticated laboratory rodents rather than wild-caught species, so few attempts have been made to correlate lifestyle and evolutionary history with auditory processing. Nonetheless, a baseline knowledge of hearing abilities in rodents will facilitate experiments on the perception of more complex, natural acoustic stimuli in the future.

Keywords Audiogram · Critical ratio · Discrimination · Frequency difference limen · Intensity difference limen · Speech perception · Ultrasonic vocalizations

4.1 Introduction

The widespread diversity in the habitats of rodents requires a matching diversity in acoustic communication systems. Since many rodents are social and quite a few are nocturnal, auditory signals play a dominant role in their lives. Rodents are known to produce warning signals to conspecifics, sonic and ultrasonic vocalizations for coordinating sexual behaviors, and pup distress calls when separated

M. L. Dent (✉) · L. A. Screven · A. Kobrina
Department of Psychology, University at Buffalo, SUNY, Buffalo, NY, USA
e-mail: mdent@buffalo.edu; laurelsc@buffalo.edu; akobrina@buffalo.edu

M. L. Dent et al. (eds.), *Rodent Bioacoustics*, Springer Handbook of Auditory Research 67, https://doi.org/10.1007/978-3-319-92495-3_4

from the mother (e.g., Okanoya and Screven, Chap. 2). Rodents are also widely preyed upon and hearing the predators in their environment is clearly advantageous. Thus, the ability to detect, discriminate, localize, and identify acoustic signals is important for the survival of all members of the rodent order. Unfortunately, most studies of hearing abilities in rodents have been conducted in just a handful of the more than 2200 species, and most tests have been administered in common laboratory-bred species. There are limits to the descriptions of hearing in rodents that are summarized in this chapter. There are, nonetheless, studies on many different aspects of hearing in rodents, both in quiet and in noise, using both simple and complex stimuli, and employing various behavioral and physiological methods.

Behavioral studies of hearing in rodents have utilized both trained and untrained animals to measure operant and reflexive responses to acoustic stimuli. Generally, operant experiments on trained animals are preferred because the rodents are reliable attending observers under stimulus control. Unfortunately, these types of experiments take longer than those using untrained subjects and, therefore, are somewhat rare in rodents relative to those in birds and other mammals (e.g., Fay 1988). One of the benefits of operant experiments is that they can be conducted on low intensity stimuli, which will generally not work for acoustic startle types of experiments (Heffner and Heffner 2001; Behrens and Klump 2015). Complex tasks such as sound classification can only be performed using conditioning paradigms. Finally, and probably the most important issue for startle experiments, sensitivity is typically much worse for acoustic startle experiments relative to conditioning experiments (reviewed in Lauer et al. 2017). Thus, to obtain accurate measures of an awake and behaving animal's ability to hear the world, operant experiments are preferable. However, for measurements of very young animals or when sensorimotor gating is of interest, it is often necessary to use acoustic startle to obtain a quick estimate of an animal's auditory acuity.

Some experiments utilized physiological measurements such as auditory brainstem responses (ABRs), cochlear microphonics (CM), or single-unit recordings in auditory nuclei. Those measurements can be even more difficult to compare to behavioral results due to issues that include the effects of anesthesia on neural responses and the limits of understanding a rodent's behavior by obtaining recordings from a single neuron and not a population of neurons. Regardless, in some instances, these are the only measurements of auditory acuity available from that species on that dimension of hearing. Where possible, comparisons are made between physiological responses and behavioral responses in the same species under similar experimental conditions.

In this chapter, auditory acuity for simple and complex stimuli is described for rodents. First, absolute sensitivity for pure tones in quiet is outlined in Sect. 4.2. Frequency selectivity measured by several techniques is described in Sect. 4.3. The discrimination of frequency and intensity in simple stimuli are characterized next (Sects. 4.4 and 4.5), followed by the resolution for temporal properties of sounds (Sect. 4.6). Finally, the perception of complex stimuli, including vocalizations and human speech sounds, by rodents is summarized in Sect. 4.7. As stated previously,

Table 4.1 Abbreviations

ABR	Auditory brainstem response
AC	Auditory cortex
CM	Cochlear microphonics
CMR	Comodulation masking release
CN	Cochlear nucleus
ERB	Equivalent rectangular bandwidths
FDL	Frequency difference limen
IC	Inferior colliculus
IDL	Intensity difference limen
NMRI	Naval Medical Research Institute
PPI	Pre-pulse inhibition
PTC	Psychophysical tuning curve
SL	Sensation level
SPL	Sound pressure level
TMTF	Temporal modulation transfer function
USV	Ultrasonic vocalization
VOT	Voice onset time

this chapter primarily focuses on behavioral studies, but examples of physiological experiments are included in most sections. All abbreviations are defined in Table 4.1.

4.2 Absolute Sensitivity

Behavioral and physiological audiograms have been reported for twenty-two rodent species (Fig. 4.1), including strains of laboratory mice (*Mus musculus*) and rats (*Rattus norvegicus*) (Fig. 4.2). In general, the range of hearing in rodents is broader than that of other mammals with some rodents hearing better at low frequencies, some hearing better at high frequencies, and some hearing better at both lower and higher frequencies than humans (e.g., Fay 1988). Peak sensitivity for wild rodents ranges from −9 dB SPL in the guinea pig (*Cavia porcellus*) (Prosen et al. 1978) to just 35 dB SPL in the naked mole rat (*Heterocephalus glaber*) (Heffner and Heffner 1993). The best frequency of hearing also shows variability across species with some rodents, like the common mole rat (*Cryptomys hottentotus*) (Muller and Burda 1989) and the Zambian mole rat (*Fukomys amatus*) (Bruckmann and Burda 1997), more sensitive to very low frequencies (0.8 kHz) and others, like the feral house mouse (Heffner and Masterton 1980), most sensitive to higher frequencies (16 kHz). In general, rodents in the Castorimorpha (pocket gophers) and Hystricognathi (mole rats, chinchillas, guinea pigs, and coruros) suborders have lower best frequencies of hearing than rodents in the Sciurognathi (groundhogs, chipmunks, fox squirrels, and prairie dogs) and Myomorpha (wild rats and mice, and hamsters) suborders. The fossorial and subterranean animals hear better at low frequencies, and rodents that live completely or partially above ground have better high-frequency hearing.

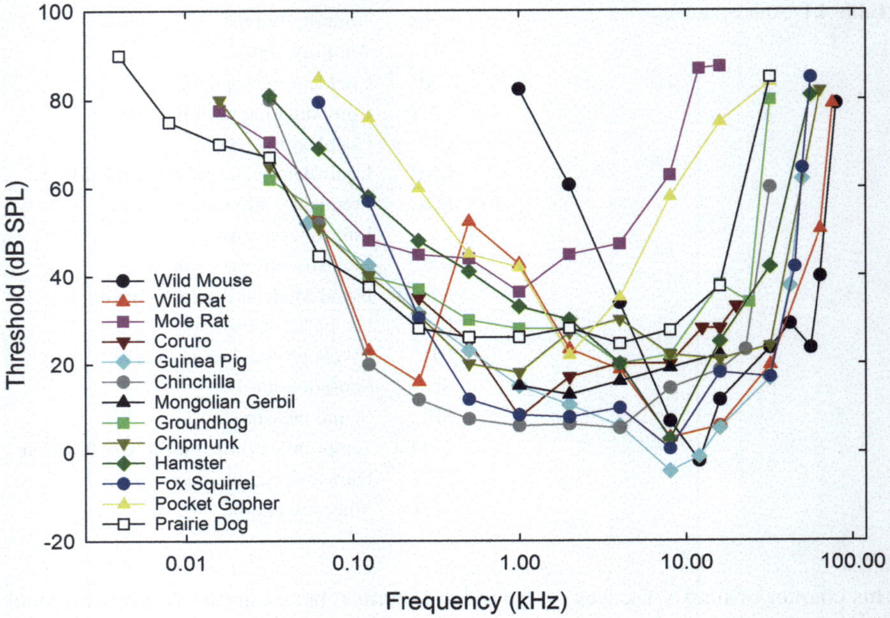

Fig. 4.1 Audiograms (means) for several rodent species: the wild mouse (*black circles*; Heffner and Masterton 1980; Heffner and Heffner 1985; Heffner et al. 2001); wild rat (*red triangles*; Heffner and Masterton 1980; Heffner and Heffner 1985); mole rat, *Cryptomys hottentotus* (*pink squares*; Muller and Burda 1989; Heffner and Heffner 1993; Bruckmann and Burda 1997); coruro, *Spalacopus cyanus* (*dark red inverted triangles*; Begall et al. 2004); guinea pig, *Cavia porcellus* (*cyan diamonds*; Heffner et al. 1971; Prosen et al. 1978; Syka and Popelar 1988); chinchilla, *Chinchilla laniger* (*grey circles*; Miller 1970; Heffner and Heffner 1991); Mongolian gerbil, *Meriones unguiculatus* (*black triangles*; Hamann et al. 2002); groundhog, *Marmota monax* (*light green squares*; Heffner et al. 2001); chipmunk, *Tamias striatus* (*dark yellow inverted triangles*; Heffner et al. 2001); hamster, *Mesocricetus auratus* (*green diamonds*; Heffner et al. 2001); fox squirrel, *Sciurus niger* (*blue circles*; Jackson et al. 1997); pocket gopher, *Geomys bursarius* (*yellow triangles*; Heffner and Heffner 1990); and prairie dog, *Cynomys ludovicianus* and *C. leucurus* (*white squares*; Heffner, R. S., et al. 1994)

There are likely specializations for underground communication in the subterranean rodents (Schleich and Francescoli, Chap. 3), as well as specializations for ultrasonic communication in air (Okanoya and Screven, Chap. 2). Given the great diversity in rodent habitats, the great diversity across rodent audiograms is no surprise.

Many audiograms have been obtained from domesticated and inbred laboratory rats and mice, using both behavioral and physiological methods (Fig. 4.2). Rats are more sensitive to lower frequencies than are mice, although peak sensitivity is about the same across the two species. There is huge variability amongst wild and laboratory mouse audiograms relative to rat audiograms. Thresholds from some strains of mice are up to 80 dB higher than in other strains at the same frequencies, although more strains of mice have been measured. The wide range of threshold sensitivity in

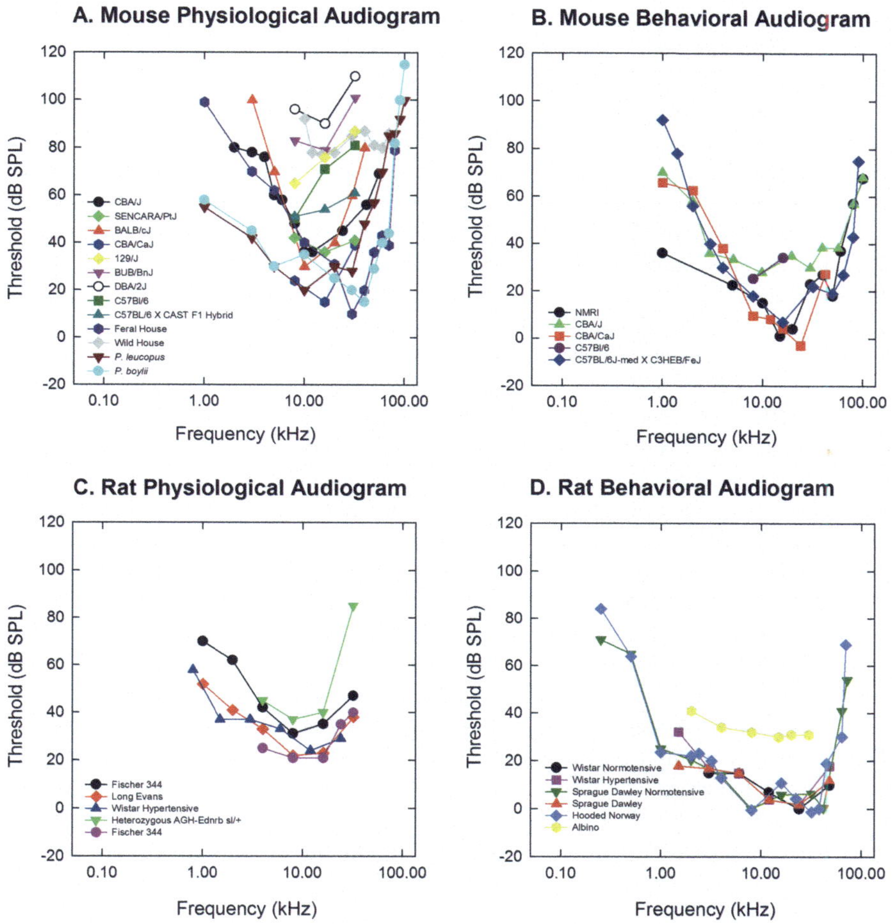

Fig. 4.2 Audiograms for mice (**a–b**) and rat strains (**c–d**) collected using physiological (**a**, **c**) or behavioral (**b**, **d**) methods. (**a**) CBA/J (*black circles*; Muller et al. 2005); SEN CARA/PtJ (*light green diamonds*; Zheng et al. 1999), BALB/cJ (*red triangles*; Ralls 1967); CBA/CaJ (*blue hexagons*, Zheng et al. 1999); 129/J (*yellow diamonds*; Zheng et al. 1999); BUB/BnJ (*pink inverted triangles*; Zheng et al. 1999); DBA/2J (*white circles*; Zheng et al. 1999); C57BL/6 (*dark green squares*; Prosen et al. 2003); C57BL/6 X CAST F1 hybrid (*dark cyan triangles*; Prosen et al. 2003); feral house mouse (*purple hexagon*; Ralls 1967); wild house mouse (*grey diamonds*; Francis 1979); *P. leucopus* (*dark red inverted triangles*; Ralls 1967); *P. boylii* (*cyan circles*; Ralls 1967). (**b**) NMRI (*black circles*; Ehret 1974), CBA/J (*green triangles*; Birch et al. 1968); CBA/CaJ (*red squares*; Radziwon et al. 2009); C57BL/6 (*purple circles*; Prosen et al. 2003); C57BL/6J-med X C3HEB/FeJ (*blue diamonds*; Koay et al. 2002). (**c**) Fischer 344 (*black circles*; Popelar et al. 2006); Long Evans (*red diamonds*; Popelar et al. 2006); Wistar hypertensive (*blue triangles*; Borg 1982); Heterozygous AGH-Ednrb sl/+ (*green inverted triangles*; Dang et al. 2011); Fischer 344 (*pink circles*; Polak et al. 2004). (**d**) Wistar normotensive (*black circles*; Borg 1982); Wistar hypertensive (*pink squares*; Borg 1982); Sprague-Dawley normotensive (*green inverted triangles*; Borg 1982); Sprague-Dawley (*red triangles*; Kelly and Masterton 1977); hooded Norway rat (*purple diamonds*; Heffner, H. E., et al. 1994); albino rat (*yellow octagons*, Hack 1971)

laboratory mice is mainly due to spontaneous and engineered genetic mutations that affect cochlear function (Ohlemiller, Chap. 7). In rats, the variability is only about 20 dB between strains.

Finally, methodology plays a large role in the hearing sensitivity that is measured, as mentioned in Sect. 4.1. Physiological audiograms can be 10–30 dB higher than behavioral audiograms and can also take on different shapes. For instance, in the CBA/J mouse strain, thresholds range from 80 dB for the ABR measurements to 58 dB for the behavioral measurements at 2 kHz. The difference is much smaller at 10 kHz, the frequency of best sensitivity for this strain of mouse, ranging from 36 dB for the ABR thresholds to 28 dB for the behavioral threshold. Thus, both the actual thresholds and the audiogram shapes differ depending on the methodology used. These differences are likely due, in part, to differences in temporal integration of the sound stimuli, since very brief sounds (~50 microseconds to 5 ms) are used to elicit ABRs and longer sounds (~200 ms) are used for behavioral detection experiments. In addition, ABRs represent synchronous neural activity across many neurons in response to the onset of a sound, whereas behavior may only require a small number of neurons responding to the onset or steady state of a sound to support detection. Finally, attentional and cortical activity is suppressed when measuring ABRs under anesthesia, yet cortical processes can enhance behavioral sound detection (Chambers et al. 2016).

In summary, rodents as a group do not generally exhibit sensitivity below 0 dB, putting them on par with humans. Their hearing in the ultrasonic range can be quite good, allowing for the perception of vocalizations in those high frequencies (Sect. 4.7.1; Okanoya and Screven, Chap. 2; Hurley and Kalcounis-Rueppell, Chap. 8). The variation in acuity may be related to lifestyle differences and evolutionary pressure, but since there are so few rodents tested to date, this remains largely unknown. Auditory acuity needs to be measured in many more rodent species in order to determine the reasons for the differences among rodents.

4.3 Frequency Selectivity

An animal's environment is rarely absolutely quiet. Thus thresholds for tones in quiet have limited usefulness for understanding both general auditory ability and acoustic communication. More likely, animals are trying to hear conspecifics and predators through some level of noise, often in the form of rain, wind, insects, and anthropogenic noise. They are also trying to discriminate between sounds of varying frequencies to distinguish between predators, offspring crying for help, or an attractive versus unattractive suitor. Several methods have been used to calculate frequency selectivity in rodents, including psychophysical tuning curves, critical ratios and critical bands, notched noise masking, tone-on-tone masking, and forward masking. Unfortunately, only laboratory rodents have been tested to date, leaving the frequency selectivity abilities of wild rodent species unexplored.

4.3.1 Critical Ratios and Critical Bands

Thresholds for hearing in noise and estimates of auditory filter bandwidth have been measured in several species of laboratory rodents using the critical ratio method, and these measures generally demonstrate similar acuity to other mammals (like cats) but worse acuity than humans (King et al. 2015). In general, across the animal kingdom, masked thresholds for pure tones embedded in white noise increase 3 dB/octave (Fay 1988). This trend is also seen in the critical ratios of rats, mice, chinchillas (*Chinchilla laniger*), and Mongolian gerbils (*Meriones unguiculatus*) (Fig. 4.3). Critical ratios range from around 20 dB at the lowest frequencies (chinchilla, Miller 1964) to about 45 dB at the highest frequencies (mouse, Ehret 1975) across these rodents, with across species' critical-ratio functions overlapping almost completely. Unfortunately, critical ratios have only been obtained for four species of rodents to date, and all measurements were taken in domesticated rodents.

In humans, critical bandwidths and critical ratios follow similar trends, with critical bands typically 2.5 times larger than critical ratios across frequencies. Critical bands are not usually measured in animals because the experiments are very time consuming, but they were measured in chinchillas by Seaton and Trahiotis (1975). The authors found that critical bands do not follow the typical trend seen in humans, deviating from critical ratios at 2 kHz and 4 kHz. Yost and Shofner (2009) later showed that critical ratios are better indicators of processing style (broadband versus narrowband) in both humans and nonhumans (using the chinchilla as a model) and do not accurately reflect critical bandwidths. Thus, the critical ratio is not necessarily a good measure of frequency selectivity, but it does indicate that

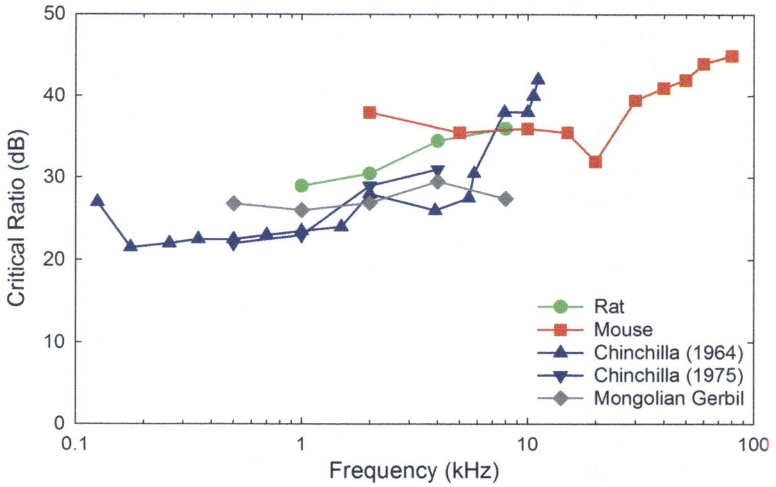

Fig. 4.3 Critical ratios for rat (*green circles*; Gourevitch 1965); mouse (*red squares*; Ehret 1975); chinchilla (*blue triangles*; Miller 1964) (*blue inverted triangles*; Seaton and Trahiotis 1975); and Mongolian gerbil (*gray diamonds*; Kittel et al. 2002)

chinchillas (and probably other rodents) are using broad bandwidths when detecting signals in noise, and humans listen in much narrower bands, showing greater acuity in these tasks (reviewed in King et al. 2015).

4.3.2 Simultaneous and Nonsimultaneous Masking

Psychophysical tuning curves (PTCs) also reflect frequency selectivity and can be obtained using *masking paradigms*. In general, PTCs are obtained using a detection task during which a fixed intensity pure tone is presented simultaneously with (i.e., *simultaneous masking*) or following (*partial forward masking* and *forward masking*) tones of varying intensities with frequencies above and below the test probe. The resulting measure is the masker intensity necessary to eliminate the perception of the test tone. Mammalian psychophysical tuning curves are generally V-shaped with steeper slopes for high frequency maskers than for low frequency maskers. Low-frequency test tones lead to wide and shallow functions, while high-frequency PTCs are sharp and narrow with flat tails in the low-frequency region. Interestingly, behaviorally obtained PTCs generally show the same pattern of results as single cell auditory nerve recordings and physiological tuning curves, revealing a close connection between the areas of activation of the masker and test tones in the auditory periphery.

In rodents, PTCs have been obtained for a variety of test frequencies only for chinchillas (McGee et al. 1976; Ryan et al. 1979). Tuning curves have been acquired for several strains of laboratory mice, including C57BL/6J and CBA/CaJ (Saunders et al. 1980; Taberner and Liberman 2005), guinea pigs (Mitchell and Fowler 1980), and chinchillas (Fig. 4.4) (Salvi et al. 1982). Overall, physiological tuning thresholds, such as ABRs and cochlear recordings, are higher than psychophysical tuning curves. Nonetheless, Salvi and colleagues (1982) found that chinchilla psychophysical tuning curves and ABR tuning curves follow similar patterns (Fig. 4.4a). Furthermore, Mitchell and Fowler (1980) showed that cochlear and auditory brainstem tuning curves were highly correlated across frequencies in guinea pigs (Fig. 4.4b). High frequency tuning curves in the C57BL/6J laboratory mouse strain exhibit the characteristic sharp and narrow PTCs with flat low-frequency tails (Saunders et al. 1980).

Fig. 4.4 (continued) and 11.2 kHz (*cyan circles*); chinchilla ABR tuning curves at 0.5 kHz (*blue squares*), 1 kHz (*pink squares*), 2 kHz (*black squares*), 4 kHz (*green squares*), 8 kHz (*red squares*), and 11.2 kHz (*cyan squares*). (**b**) ABR tuning curves for guinea pig (Mitchell and Fowler 1980) for 0.5 kHz (*red diamonds*), 1 kHz (*dark green diamonds*), 2 kHz (*blue diamonds*), 6 kHz (*pink diamonds*), 8 kHz (*cyan diamonds*); ABR tuning curves for mouse at 5 kHz (*green hexagons*), 16 kHz (*black hexagons*), and 20 kHz (*gray hexagons*). (**c**) ABR Q10dB values as a function of frequency for chinchilla (*red squares*; Salvi et al. 1982), guinea pig (*blue diamonds*; Mitchell and Fowler 1980), and mouse (*black hexagons*; Saunders et al. 1980); psychophysical Q10dB values for chinchilla (*yellow triangles*; McGee et al. 1976) (*light green circles*; Salvi et al. 1982)

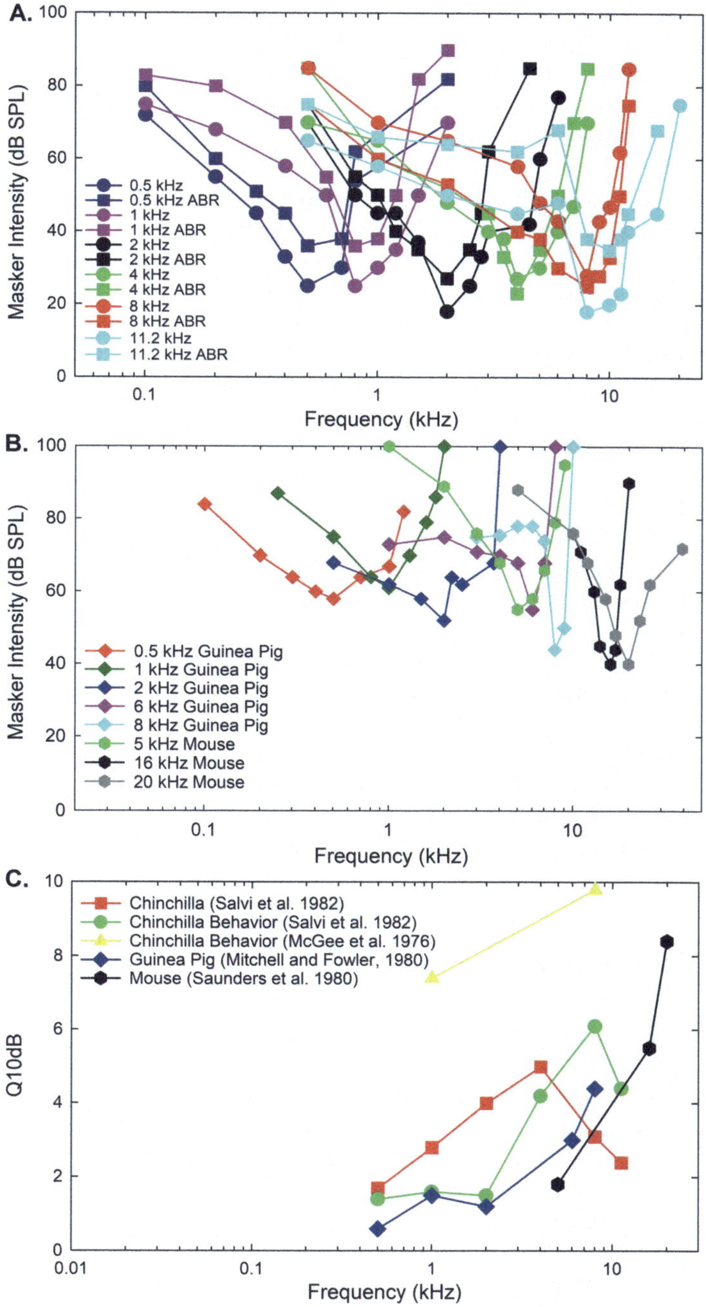

Fig. 4.4 Psychophysical tuning curves. (**a**) Chinchilla data from Salvi et al. (1982): 0.5 kHz (*dark blue circles*), 1 kHz (*pink circles*), 2 kHz (*black circles*), 4 kHz (*green circles*), 8 kHz (*red circles*),

The *Q10dB* is a measure of quantitatively assessing the frequency selectivity of tuning curves, defined as the center frequency of the tuning curve divided by the bandwidth 10 dB above the minimum threshold. Generally, at low frequencies the Q10dB values are low, indicating broad tuning, with progressively increasing Q10dB values at higher frequency PTCs (Fig. 4.4c). Across species, Q10dB values increase with frequency, although the values vary across rodents and methodologies. Indeed, Salvi and colleagues (1982) discovered that ABR Q10dB values were higher than behaviorally obtained Q10dBs in chinchillas for low frequencies, suggesting sharper tuning with ABR measures than with behavioral measures. McGee and colleagues (1976) showed sharper tuning for 1 kHz and 8 kHz test tones in chinchillas than were discovered by Salvi and colleagues (1982), suggesting a degree of within species variability in PTCs. Interestingly, Mitchell and Fowler (1980) did not find any differences between cochlear and brainstem Q10dBs in guinea pigs, and mouse Q10dB values indicated a linear trend similar to the other rodents (Saunders et al. 1980).

McGee and colleagues (1976) compared simultaneous and forward masking in chinchilla PTCs for 1 kHz and 8 kHz tones. The forward masking results were slightly lower than those for simultaneous masking with no significant differences between tuning curves. Long and Miller (1981) used simultaneous masking techniques for 1 kHz and 1.5 kHz maskers and postulated that chinchillas may be perceiving and responding to combination tones, leading to lower thresholds in certain areas of the tuning curves. These results suggest that, when measured using similar methodologies, chinchillas may have much poorer frequency selectivity than humans.

Finally, forward masking tuning curves have been found to be more comparable to tuning curves from single unit recordings than simultaneous masking results (Cheatham et al. 2004). Unfortunately, animal model types and equipment may limit the forward masking technique applicability.

4.3.3 Notched-Noise Masking

Notched-noise masking is another technique used to measure frequency resolution and auditory filter shapes in animals. In this procedure, frequency resolution is obtained based on the presentation of noise in which a notch of a certain bandwidth has been removed and the simultaneous presentation of a pure tone test probe. Notched noise thresholds have been measured in a few species of rodents to better estimate the frequency resolving abilities in these animals. In chinchillas, the obtained filters are broader in bandwidth than those in humans (Halpern and Dallos 1986). May and colleagues (2006) used operant conditioning procedures with positive reinforcement to assess auditory filter shapes in adult CBA/CaJ mice for 8 kHz, 11.2 kHz, and 16 kHz tones. These mice have auditory filtering effects that are similar to other mammalian species, including humans.

Lina and Lauer (2013) expanded on the findings from May and colleagues (2006), using the ABR as an alternative to behavioral techniques, to assess filter bandwidth in this strain of laboratory mice. The results revealed that, similar to behavioral results, ABR derived filter bandwidths broadened with increasing center frequency. Mouse frequency tuning is sharper at frequencies above 8 kHz as compared to humans. Overall, the ABR-derived *equivalent rectangular bandwidths* (ERBs) were similar to, but slightly lower than, behaviorally derived ERBs in mice and in humans.

In conclusion, frequency tuning as measured through notched noise, critical bands and critical ratios, and simultaneous and nonsimultaneous masking techniques in rodents varies depending on the species and methodology used. Unfortunately, little is currently known about frequency selectivity in rodents as a group, since only a few species have been tested.

4.4 Frequency Discrimination

Animals in their natural environment are inundated by complex sounds that must be divided into their various components and deciphered. Successful parsing of sounds into separate auditory objects is aided by frequency discrimination. Frequency discrimination is often investigated by measuring *frequency difference limens* (FDLs), or *just noticeable differences* in frequency. Researchers have been able to behaviorally measure frequency discrimination in various rodent species, including mice, rats, guinea pigs, chinchillas, and Mongolian gérbils, using operant conditioning methods (Fig. 4.5). Generally, there is a great deal of overlap across and within animal species, and FDLs follow the same general trend in all rodents tested to date.

FDLs have been measured in mice using various methodologies, and the results vary across strains. Heffner and Masterton (1980) found that the feral house mouse had a slightly shallower than expected FDL function. Kulig and Willott (1984) measured FDLs in two strains of laboratory mice. The DBA/2 mice could discriminate differences in frequency at a higher resolution than the C57BL/6 mice at 12 kHz and 16 kHz, but C57BL/6 mice had better frequency selectivity at 8 kHz. The authors believed that the differences observed between these two strains of mice were the result of differences in auditory neuron fibers. Radziwon and Dent (2014) measured FDLs at two tone intensities in CBA/CaJ mice. They found that these mice performed similarly under both 10 dB and 30 dB sensation levels (SL), and their FDLs were similar to thresholds measured in other rodents.

Frequency difference limens have also been measured in multiple species of rats. Historically, researchers believed that rats were very poor at frequency discrimination (e.g. Kelly 1970; Fay 1974). Furthermore, it was assumed in many species that, despite the ability to hear at high frequencies, animals had very poor frequency resolution at those frequencies. Heffner and Masterton (1980) observed that relative frequency discrimination in rats was as good at high frequencies as at low frequencies, indicating that rats are no worse at high-frequency than low-frequency differ-

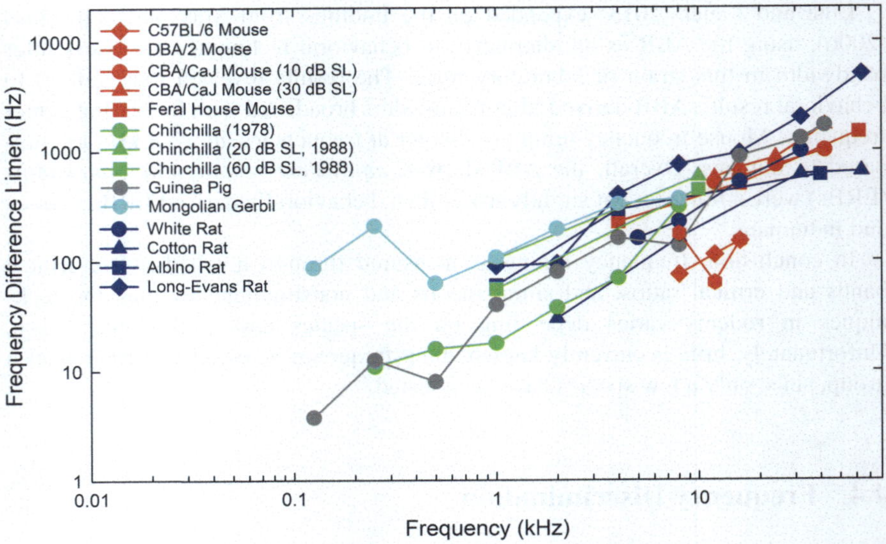

Fig. 4.5 Frequency difference limens for several rodent species: the CBA/CaJ mouse (*red circles*, 10 dB sensation level (SL); *red triangles*, 30 dB SL; Radziwon and Dent 2014); feral house mouse (*red squares*; Heffner and Masterton 1980); C57BL/6 mouse (*red diamonds*; Kulig and Willott 1984); DBA/2 mouse (*red hexagons*; Kulig and Willott 1984); white rat (*blue circles*; Talwar and Gerstein 1998), cotton rat (*blue triangles*; Heffner and Masterton 1980); albino rat (*blue squares*; Kelly 1970); Long-Evans rat (*blue diamonds*; Syka et al. 1996); chinchilla (*green circles*; Nelson and Kiester 1978) (*green triangles*, 20 dB SL; *green squares*, 60 dB SL; Prosen et al. 1988); guinea pig (*grey circles*; Heffner et al. 1971); Mongolian gerbil (*cyan circles*; Sinnott et al. 1992)

ence discrimination. Kelly (1970) and Syka and colleagues (1996) investigated FDLs in albino and Long-Evans rats, respectively, and discovered that frequency discrimination thresholds were much poorer in rats when compared to other mammals. However, Talwar and Gerstein (1998) showed that frequency discrimination could be much more sensitive in some rats than previous findings had indicated.

Heffner and colleagues (1971) found that the guinea pigs' frequency discrimination function had a shallower slope at low and high frequencies and was steeper for mid-frequencies, but overall, the results were comparable to other rodents. In chinchillas, FDLs were higher than those seen in other mammals but similar to other rodent species (Nelson and Kiester 1978; Long and Clark 1984). Furthermore, the mechanism of frequency difference detection was investigated by Prosen and colleagues (1988), who monaurally damaged the organ of Corti and observed no changes in measured FDLs at either 1 kHz or 10 kHz despite an absolute threshold shift at 10 kHz. Mongolian gerbils have the shallowest FDLs measured so far and much higher FDLs at the lowest frequencies relative to other rodents (Sinnott et al. 1992). Klinge and Klump (2008) later confirmed these high thresholds at low frequencies and suggested that the short cochlea paired with the large frequency range of hearing is to blame.

In summary, different species of rodents have similar frequency difference thresholds, with some variation in ability. It is widely agreed upon that the performance of rodents is worse, however, when compared to birds, other mammals, and humans. Again, frequency discrimination abilities in nondomestic, nonlaboratory rodents are needed to confirm if there are commonalities across all animals of this order.

4.5 Intensity Discrimination

The ability to utilize and discriminate all aspects of sound is critical for survival. Intensity is an important auditory cue that plays a role in distance perception and may be used for conspecific and predator ranging (Naguib 1997). The ability to discriminate intensity levels across frequencies and background levels is measured through *intensity difference limens* (IDLs) by determining the *just noticeable difference* in loudness of the same sound. Thus far, IDLs have been determined for mice, rats, chinchillas, and Mongolian gerbils (Fig. 4.6).

In mice, IDLs have been measured in three separate strains using operant conditioning procedures. Ehret (1975) discovered that IDLs in Naval Medical Research Institute (NMRI) mice significantly decreased with increasing sensation levels for all frequencies except 10 kHz. Behrens and Klump (2015) found that C57BL/6J mice show increased sensitivity for IDL detection from low to high background levels, similar to the results from Ehret (1975). Finally, Kobrina and colleagues (2018) measured IDLs in the CBA/CaJ mouse strain. The CBA/CaJ mice could discriminate intensities across two background levels, with the 10 dB SL thresholds for 16 kHz and 42 kHz tones significantly higher than thresholds for those same tones at 30 dB SL. Otherwise, thresholds did not differ across background intensity. The CBA/CaJ mice showed higher IDLs than in other mouse strains and rodent species, except for Mongolian gerbils (Fig. 4.6), but the reason for this deviation is not yet known.

Intensity difference limens also have been studied in two rat strains: Wistar albino rats and Long–Evans pigmented rats (Fig. 4.6). Hack (1971) found that albino rats were reliably able to detect very small intensity differences at 20 dB and 40 dB SL. Similar to other mammals, IDLs in this rat strain decreased with increasing sensation levels. Syka and colleagues (1996) found that pigmented hooded rats could discriminate upward and downward intensity shifts at 50 dB SL. IDLs in these rats were frequency independent, but direction of the shift (upward versus downward) had a significant effect on results. The thresholds for upward intensity shifts were significantly lower than those of downward intensity shifts. Downward intensity shifts are typically not measured in comparative psychoacoustic experiments, thus it is unknown if this trend is typical for rodents and other animals.

The only other measured IDLs in rodents are in chinchillas and Mongolian gerbils (Fig. 4.6). Chinchilla IDLs were similar to those from rats and mice, decreasing with increasing intensity and relatively stable across frequency (Saunders et al. 1987). In Mongolian gerbils, thresholds at the lowest intensities were very high and

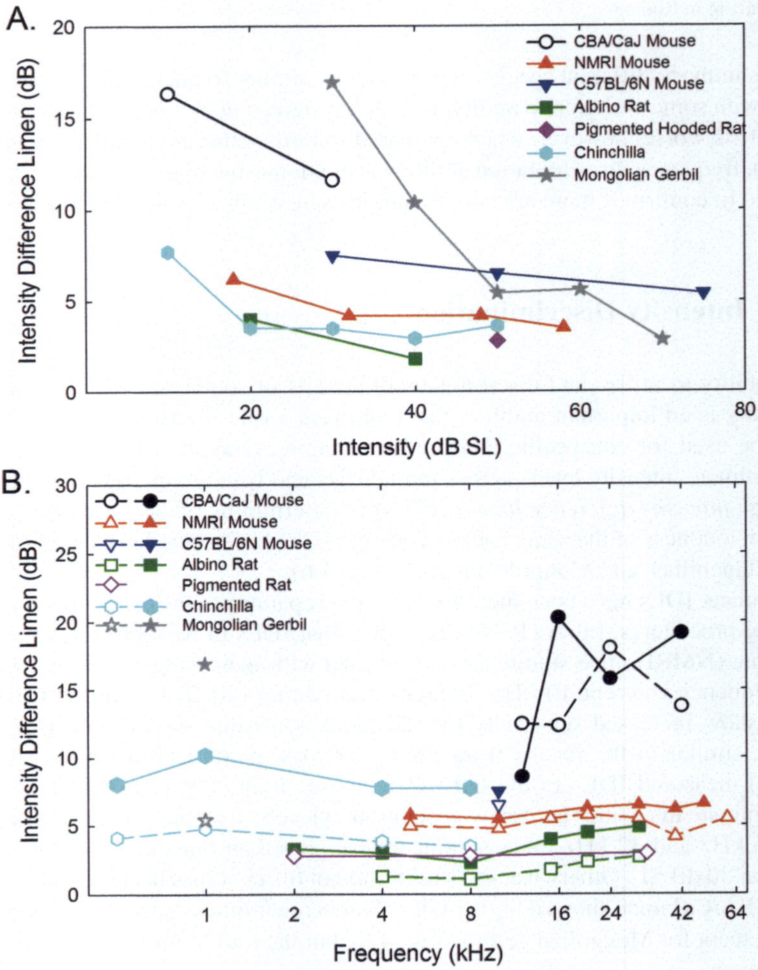

Fig. 4.6 Intensity difference limens (IDL) for a variety of rodents as a function of background sound level (**A**) and frequency (**B**). (**A**) Mean IDL values across intensities for several rodent species using operant conditioning procedures with positive reinforcement; mean IDLs for 16 kHz tones, 100 ms in duration for CBA/CaJ mice (*white circles*; Kobrina et al. 2018); NMRI mouse mean IDLs for 100 ms, 15 kHz tone bursts (*red triangles*; Ehret 1975); mean IDLs for the C57BL/6N mouse for 100 ms, 10 kHz tones (*blue inverted triangles*; Behrens and Klump 2015); albino rat mean IDLs for 1.2 s, 15 kHz tones (*green squares*; Hack 1971); pigmented rat mean IDLs for 1.0 s, 8 kHz tones (*pink diamonds*; Syka et al. 1996); chinchilla mean IDLs for 500 ms, 8 kHz tone bursts (*cyan hexagons*; Saunders et al. 1987); Mongolian gerbil IDLs for 400 ms, 1 kHz tones (*gray stars*; Sinnott et al. 1992). (**B**) Mean IDL values across frequencies for several rodent species using operant conditioning procedures with positive reinforcement: CBA/CaJ mouse mean IDLs for 100 ms tones at 10 dB sensation level (SL) (*black circles*) and 30 dB SL (*white circles*) (Kobrina et al. 2018); NMRI mouse mean IDLs at 18 dB SL (*red triangles*) and 32 dB SL (*white triangles*) for 100 ms tone bursts (Ehret 1975); mean IDLs for the C57BL/6N mouse for 100 ms tones at 30 dB SL (*blue inverted triangles*) and 50 dB SL (*white inverted triangles*) (Behrens and Klump 2015); albino rat mean IDLs for 1.2 s tones at 20 dB SL (*green squares*) and 40 dB SL (*white squares*) (Hack 1971); pigmented hooded rat mean IDLs for 1.0 s tones at 50 dB SL (*white diamonds*; Syka et al. 1996); chinchilla mean IDLs at 10 dB SL (*light blue hexagons*) and 30 dB SL (*white hexagons*) for 500 ms tone bursts (Saunders et al. 1987); Mongolian gerbil IDLs at 30 dB SL (*gray stars*) and 50 dB SL (*white stars*) for 400 ms tones (Sinnott et al. 1992)

similar to IDLs in the CBA/CaJ mouse (Sinnott et al. 1992). Overall, IDLs in rodents are somewhat higher than those seen in other mammals (e.g., Fay 1988). Saunders et al. (1987) suggest that this difference reflects differences in frequency selectivity in rodents compared to humans and other mammals (discussed in Sect. 4.3).

4.6 Temporal Resolution

Temporal processing is an important feature of many animals' auditory systems that allows them to identify and understand complex sounds in their environment. Temporal processing can be measured across different scales, from the detection of short versus long tones to the discrimination of amplitude modulated sounds with different modulation rates. Combined, these measures describe the ability of rodents to distinguish between complex auditory objects and to comprehend complex acoustic signals used for communication. Without adequate temporal resolution, neither is possible.

4.6.1 Temporal Summation

Temporal summation is a phenomenon in which an increase in stimulus length results in a decrease of the absolute hearing threshold for that stimulus. For auditory temporal summation, thresholds decrease about 3 dB as the duration of the stimulus doubles (Viemeister 1996). Temporal summation has been investigated in only three species of rodents. Ehret (1976) used a wide variety of durations, ranging from 1 ms to 3 s, and frequencies from 1 kHz to 120 kHz to determine if mice would follow the expected pattern of showing decreased thresholds for longer stimuli. Across all frequencies, mice had lower thresholds as the stimuli increased in duration up to 1000 ms. This suggests that the additional energy from longer duration sounds helps animals to detect these sounds more easily. A similar pattern was found in chinchillas. Henderson (1969) discovered that chinchillas show a decrease in thresholds for pure tones at 2 kHz by as much as 12 dB when duration was increased from 25 ms to 750 ms. Finally, Gleich and colleagues (2007) obtained thresholds for 2 kHz pure tones ranging from 10 ms to 1000 ms in Mongolian gerbils. Thresholds improved for tone durations up to 300 ms, making their temporal integration functions like those from other animals tested to date.

The results from these studies provide evidence that the auditory system integrates information across the duration of a sound and uses this to improve signal detection across a wide variety of frequencies. Physiological responses in the auditory periphery to stimuli differing in duration can be comparable to behavior (Eddins et al. 1998), and can account for the detection and discrimination of natural vocalizations (Brand et al. 2000).

4.6.2 Gap Detection

Another commonly used method for assessing temporal processing in animals is measuring gap detection thresholds. These thresholds are obtained through both behavioral and physiological techniques, and they indicate the minimum silent interval embedded within a stimulus that an animal can detect. Gap detection has been investigated in several species of rodents to determine how well these animals are able to perceive temporal changes, a cue that is likely important for processing environmental and conspecific stimuli. There is a general trend that gap detection thresholds are the best (smallest) when the stimuli presented are more intense, and gap detection thresholds are worse (larger) at lower intensities (Fig. 4.7). Additionally, it is believed that high-frequency hearing plays a role in how well animals can detect intervals of silence.

Behavioral methods that have been employed to investigate gap detection vary from simple acoustic startle reflex analysis to more complex operant conditioning paradigms. One paradigm for studying gap detection has been *pre-pulse inhibition* (PPI), where detection of a period of silence enables the animal to suppress their acoustic startle reflex during the presentation of a loud sound immediately following the silent period. Radziwon and colleagues (2015) used a PPI procedure in rats and found that gap detection thresholds were between 2 ms and 3 ms for loud broadband sounds. Thresholds increased to 4.3 ms for the quietest broadband noise.

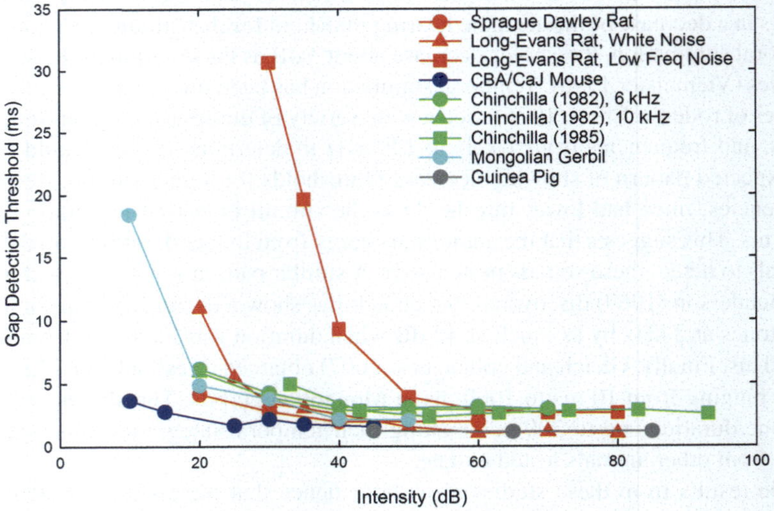

Fig. 4.7 Gap detection thresholds as a function of intensity for Sprague-Dawley rats (*red circle*; Radziwon et al. 2015); Long-Evans rats for white noise (*red triangles*) and low frequency noise (*red squares*) (Syka et al. 2002); CBA/CaJ mice (*blue circle*; Radziwon et al. 2009); chinchillas at 6 kHz (*green circles*) and 10 kHz (*green triangles*) (Giraudi-Perry et al. 1982) and for broadband noise (*green squares*; Salvi and Arehole 1985); Mongolian gerbils (*cyan circles*; Wagner et al. 2003) and guinea pigs (*gray circle*; Feng et al. 2007)

Additionally, as the frequency range of the stimuli was decreased, gap detection worsened. Walton and colleagues (1997) discovered that mice display gap detection thresholds between 1 ms and 3 ms using the PPI paradigm. PPI has been used to measure gap detection in a variety of rodents, including Mongolian gerbils and mice (e.g., Gaese et al. 2009; reviewed in Lauer et al. 2017).

Operant conditioning paradigms also have been used to determine gap detection acuity. Syka and colleagues (2002) discovered that gap detection thresholds in rats were about 1.6 ms for loud, broadband white noise stimuli, slightly smaller than the thresholds measured using PPI. Thresholds were larger for low-frequency noise, which was in agreement with other measures of gap detection. Further, Radziwon and colleagues (2009) measured gap detection thresholds of 1.6 ms for high intensity sound and 3.4 ms for low intensity sound. In chinchillas, Salvi and Arehole (1985) found gap detection thresholds of 2.6 ms and 5 ms for high and low intensity stimuli, respectively. Finally, in Mongolian gerbils, gap detection thresholds for high intensity noise ranged from 2.1 to 3 ms and increased to 5–6 ms for lower intensity stimuli (Wagner et al. 2003).

Physiological measurements are a widespread method of measuring gap detection thresholds. For example, Syka and colleagues (2002) used auditory cortex (AC) lesions to determine temporal processing in rats. The authors found that following ablation of the AC, gap detection thresholds increased from about 1.6 ms to 4 ms, indicating that the auditory cortex does play a role in perceiving temporal information. This inflated threshold gradually decreased but remained higher than the pre-ablation threshold two months later.

The inferior colliculus (IC) has also been identified as an area likely to be involved in processing of auditory temporal information. Walton and colleagues (1997) obtained gap detection thresholds from the central nucleus of the IC and found that thresholds were equivalent to those measured using PPI (1–3 ms). Lastly, Wang and colleagues (2006) measured responses from both the AC and IC in guinea pigs and found that gap detection thresholds differed in the two regions. In the IC, the thresholds were 1–2 ms, whereas in the AC, thresholds were 2–4 ms, depending on the frequency and intensity of the stimuli. These physiological measurements of gap detection thresholds obtained through recordings throughout the central auditory system agree with the behavioral measurements.

The results of both behavioral and physiological studies suggest a relationship between high frequency hearing and gap detection ability. Radziwon and colleagues (2009) proposed that mice have the smallest measured gap detection thresholds of all rodents tested to date because mice have the best high-frequency hearing. The rank order of the upper limit for high-frequency hearing is mirrored by gap detection ability: mice have the best gap detection and the best high frequency hearing ability, chinchillas have the worst gap detection and worst high frequency hearing, and rats and Mongolian gerbils fall in the middle. This assertion is supported by the findings of Salvi and Arehole (1985), who discovered that the high frequency content of a stimulus is the most important characteristic for gap resolution.

Gap detection is an important measurement for determining temporal processing ability. The resolution of gap detection thresholds in rodents varies, depending upon the frequency and intensity information present in the sound stimulus: broader frequency ranges and higher intensities produce the smallest thresholds. Measurements of the temporal processing power of rodents provide information for how small temporal changes, which are present in environmental noise and in calls from conspecifics, can be used for deciphering acoustic stimuli.

4.6.3 Perception of Sweeps

Frequency modulation is an important characteristic of many vocalizations by rodents. It is therefore critical that animals can perceive changes in frequency modulation to distinguish among calls. The mammalian brain has a considerable number of neurons that are finely tuned to changes in the direction of frequency modulation, including in the rat auditory cortex (e.g., Gaese and Ostwald 1995) and IC (Felsheim and Ostwald 1996). This is also true for the mouse auditory cortex (Honma et al. 2013). A few researchers have utilized behavioral methods and operant conditioning paradigms to ask animals directly how they perceive frequency-modulated sweeps. Gaese and colleagues (2006) tested perception of the direction of frequency-modulated sweeps and found that rats rely on both the duration as well as the frequency range of the sweep to distinguish between them. Sweeps that were longer in duration and had a wider frequency range were easier to discriminate than shorter sweeps or those that had a smaller frequency range. The authors believe that this could be the result of differences in the audibility of the parameters of the sweeps when they are longer or have a wider bandwidth, differences in the selectivity of neurons at the frequencies tested, or the unavailability of phase locking at frequencies above 4 kHz.

Screven and Dent (2016) investigated discrimination of frequency-modulated sweeps in CBA/CaJ mice. The authors determined that, like rats, discrimination performance between upward and downward frequency-modulated sweeps improved for sweeps that were longer in duration or wider in frequency range. While these results agree with what has been found in rats, the results were surprising because the mice were unable to discriminate between sweeps that were like their natural vocalizations with regard to duration and frequency range. This indicates that the upsweep and downsweep vocalizations produced by the mice may not be perceived as different by the mice.

Finally, frequency-modulated sweep discrimination was demonstrated by Wetzel and colleagues (1998b) in Mongolian gerbils. Mongolian gerbils showed the same pattern of increased performance for sweeps starting at a lower frequency than those starting at a higher frequency range, which the authors attributed to the activation of low-frequency neurons because there are comparatively more neurons that are tuned to lower frequencies than to higher frequencies. Discrimination ability appears to be at least partially reliant on the auditory cortex, as performance decreases following ablation of the auditory cortex (Wetzel et al. 1998a). The results of frequency modulation discrimination experiments indicate that this is an important characteristic that rodents likely rely upon when perceiving vocalizations from conspecifics. The parameters of the sweeps, including duration and frequency range, clearly play a critical role in their discriminability in all rodents that have been tested so far.

4.6.4 Modulation Transfer Functions

The sensitivity of the auditory system to temporal changes is often measured using *temporal modulation transfer functions* (TMTFs), where sensitivity to sinusoidally amplitude-modulated noise is measured at various rates of frequency modulation. In general, this process is easier at low frequency-modulation rates and becomes more difficult for animals as frequency modulation rate increases. The TMTF measurements can be taken either behaviorally or physiologically.

Behavioral methods have been used to examine TMTFs in two species of rodents: rats and chinchillas. Kelly and colleagues (2006) measured TMTFs in rats behaviorally and observed that, like other mammals, rats experienced the lowest thresholds at low modulation rates and became increasingly worse as the frequency modulation rate increased. The rats lost the ability to perceive any modulation for rates above 2 kHz, which the authors postulated was due to the inability of the rats' auditory nerve to synchronize to the amplitude modulation at high rates. Although synchronization of firing is best for frequencies up to 100 Hz, there is a certain degree of synchronization up to 250 Hz; these results indicate that there is a mismatch between what would be inferred about TMTF thresholds measured physiologically compared to what was observed behaviorally.

Measurements of TMTFs were conducted in chinchillas by Salvi and colleagues (1982). Similar to rats, thresholds were best for modulation rates up to 30 Hz, above which detection became poorer. At higher rates, the temporal fine structure of the sound envelope became smeared, leading to a decreased ability to detect the amplitude modulation rate. Further, the authors related the chinchillas' behavioral responses to low-pass filters at low frequencies. These behavioral studies in just two species of rodents generally show similar trends to other animals.

The majority of the modulation transfer function measurements in rodents have used physiological methods, recording from various structures along the auditory pathway. Palombi and colleagues (2001) found a slightly higher proportion of neurons that responded to sinusoidally amplitude-modulated noise in the central nucleus of the IC as compared to the external nucleus of the IC in rats. Additionally, Krishna and Semple (2000) recorded from the IC of Mongolian gerbils and found that neurons showed characteristics of a low-pass filter for low modulation frequencies with the majority of the neurons showing no synchrony for modulation rates over 300 Hz.

Recordings from the cochlear nucleus (CN) of the Mongolian gerbil revealed that cells in the ventral CN that have precise onset timing show the greatest firing synchrony to amplitude modulation. Schulze and Langner (1997) discovered that there are two populations of neurons within the auditory system that respond to sinusoidally amplitude-modulated noise in Mongolian gerbils. One population encodes temporal information for low modulation rates through synchrony of neural discharges, while the other population encodes information about high modulation rates through nonsynchronous discharges.

Cooke and colleagues (2007) examined neural firing in the auditory cortex of rats in response to amplitude-modulated noise. They believed that rats would show similar neural activity patterns as Mongolian gerbils because synchrony of neural firing to the modulation rate is impossible once modulation frequency becomes too high. After ablating the auditory cortex, the rats experienced deficits in perceiving amplitude modulation, particularly at high modulation rates. Cooke and colleagues (2007) hypothesized that subcortical auditory structures preserved sensitivity at the lower frequencies, but that the cortex was important for detecting amplitude modulation at high modulation rates.

Temporal variations in natural complex sounds in the environment likely contain critical information. It is therefore likely that animals have intricate systems to help them understand this information so they may respond appropriately. Using physiological and behavioral methods, researchers have attempted to discover how rodents perceive and encode this information throughout the auditory system. In general, the ability of rodents to perceive temporal information is fine-tuned at low modulation rates and becomes increasingly worse as the rate gets higher, until they are no longer able to detect the modulation at all. Researchers believe this is because of the upper limit of the rodent auditory system to fire in response to high modulation frequencies, a process common across the animal kingdom.

4.6.5 Duration Discrimination

Although animals are finely tuned to many of the temporal properties of a stimulus, duration discrimination has not yet been thoroughly explored in rodents. Klink and Klump (2004) examined duration discrimination in the NMRI mouse using tone durations of 50 ms, 100 ms, and 200 ms. The authors discovered that for short tones, the mice could discriminate relatively small differences in durations. As the reference tones became longer, the mice needed larger differences in duration to be able to discriminate the difference. Brand and colleagues (2000) found that 55% of cells in the nonalbino mouse IC were sensitive to duration, similar to other animals.

Kelly and colleagues (2006) measured duration discrimination in rats to determine if rats show a similar pattern as mice. Previous duration discrimination thresholds in rats indicated that for durations greater than 500 ms, their performance became elevated beyond what would be expected if rats were discriminating similarly to mice (Church et al. 1976). Kelly and colleagues (2006) showed that rats exhibited discrimination thresholds that were comparable to those from mice, contradicting the results of Church and colleagues (1976), although strain differences are a possible factor.

Kelly and colleagues (2006) also discovered that duration discrimination was contingent upon the loudness of the stimuli that were used: louder stimuli showed improved thresholds compared to quieter stimuli. To test this, the authors used stimuli that were the same duration and were presented at either 50 dB or 80 dB SPL. The rats showed better duration discrimination for the 80 dB stimuli than for the 50 dB stimuli. Temporal summation, where longer stimuli are perceived as louder, is a likely factor in the improved duration discrimination thresholds for more intense stimuli (Sect. 4.6.1).

The results of the duration discrimination and other temporal processing experiments contribute to our understanding of how rodents can process small changes in the timing of stimuli. Evolutionarily, this capability is a potentially critical component of understanding natural sounds in the environment.

4.7 Complex Sound Perception

The detection and discrimination of simple acoustic stimuli, summarized in Sects. 4.2, 4.3, 4.4, 4.5, and 4.6 for rodents, demonstrate an animal's auditory acuity and potential for acoustic communication. Without the ability to distinguish stimuli that are changing in location (Lauer, Engel, and Schrode, Chap. 5), intensity, frequency, or over time, animals would not be able to use those cues to differentiate more complex stimuli. As discussed in other chapters (Okanoya and Screven, Chap. 2; Schleich and Francescoli, Chap. 3), rodents produce signals both in air and seismically. To use these signals for communication purposes, rodents must be able to detect them. To use these signals to understand specific information about a situation, such as a predator type and location, rodents must be able to discriminate and categorize the signals (Hurley and Kalcounis-Rueppell, Chap. 8). Laboratory and field studies have demonstrated the potential for discrimination and categorization acuity in rodents. Further, studies on the perception of human speech by rodents may increase the utility of these animals as models for human communication and auditory disorders (see Sect. 4.7.2). Finally, studies of auditory scene analysis add to what is known about the complex auditory world of rodents, highlighting how individual auditory objects are assigned within the cacophony of sounds an animal hears.

4.7.1 Vocalizations

Animals produce vocalizations in multiple circumstances, including social and agonistic interactions. Rodent calls during these situations may be context specific, although this has not been shown in all species. For vocalizations to carry specific meanings, it is critical that animals discriminate among the various vocalizations in their repertoire. If animals are unable to tell the difference between these sounds, it is unlikely that the animals are using them in a way to communicate specific things about a situation, environment, or condition, because all vocalizations would be perceived as being the same. The encoding of vocalizations by rodents has been studied using a variety of behavioral and physiological techniques in both the field and the laboratory.

The discrimination of vocalizations has been extensively studied in juvenile and adult mice. Mouse pups produce characteristic ultrasonic vocalizations (USVs) that have been previously shown to elicit approach and retrieval behavior from maternal females (e.g. Noriot 1972; Smith 1976). Ehret and Haack (1981) examined how lactating female NMRI mice perceive natural pup calls compared to other ultrasonic sounds. Females placed narrowband noise and pup USVs within the same category, indicating that these mice rely on the frequency bands of calls more than the frequency or amplitude modulation seen within natural USVs. Following these findings, the response of maternal female mice to natural pup USVs and several synthetic noise stimuli was investigated in NMRI mice (Ehret and Haack 1982). Female mice show a similar preference for ultrasonic pure tones, narrowband noise, and ultrasonic pup calls. These results support previous findings that mice do not require the intricate frequency sweeps and amplitude modulation seen in mouse calls to elicit a response.

Although the female mice in the studies above responded to multiple types of acoustic stimuli in a similar manner, other studies have found that mice at least have the capacity to detect and discriminate between ultrasonic vocalizations, suggesting that different calls could be used for communicating different details about the mouse's situation. CBA/CaJ mice can detect ultrasonic vocalizations at a lower intensity than they can pure tones of similar frequencies, which is not surprising given the spectrotemporal complexity of those vocalizations (Kobrina and Dent 2016). Neilans and colleagues (2014) illustrated that discrimination of adult USVs depends upon the spectrotemporal complexity of the calls (Fig. 4.8). When calls were spectrotemporally similar, mice were less able to discriminate between them, whereas if calls were spectrotemporally dissimilar, mice showed increased discrimination performance. Discrimination ability was also measured for manipulated USVs by Neilans and colleagues (2014). They halved or doubled the duration of USVs and removed the frequency modulation in the calls, which are manipulations that previously elicited similar single unit responses in the IC (Holmstrom et al. 2010). These manipulated calls were found to be difficult to behaviorally discriminate from the natural USVs, providing behavioral evidence that supports Holmstrom and colleagues' findings that these calls are represented similarly in the central nervous system.

Finally, to determine if certain portions of mouse USVs contain more information than other portions, mice were trained to discriminate between partial and whole vocalizations (Holfoth et al. 2014). The researchers discovered that mice struggled to discriminate between the first one-third of a call and the whole call, but discrimination improved when presented with either of the other thirds of the call (Fig. 4.9). These findings indicate that the beginning of a call carries the most information or is the most salient, since the mice perceived these partial calls as the same as whole calls much like humans do with human speech sounds. Overall, these limited results on call perception by mice suggest that they at least have the potential

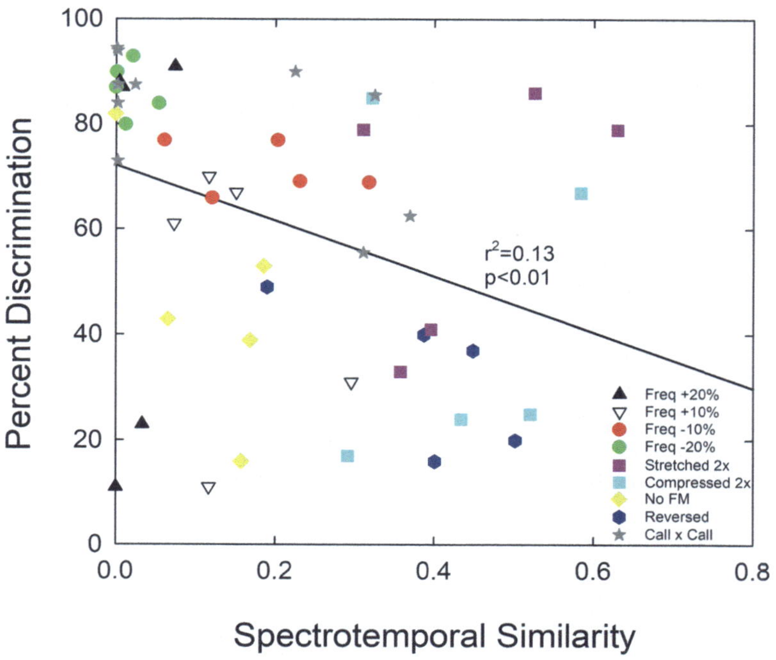

Fig. 4.8 Discrimination of ultrasonic vocalizations as a function of spectrotemporal similarity. Calls were increased or decreased in frequency, frequency modulation was removed, calls were reversed, call duration was doubled or halved, and unmanipulated calls were tested against each other. (Reproduced from Neilans et al. (2014) with permission)

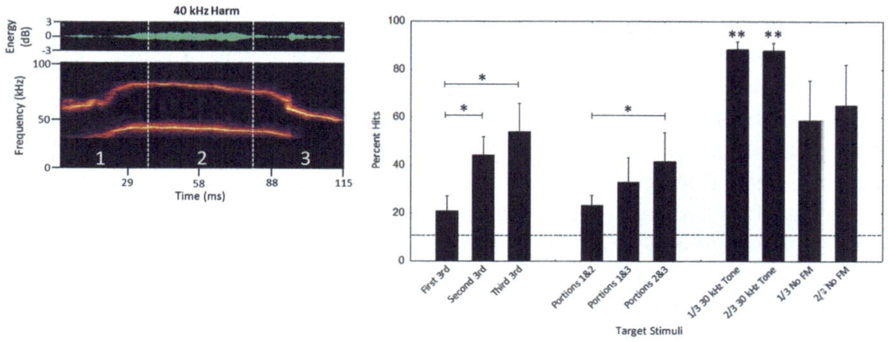

Fig. 4.9 Spectrogram and time waveform of a "40 kHz Harmonic" call from a CBA/CaJ mouse (*left*). Discrimination of the first third, second third, and last third of that call from the whole call (*right figure, left three bars*). Discrimination of two portions of that call from the whole call (*right figure, middle three bars*). Discrimination of shorter pure tones or a no frequency modulation (no FM) version of the call from the whole call (*right figure, right four bars*). (Reproduced from Holfoth et al. (2014) and used with permission of AIP Publishing)

for hearing different categories of vocalizations and, perhaps, the different categories could be used to convey distinct information to receivers.

One of the most critical classes of vocalizations that animals emit are alarm calls. These calls serve to alert conspecifics to potential dangers in the environment, helping them to avoid predation. Yellow-bellied marmots (*Marmota flaviventris*) produce alarm calls and can communicate the risk of a particular predator or situation by varying both the number and rate of calls that they produce (Blumstein and Armitage 1997). These animals also are one of a very short list of animals that can recognize individuals based on their alarm calls, as illustrated by Blumstein and Daniel (2004). Blumstein and Daniel (2004) theorized that the ability to identify specific callers could allow receivers to discriminate between conspecifics and heterospecifics if calls from conspecifics are considered more reliable. Further, the authors determined that the marmots could distinguish adult callers from juvenile callers, and, somewhat surprisingly, adults treated calls from juveniles as more salient than vocalizations from adults. It is possible that the parents placed extra importance on calls from juveniles so that they could provide the youngsters with extra protection.

Yellow-bellied marmots and golden-mantled ground squirrels (*Callospermophilus lateralis*) are able to discriminate and understand the alarm calls of other species. Shriner (1998) discovered, using alarm call playbacks, that both yellow-bellied marmots and golden-mantled ground squirrels responded to the calls of the other species, even in the absence of an actual predator. This provides concrete evidence that these animals are not simply responding to the sight of a predator at the same time as the other species produces the alarm call. Blumstein and Armitage (1997) reported that yellow-bellied marmots also respond to the alarm calls of rock squirrels, although the calls of these two species are similar, unlike the vocalizations of the golden-mantled ground squirrel.

Koeppl and colleagues (1978) identified Richardson's ground squirrels (*Spermophilus richardsonii*) as a species that produces individual-specific vocalizations. It is likely that these individually identifiable calls serve a purpose for squirrels, as these animals tend to habituate to repeated production of alarm calls. Perception of multiple individuals producing an alarm call could lead to increased vigilance given Hare's (1998) observation that only one individual producing these calls resulted in decreased vigilance. Finally, Nakano and colleagues (2013) found that degus (*Octodon degus*) could discriminate between the alarm calls of juvenile and adult conspecifics and that they use this information to preferentially respond more frequently to the calls of adults than to those of juveniles. The authors hypothesized that this is because adults are considered more reliable than young animals and, therefore, they preferentially respond to the calls of these animals in order to increase time spent pursuing other activities. These results, taken together, indicate that rodents can produce and understand complex, diverse, and individualistic vocalizations and utilize these calls to evade predators.

In summary, a wide variety of rodents produce vocalizations that have widespread uses, including maternal, social, and evasive behavior. Accurate perception of these vocalizations is critical to the survival of these animals. Furthermore, some animals respond even to the calls of other species within their habitat, indicating that rodents are capable of learning to associate vocalizations they do not produce with the appropriate behavioral response.

4.7.2 Human Speech

To understand the evolution of human speech, many researchers have looked at nonhuman animals and investigated if these animals are able to process speech or speech-like stimuli similarly to humans. By doing this, researchers can explore the physiological and anatomical frameworks required for speech perception using methodologies that would not be available when studying humans. Rodents have emerged as an important model for human speech because many rodents perceive phonetic boundaries and discriminate speech sounds similarly to humans.

One critical feature of understanding speech is the perception of voice onset time (VOT). According to the *phoneme boundary effect*, VOTs create a phonetic boundary that allows listeners to classify speech sounds into their appropriate perceptual category (Wood 1976). A common example of VOT is the boundary between /ba/ and /pa/, with short VOTs categorized as /ba/ and longer VOTs categorized as /pa/. There is a sharp boundary in humans that divides the perception of two syllables into the two distinct categories. Phonetic boundaries were first investigated in chinchillas by Kuhl and Miller (1975) on a classification task using /ta/ and /da/ syllables as stimuli. Chinchillas can not only discriminate between these two syllables, but they also demonstrate phonetic boundaries at the same locations as humans.

The similarity of phonetic boundaries between chinchillas and humans was further supported by Ohlemiller and colleagues (1999) who found that the classification of stop consonants also showed a similar boundary. In chinchillas, VOT difference limens show the greatest sensitivity to changes in VOT that occur around their perceptual boundary (Kuhl 1981), indicating that at least some of the framework underlying speech perception is not a result of experience or production of speech sounds. The auditory processing of VOTs in Mongolian gerbils (Sinnott and Mosteller 2001) has also been found to be comparable to humans, and the ability does not appear to degrade with age (Sinnott and Mosqueda 2003).

Another important feature required for understanding speech is the ability to discriminate between various speech sounds. If humans were unable to tell the difference between words, individual words would not carry context-specific or individual meaning. This capability to discriminate between speech signals was found in rats by Floody and Kilgard (2007). Those results agree with other reports of the capacity of animals to understand speech sounds (e.g., Sinnott and Mosteller 2001; Brown and Sinnott 2006).

The discrimination of speech sounds is dependent on processing in the primary auditory cortex. Floody and colleagues (2010) discovered that discrimination of /bæ/ and /pæ/ syllables was negatively affected in rats following damage to the anteroventral quadrant of the right hemisphere of the primary auditory cortex. The deficits seen in speech discrimination following partial ablation of the primary auditory cortex suggest that this area is required for complex temporal processing (e.g., VOT) but not for simpler temporal cues. Those findings were further supported by Porter and colleagues (2011) who found that the primary auditory cortex was necessary for rats to discriminate truncated speech sounds. This indicates that the primary auditory cortex is required for perception of the acoustic transitions at the beginning of speech sounds.

Finally, humans can discriminate degraded speech sounds, such as *vocoded speech*, with up to 90% accuracy (e.g., Shannon et al. 1995; Xu et al. 2005). Vocoded speech breaks the speech signal apart into different frequency bands (discussed in Shannon et al. 1995) and sounds similar to speech heard by humans using cochlear implants. Ranasinghe and colleagues (2012) discovered that rats are able to process degraded speech sounds as well as normal speech, indicating speech perception in rats is comparable to humans. In contrast, Shofner (2014) investigated the perception of noise vocoded speech in chinchillas using a stimulus generalization paradigm and reported that chinchillas are unable to understand degraded speech sounds. The author argued that speech perception relies on top–down processing because of experience with speech, and therefore chinchillas have inadequate top–down processing of these signals. This does not account for the ability of rats to discriminate vocoded speech sounds, using certain techniques, however.

In summary, rodents show the capacity to detect, discriminate, and categorize various characteristics of speech sounds. This illustrates that rodents are an appropriate model for investigating the mechanisms that contribute to human speech perception, although this perceptual ability is still not fully understood.

4.7.3 Auditory Scene Analysis

Experiments on the perception of complex signals mentioned throughout this chapter typically involve just one signal, or perhaps a signal and a masker, presented to the animal. In the real world, animals are regularly inundated with many sounds, often overlapping in time, frequency, and/or location. Yet, they can parse auditory objects in similar ways as humans. These studies, which attempt to determine the *auditory scene* of an animal, can take the physiological form or the behavioral form. Physiological correlates to auditory scene analysis have been measured in rats from multiple points along the auditory pathway (Yao et al. 2015) and in the auditory cortex of guinea pigs (Scholes et al. 2015).

Behavioral evidence of auditory scene analysis in rodents is limited to one study on Mongolian gerbils. Kobayasi and colleagues (2012) measured an auditory phenomenon known as *induction*. Humans and other animals presented with a tone with a silent gap embedded in the middle perceive that gap. If, however, a noise burst is presented during the gap that includes the frequency range of the missing tone, listeners report an ungapped tone. Kobayasi and colleagues presented Mongolian gerbils with gapped tones where the gap was filled with noise bursts of various frequencies, durations, and intensities. The Mongolian gerbils could hear the gapped stimuli as complete, but only under conditions that led to induction in humans. If the noise was of the wrong frequency, duration, or intensity, the Mongolian gerbils could still detect the gap. Noto and colleagues (2016) found physiological correlates to induction in the auditory cortex of rats. The lone behavioral study along with the few physiological studies on the perception of the auditory scene in just three species of rodents suggest that their auditory worlds are likely organized into separate auditory objects in the same way as in humans and other animals.

4.7.4 Comodulation Masking Release

Comodulation masking release (CMR) is the improvement in threshold for a pure tone when maskers are amplitude modulated relative to when they are unmodulated or asynchronously modulated. The CMR is thought to be analogous to the experience of auditory grouping, whereby individual auditory objects are formed by sounds changing in amplitude at the same pace. Quantitatively, CMR is an indirect measure of frequency selectivity because maskers have differential effects on tone detection depending on whether they are within the same auditory filter or outside the frequency range of that filter (Hall and Grose 1990). Gleich and colleagues (2007) found evidence for CMR in Mongolian gerbils, whose thresholds for detecting long tones in noise were better when the noise was amplitude modulated than when it was unmodulated. The same was not true for short duration tones. Gleich and colleagues (2007) suggested that the troughs created by modulating the masker aided in detecting those long tones, while the short tones got washed out in the same peaks when the maskers were modulated.

Another method used to measure CMR is the *flanker method*. In this paradigm, tone detection is measured in a narrow band of noise surrounding the tone and with off-frequency narrowband maskers at higher and lower frequencies. Those maskers are amplitude modulated in synchrony with one another or asynchronously. When the maskers are coherently modulated, tone thresholds are lower than when they are asynchronously modulated. In humans, CMR occurs both within channels, when flanking maskers are similar in frequency to the pure tone, and across channels, when the flankers are located at frequencies well above and/or below the tone (Hall and Grose 1990). In humans, CMR tends to be larger for within-channel masking conditions than for across-channel masking conditions (e.g., Schooneveldt and Moore 1987). Thus, CMR is thought to reflect the bandwidth of the subjects' auditory filters because the improvement in tone detection thresholds differs between the within-channel and across-channel masking conditions. Klink and colleagues (2010) found CMR in NMRI mice within channels, but not across channels, differing from the results in humans. Further, the larger auditory filter size of mice relative to humans (Sect. 4.3.1) allowed for better within-channel tone detection in larger masker frequency separations for the mice relative to humans. Klink and colleagues did not find CMR for flanking maskers at large frequency separations, suggesting no across-channel benefits to correlated maskers in the mice. The lack of across-channel CMR was hypothesized to be caused by a problem with phase locking to composite maskers with high-frequency sound envelopes or an inability to detect modulation of maskers at high modulation rates.

Neural correlates to behavioral CMR have been measured in the ventral cochlear nucleus of guinea pigs (Pressnitzer et al. 2001). Here, across-channel processing was found, such that some single units responded less to maskers when comodulated flanking bands were added to an on-frequency masker. These qualitative results were similar to those found in humans, despite coming from single units in anesthetized animals at a fairly early stage of the auditory system. Pressnitzer and colleagues noted that the CMR thresholds by behaving animals likely involved population responses of auditory system neurons.

Finally, measurements of CMR stimuli in the auditory cortex of rats were taken by Hershenhoren and Nelken (2016). Across-channel integration in the auditory cortex likely led to the high masked thresholds seen to both modulated and unmodulated maskers relative to the differentiated responses to the two masker types in the ventral cochlear nucleus used by Pressnitzer and colleagues (2001). There was a difference in the temporal structure of responses to modulated and unmodulated maskers in the cortex, however, with lower thresholds when detecting a tone in a modulated masker than an unmodulated masker. Hershenhoren and Nelken (2016) suggested that the neurons locking to the envelope of the noise in the unmodulated masker disrupted the detection of the tone. When the noise was modulated, suppression of envelope locking was found. Together, these three CMR studies in rodents highlight the use of temporal and frequency interactions in detecting objects in fluctuating noise and reveal yet another way to indirectly measure auditory filter size in rodents.

4.8 Summary

The experiments described in this chapter characterize various aspects of hearing in a few species of rodents as determined using various methodologies. Rodents have a broad bandwidth of sensitivity for pure tones, in general, and can distinguish between sounds changing in frequency, time, and intensity. They can detect and discriminate between complex sounds, including natural vocalizations and human speech.

Missing from these experiments are a very large number of rodent species that remain untested—virtually nothing is known about hearing in thousands of species. It is possible that the specializations seen in the audiograms across suborders also occur for temporal processing and frequency selectivity. It is possible that the untested rodent species have completely different auditory acuity abilities that may be related to their lifestyles. For instance, the purported echolocation abilities of the arboreal Vietnamese pygmy dormouse (*Typhlomys chapensis*), described by Panyutina and colleagues (2016), would require more sensitive auditory processing than in other rodents in order to determine the characteristics of the environment, such as that measured in bats. Yet, virtually nothing is known about hearing in this rodent. Further, this is unlikely to be the only echolocating rodent to be discovered. In 1950, Frank Beach warned psychologists of the repercussions of limiting scientific research to rats. Unfortunately, that is almost exactly what researchers interested in acoustic communication of rodents have done. To truly understand hearing and communication in rodents, scientists must go beyond the five domesticated laboratory rodents included in almost every section of this chapter and in almost every chapter of this book.

Acknowledgments The work described here was supported by NIH R03DC009483 and R01DC012302 (Dent). This work would not have been possible if not for the significant contributions of Dr. Kelly Radziwon and numerous undergraduate and graduate students in the Dent Laboratory. Thanks to Dr. Amanda Lauer for helpful comments on this chapter.

Compliance with Ethics Requirements Micheal Dent declares that she has no conflict of interest.

Laurel Screven declares that she has no conflict of interest.

Anastasiya Kobrina declares that she has no conflict of interest.

References

Beach, F. A. (1950). The snark was a boojum. *American Psychologist*, 5(4), 115–124.

Begall, S., Burda, H., & Schneider, B. (2004). Hearing in coruros (*Spalacopus cyanus*): Special audiogram features of a subterranean rodent. *Journal of Comparative Physiology A: Neuroethology, Sensory, Neural, and Behavioral Physiology*, 190(11), 963–969.

Behrens, D., & Klump, G. M. (2015). Comparison of the sensitivity of prepulse inhibition of the startle reflex and operant conditioning in an auditory intensity difference limen paradigm. *Hearing Research*, 321(1), 35–44.

Birch, L., Warfield, D., Ruben, R., & Mikaelian, D. (1968). Behavioral measurements of pure tone thresholds in normal CBA-J mice. *Journal of Auditory Research*, 8(1), 459–468.

Blumstein, D. T., & Armitage, K. B. (1997). Does sociality drive the evolution of communicative complexity? A comparative test with ground-dwelling sciurid alarm calls. *The American Naturalist*, 150(2), 179–200.

Blumstein, D. T., & Daniel, J. C. (2004). Yellow-bellied marmots discriminate between the alarm calls of individuals and are more responsive to calls from juveniles. *Animal Behaviour*, 68(6), 1257–1265.

Borg, E. (1982). Auditory thresholds in rats of different age and strain. A behavioral and electrophysiological study. *Hearing Research*, 8(2), 101–115.

Brand, A., Urban, R., & Grothe, B. (2000). Duration tuning in the mouse auditory midbrain. *Journal of Neurophysiology*, 84(4), 1790–1799.

Brown, C. H., & Sinnott, J. M. (2006). Cross-species comparisons of vocal perception. In S. Greenberg & W. A. Ainsworth (Eds.), *Listening to speech: An auditory perspective* (pp. 183–201). Mahwah, NJ: Lawrence Erlbaum Associates.

Bruckmann, G., & Burda, H. (1997). Hearing in blind subterranean Zambian mole-rats (*Cryptomys sp.*): Collective behavioural audiogram in a highly social rodent. *Journal of Comparative Physiology A: Sensory, Neural, and Behavioral Physiology*, 181(1), 83–88.

Chambers, A. R., Resnik, J., Yuan, Y., Whitton, J. P., et al. (2016). Central gain restores auditory processing following near-complete cochlear denervation. *Neuron*, 89(4), 867–879.

Cheatham, M., Huynh, K., Gao, J., Zuo, J., & Dallos, P. (2004). Cochlear function in Prestin knockout mice. *The Journal of Physiology*, 560(3), 821–830.

Church, R. M., Getty, D. J., & Lerner, N. D. (1976). Duration discrimination by rats. *Journal of Experimental Psychology: Animal Behavior Processes*, 2(4), 303. https://doi.org/10.1037/0097-7403.2.4.303

Cooke, J. E., Zhang, H., & Kelly, J. B. (2007). Detection of sinusoidal amplitude modulated sounds: Deficits after bilateral lesions of auditory cortex in the rat. *Hearing Research*, 231(1), 90–99.

Dang, R., Torigoe, D., Suzuki, S., Kikkawa, Y., et al. (2011). Genetic background strongly modifies the severity of symptoms of Hirschsprung disease, but not hearing loss in rats carrying Ednrb sl mutations. *PLoS One*, 6(9), e24086.

Eddins, A. C., Salvi, R. J., Wang, J., & Powers, N. L. (1998). Threshold-duration functions of chinchilla auditory nerve fibers. *Hearing Research*, 119(1), 135–141.

Ehret, G. (1974). Age-dependent hearing loss in normal hearing mice. *Naturwissenschaften*, 61(11), 506–507.

Ehret, G. (1975). Frequency and intensity difference limens and nonlinearities in the ear of the housemouse (*Mus musculus*). *Journal of Comparative Physiology*, 102(4), 321–336.

Ehret, G. (1976). Temporal auditory summation for pure tones and white noise in the house mouse (*Mus musculus*). *The Journal of the Acoustical Society of America*, 59(6), 1421–1427.

Ehret, G., & Haack, B. (1981). Categorical perception of mouse pup ultrasound by lactating females. *Naturwissenschaften*, 68(4), 208–209.

Ehret, G., & Haack, B. (1982). Ultrasound recognition in the house mouse: Key-stimulus configuration and recognition mechanism. *Journal of Comparative Physiology*, 148(2), 245–251.

Fay, R. R. (1974). Auditory frequency discrimination in vertebrates. *The Journal of the Acoustical Society of America*, 56(1), 206–209.

Fay, R. R. (1988). *Hearing in vertebrates: A psychophysics databook*. Winnetka, IL: Hill-Fay Associates.

Felsheim, C., & Ostwald, J. (1996). Responses to exponential frequency modulations in the rat inferior colliculus. *Hearing Research*, 98(1), 137–151.

Feng, Y., Wang, J., & Yin, S. (2007). General anesthesia changes gap-evoked auditory responses in guinea pigs. *Acta Oto-Laryngologica*, 127(2), 143–148.

Floody, O. R., & Kilgard, M. P. (2007). Differential reductions in acoustic startle document the discrimination of speech sounds in rats. *The Journal of the Acoustical Society of America*, 122(4), 1884–1887.

Floody, O. R., Ouda, L., Porter, B. A., & Kilgard, M. P. (2010). Effects of damage to auditory cortex on the discrimination of speech sounds by rats. *Physiology and Behavior*, 101(2), 260–268. https://doi.org/10.1016/j.physbeh.2010.05.009

Francis, R. L. (1979). The preyer reflex audiogram of several rodents, and its relation to the "absolute" audiogram in the rat. *Journal of Auditory Research*, 19(3), 217–233.

Gaese, B. H., & Ostwald, J. (1995). Temporal coding of amplitude and frequency modulation in the rat auditory cortex. *European Journal of Neuroscience*, 7(3), 438–450.

Gaese, B. H., King, I., Felsheim, C., Ostwald, J., & von der Behrens, W. (2006). Discrimination of direction in fast frequency-modulated tones by rats. *Journal of the Association for Research in Otolaryngology*, 7(1), 48–58.

Gaese, B. H., Nowotny, M., & Pilz, P. K. (2009). Acoustic startle and prepulse inhibition in the Mongolian gerbil. *Physiology and Behavior*, 98(4), 460–466.

Giraudi-Perry, D., Salvi, R., & Henderson, D. (1982). Gap detection in hearing-impaired chinchillas. *The Journal of the Acoustical Society of America*, 72(5), 1387–1393.

Gleich, O., Kittel, M. C., Klump, G. M., & Strutz, J. (2007). Temporal integration in the gerbil: The effects of age, hearing loss and temporally unmodulated and modulated speech-like masker noises. *Hearing Research*, 224(1), 101–114.

Gourevitch, G. (1965). Auditory masking in the rat. *The Journal of the Acoustical Society of America*, 37(3), 439–443.

Hack, M. H. (1971). Auditory intensity discrimination in the rat. *Journal of Comparative and Physiological Psychology*, 74(2), 315–318.

Hall III, J. W., & Grose, J. H. (1990). Comodulation masking release and auditory grouping. *The Journal of the Acoustical Society of America*, 88(1), 119–125.

Halpern, D. L., & Dallos, P. (1986). Auditory filter shapes in the chinchilla. *The Journal of the Acoustical Society of America*, 80(3), 765–775.

Hamann, I., Gleich, O., Klump, G. M., Kittel, M. C., et al. (2002). Behavioral and evoked-potential thresholds in young and old Mongolian gerbils (*Meriones unguiculatus*). *Hearing Research*, 171(1-2), 82–95.

Hare, J. F. (1998). Juvenile Richardson's ground squirrels discriminate among individual alarm callers. *Animal Behaviour*, 55(2), 451–460.

Heffner, H., & Masterton, B. (1980). Hearing in glires: Domestic rabbit, cotton rat, feral house mouse, and kangaroo rat. *The Journal of the Acoustical Society of America*, 68(6), 1584–1599.

Heffner, H. E., & Heffner, R. S. (1985). Hearing in two cricetid rodents: Wood rat (*Neotoma floridana*) and grasshopper mouse (*Onychomys leucogaster*). *Journal of Comparative Psychology*, 99(3), 275–288.

Heffner, H. E., & Heffner, R. S. (2001). Behavioral assessment of hearing in mice. In J. F. Willott (Ed.), *Handbook of mouse auditory research* (pp. 19–30). New York: CRC Press.

Heffner, H. E., Heffner, R. S., Contos, C., & Ott, T. (1994). Audiogram of the hooded Norway rat. *Hearing Research*, 73(2), 244–247.

Heffner, R., Heffner, H., & Masterton, B. (1971). Behavioral measurements of absolute and frequency-difference thresholds in guinea pig. *The Journal of the Acoustical Society of America*, 49(6), 1888–1895.

Heffner, R. S., & Heffner, H. E. (1990). Vestigial hearing in a fossorial mammal, the pocket gopher (*Geomys bursarius*). *Hearing Research*, 46(3), 239–252.

Heffner, R. S., & Heffner, H. E. (1991). Behavioral hearing range of the chinchilla. *Hearing Research*, 52(1), 13–16.

Heffner, R. S., & Heffner, H. E. (1993). Degenerate hearing and sound localization in naked mole rats (*Heterocephalus glaber*), with an overview of central auditory structures. *The Journal of Comparative Neurology*, 331(3), 418–433.

Heffner, R. S., Heffner, H. E., Contos, C., & Kearns, D. (1994). Hearing in prairie dogs—transition between surface and subterranean rodents. *Hearing Research*, 73(2), 185–189.

Heffner, R. S., Koay, G., & Heffner, H. E. (2001). Audiograms of five species of rodents: Implications for the evolution of hearing and the perception of pitch. *Hearing Research*, 157(1-2), 138–152.

Henderson, D. (1969). Temporal summation of acoustic signals by the chinchilla. *The Journal of the Acoustical Society of America*, 46(2), 474–475.

Hershenhoren, I., & Nelken, I. (2016). Detection of tones masked by fluctuating noise in rat auditory cortex. *Cerebral Cortex*, 27(11), 5130–5143.

Holfoth, D. P., Neilans, E. G., & Dent, M. L. (2014). Discrimination of partial from whole ultrasonic vocalizations using a go/no-go task in mice. *The Journal of the Acoustical Society of America*, 136(6), 3401–3409.

Holmstrom, L. A., Eeuwes, L. B. M., Roberts, P. D., & Portfors, C. V. (2010). Efficient encoding of vocalizations in the auditory midbrain. *The Journal of Neuroscience*, 30(3), 802–819.

Honma, Y., Tsukano, H., Horie, M., Ohshima, S., et al. (2013). Auditory cortical areas activated by slow frequency-modulated sounds in mice. *PLoS One*, 8(7), e68113.

Jackson, L. L., Heffner, H. E., & Heffner, R. S. (1997). Audiogram of the fox squirrel (*Sciurus niger*). *Journal of Comparative and Physiological Psychology*, 111(1), 100–104.

Kelly, J. B. (1970). *The effects of lateral lemniscal and neocortical lesions on auditory absolute thresholds and frequency difference thresholds of the rat*. Ph.D. dissertation, Vanderbilt University, Nashville, TN. ProQuest Information & Learning

Kelly, J. B., & Masterton, B. (1977). Auditory sensitivity of the albino rat. *Journal of Comparative and Physiological Psychology*, 91(4), 930–936.

Kelly, J. B., Cooke, J. E., Gilbride, P. C., Mitchell, C., & Zhang, H. (2006). Behavioral limits of auditory temporal resolution in the rat: Amplitude modulation and duration discrimination. *Journal of Comparative Psychology*, 120(2), 98–105.

King, J., Insanally, M., Jin, M., Martins, A. R. O., et al. (2015). Rodent auditory perception: Critical band limitations and plasticity. *Neuroscience*, 296(1), 55–65.

Kittel, M., Wagner, E., & Klump, G. M. (2002). An estimate of the auditory-filter bandwidth in the Mongolian gerbil. *Hearing Research*, 164(1), 69–76.

Klinge, A., & Klump, G. M. (2008). Frequency difference limens of pure tones and harmonics within complex stimuli in Mongolian gerbils and humans. *The Journal of the Acoustical Society of America*, 125(1), 304–314.

Klink, K. B., & Klump, G. M. (2004). Duration discrimination in the mouse (*Mus musculus*). *Journal of Comparative Physiology A: Neuroethology, Sensory, Neural, and Behavioral Physiology*, 190(12), 1039–1046.

Klink, K. B., Dierker, H., Beutelmann, R., & Klump, G. M. (2010). Comodulation masking release determined in the mouse (*Mus musculus*) using a flanking-band paradigm. *Journal of the Association for Research in Otolaryngology*, 11(1), 79–88.

Koay, G., Heffner, R. S., & Heffner, H. E. (2002). Behavioral audiograms of homozygous med J mutant mice with sodium channel deficiency and unaffected controls. *Hearing Research*, 171(1), 111–118.

Kobayasi, K. I., Usami, A., & Riquimaroux, H. (2012). Behavioral evidence for auditory induction in a species of rodent: Mongolian gerbil (*Meriones unguiculatus*). *The Journal of the Acoustical Society of America*, 132(6), 4063–4068.

Kobrina, A., & Dent, M. L. (2016). The effects of aging and sex on detection of ultrasonic vocalizations by adult CBA/CaJ mice (*Mus musculus*). *Hearing Research*, 341(1), 119–129.

Kobrina, A., Toal, K., & Dent, M. L. (2018). Intensity difference limens in adult CBA/CaJ mice (*Mus musculus*). *Behavioural Processes*, 148(1), 46–48.

Koeppl, J. W., Hoffmann, R. S., & Nadler, C. F. (1978). Pattern analysis of acoustical behavior in four species of ground squirrels. *Journal of Mammalogy*, 59(4), 677–696.

Krishna, B. S., & Semple, M. N. (2000). Auditory temporal processing: Responses to sinusoidally amplitude-modulated tones in the inferior colliculus. *Journal of Neurophysiology*, 84(1), 255–273.

Kuhl, P. K. (1981). Discrimination of speech by nonhuman animals: Basic auditory sensitivities conducive to the perception of speech-sound categories. *The Journal of the Acoustical Society of America*, 70(2), 240–249.

Kuhl, P. K., & Miller, J. D. (1975). Speech perception by the chinchilla: Voiced-voiceless distinction in alveolar plosive consonants. *Science*, 190(4209), 69–72.

Kulig, J., & Willott, J. F. (1984). Frequency difference limens of C57BL/6 and DBA/2 mice: Relationship to auditory neuronal response properties and hearing impairment. *Hearing Research*, 16(2), 169–174.

Lauer, A. M., Behrens, D., & Klump, G. (2017). Acoustic startle modification as a tool for evaluating auditory function of the mouse: Progress, pitfalls, and potential. *Neuroscience and Biobehavioral Reviews*, 77, 194–208.

Lina, I. A., & Lauer, A. M. (2013). Rapid measurement of auditory filter shape in mice using the auditory brainstem response and notched noise. *Hearing Research*, 298(1-2), 73–79.

Long, G. R., & Clark, W. W. (1984). Detection of frequency and rate modulation by the chinchilla. *The Journal of the Acoustical Society of America*, 75(4), 1184–1190.

Long, G. R., & Miller, J. D. (1981). Tone-on-tone masking in the chinchilla. *Hearing Research*, 4(3), 279–285.

May, B. J., Kimar, S., & Prosen, C. A. (2006). Auditory filter shapes of CBA/CaJ mice: Behavioral assessments. *The Journal of the Acoustical Society of America*, 120(1), 321–330.

McGee, T., Ryan, A., & Dallos, P. (1976). Psychophysical tuning curves of chinchillas. *The Journal of the Acoustical Society of America*, 60(5), 1146–1150.
Miller, J. D. (1964). Auditory sensitivity of the chinchilla in quiet and in noise. *The Journal of the Acoustical Society of America*, 36(10), 2010.
Miller, J. D. (1970). Audibility curve of the chinchilla. *The Journal of the Acoustical Society of America*, 48(2), 513–523.
Mitchell, C., & Fowler, C. (1980). Tuning curves of cochlear and brainstem responses in the guinea pig. *The Journal of the Acoustical Society of America*, 68(3), 896–900.
Muller, M., & Burda, H. (1989). Restricted hearing range in a subterranean rodent, *Cryptomys hottentotus*. *Naturwissenschaften*, 76(3), 134–135.
Muller, M., von Hunerbein, K., Hoidis, S., & Smolders, J. W. (2005). A physiological place-frequency map of the cochlea in the CBA/J mouse. *Hearing Research*, 202(1-2), 63–73.
Naguib, M. (1997). Use of song amplitude for ranging in Carolina wrens, *Thryothorus ludovicianus*. *Ethology*, 103(9), 723–731.
Nakano, R., Nakagawa, R., Tokimoto, N., & Okanoya, K. (2013). Alarm call discrimination in a social rodent: Adult but not juvenile degu calls induce high vigilance. *Journal of Ethology*, 31(2), 115–121.
Neilans, E. G., Holfoth, D. P., Radziwon, K. E., Portfors, C. V., & Dent, M. L. (2014). Discrimination of ultrasonic vocalizations by CBA/CaJ mice is related to spectrotemporal dissimilarity of vocalizations. *PLoS One*, 9(1), e85405.
Nelson, D. A., & Kiester, T. E. (1978). Frequency discrimination in the chinchilla. *The Journal of the Acoustical Society of America*, 64(1), 114–128.
Noriot, E. (1972). Ultrasounds and maternal behavior in small rodents. *Developmental Psychobiology*, 5(4), 371–387.
Noto, M., Nishikawa, J., Tateno, T. (2016). An analysis of nonlinear dynamics underlying neural activity related to auditory induction in the rat auditory cortex. *Neuroscience*, 318, 58–83.
Ohlemiller, K. K., Jones, L. B., Heidbreder, A. F., Clark, W. W., & Miller, J. D. (1999). Voicing judgments by chinchillas trained with a reward paradigm. *Behavioural Brain Research*, 100(1-2), 185–195.
Palombi, P. S., Backoff, P. M., & Caspary, D. M. (2001). Responses of young and aged rat inferior colliculus neurons to sinusoidally amplitude modulated stimuli. *Hearing Research*, 153(1), 174–180.
Panyutina, A. A., Kuznetsov, A. N., Volodin, I. A., Abramov, A. V., & Soldatova, I. B. (2016). A blind climber: The first evidence of ultrasonic echolocation in arboreal mammals. *Integrative Zoology*, 12(2), 172–184.
Polak, M., Eshraghi, A. A., Nehme, O., Ahsan, S., et al. (2004). Evaluation of hearing and auditory nerve function by combining ABR, DPOAE and eABR tests into a single recording session. *Journal of Neuroscience Methods*, 134(2), 141–149.
Popelar, J., Groh, D., Pelánová, J., Canlon, B., & Syka, J. (2006). Age-related changes in cochlear and brainstem auditory functions in Fischer 344 rats. *Neurobiology of Aging*, 27(3), 490–500.
Porter, B. A., Rosenthal, T. R., Ranasinghe, K. G., & Kilgard, M. P. (2011). Discrimination of brief speech sounds is impaired in rats with auditory cortex lesions. *Behavioural Brain Research*, 219(1), 68–74.
Pressnitzer, D., Meddis, R., Delahaye, R., & Winter, I. M. (2001). Physiological correlates of comodulation masking release in the mammalian ventral cochlear nucleus. *The Journal of Neuroscience*, 21(16), 6377–6386.
Prosen, C. A., Petersen, M. R., Moody, D. B., & Stebbins, W. C. (1978). Auditory thresholds and kanamycin-induced hearing loss in the guinea pig assessed by a positive reinforcement procedure. *The Journal of the Acoustical Society of America*, 63(2), 559–566.
Prosen, C. A., Halpern, D. L., & Dallos, P. (1988). Frequency difference limens in normal and sensorineural hearing impaired chinchillas. *The Journal of the Acoustical Society of America*, 85(3), 1302–1313.

Prosen, C. A., Dore, D. J., & May, B. J. (2003). The functional age of hearing loss in a mouse model of presbycusis. I. Behavioral assessments. *Hearing Research*, 183(1), 44–56.

Radziwon, K. E., & Dent, M. L. (2014). Frequency difference limens and auditory cue trading in CBA/CaJ mice. *Behavioural Processes*, 106(1), 74–76.

Radziwon, K. E., June, K. M., Stolzberg, D. J., Xu-Friedman, M. A., et al. (2009). Behaviorally measured audiograms and gap detection thresholds in CBA/CaJ mice. *Journal of Comparative Physiology A: Neuroethology, Sensory, Neural, and Behavioral Physiology*, 195(10), 961–969.

Radziwon, K. E., Stolzberg, D. J., Urban, M. E., Bowler, R. A., & Salvi, R. J. (2015). Salicylate-induced hearing loss and gap detection deficits in rats. *Frontiers in Neurology*, 6, 31.

Ralls, K. (1967). Auditory sensitivity in mice: *Peromyscus* and *Mus musculus*. *Animal Behaviour*, 15(1), 123–128.

Ranasinghe, K. G., Vrana, W. A., Matney, C. J., & Kilgard, M. P. (2012). Neural mechanisms supporting robust discrimination of spectrally and temporally degraded speech. *Journal of the Association for Research in Otolaryngology*, 13(4), 527–542.

Ryan, A., Dallos, P., & McGee, T. (1979). Psychophysical tuning curves and auditory thresholds after hair cell damage in the chinchilla. *The Journal of the Acoustical Society of America*, 66(2), 370–378.

Salvi, R. J., & Arehole, S. (1985). Gap detection in chinchillas with temporary high-frequency hearing loss. *The Journal of the Acoustical Society of America*, 77(3), 1173–1177.

Salvi, R., Ahroon, W., Perry, J., Gunnarson, A., & Henderson, D. (1982). Comparison of psychophysical and evoked-potential tuning curves in the chinchilla. *American Journal of Otolaryngology*, 3(6), 408–416.

Saunders, J. C., Dolgin, K. G., & Lowry, L. D. (1980). The maturation of frequency selectivity in C57BL/6J mice studied with auditory evoked response tuning curves. *Brain Research*, 187(1), 69–79.

Saunders, S. S., Shivapuja, B. G., & Salvi, R. J. (1987). Auditory intensity discrimination in the chinchilla. *The Journal of the Acoustical Society of America*, 82(5), 1604–1607.

Scholes, C., Palmer, A. R., & Sumner, C. J. (2015). Stream segregation in the anesthetized auditory cortex. *Hearing Research*, 328(1), 48–58.

Schooneveldt, G. P., & Moore, B. C. (1987). Comodulation masking release (CMR): Effects of signal frequency, flanking-band frequency, masker bandwidth, flanking-band level, and monotonic versus dichotic presentation of the flanking band. *The Journal of the Acoustical Society of America*, 82(6), 1944–1956.

Schulze, H., & Langner, G. (1997). Representation of periodicity pitch in the primary auditory cortex of the Mongolian gerbil. *Acta Oto-Laryngologica*, 117(Suppl. 532), 89–95.

Screven, L. A., & Dent, M. L. (2016). Discrimination of frequency-modulated sweeps by laboratory mice. *The Journal of the Acoustical Society of America*, 137(4), 1481–1487.

Seaton, W. H., & Trahiotis, C. (1975). Comparison of critical ratios and critical bands in the monaural chinchilla. *The Journal of the Acoustical Society of America*, 57(1), 193–199.

Shannon, R. V., Zheng, F. G., Kamath, V., Wygonski, J., & Ekelid, M. (1995). Speech recognition with primarily temporal cues. *Science*, 270(5234), 303–307.

Shofner, W. P. (2014). Perception of degraded speech sounds differs in chinchilla and human listeners. *The Journal of the Acoustical Society of America*, 135(4), 2065–2077.

Shriner, W. M. (1998). Yellow-bellied marmot and golden-mantled ground squirrel responses to heterospecific alarm calls. *Animal Behaviour*, 55(3), 529–536.

Sinnott, J. M., & Mosteller, K. W. (2001). A comparative assessment of speech sound discrimination by the Mongolian gerbil. *The Journal of the Acoustical Society of America*, 110(4), 1729–1732.

Sinnott, J. M., & Mosqueda, S. B. (2003). Effects of aging on speech sound discrimination in the Mongolian gerbil. *Ear and Hearing*, 24(1), 30–37.

Sinnott, J. M., Brown, C. H., & Brown, F. E. (1992). Frequency and intensity discrimination in Mongolian gerbils, African monkeys and humans. *Hearing Research*, 59(2), 205–212.

Smith, J. C. (1976). Responses of adult mice to models of infant calls. *Journal of Comparative and Physiological Psychology*, 90(12), 1105–1115.
Syka, J., & Popelar, J. (1988). Hearing threshold shifts from prolonged exposure to noise in guinea pigs. *Hearing Research*, 3(3), 205–213.
Syka, J., Raybalko, N., Brozek, G., & Jilek, M. (1996). Auditory frequency and intensity discrimination in pigmented rats. *Hearing Research*, 100(1-2), 107–113.
Syka, J., Rybalko, N., Mazelova, J., & Druga, R. (2002). Gap detection threshold in the rat before and after auditory cortex ablation. *Hearing Research*, 172(1-2), 151–159.
Taberner, A. M., & Liberman, M. C. (2005). Response properties of single auditory nerve fibers in the mouse. *Journal of Neurophysiology*, 93(1), 557–569.
Talwar, S. K., & Gerstein, G. L. (1998). Auditory frequency discrimination in the white rat. *Hearing Research*, 126(1-2), 135–150.
Viemeister, N. (1996). Auditory temporal integration: What is being accumulated? *Current Directions in Psychological Science*, 5(1), 28–32.
Wagner, E., Klump, G. M., & Hamann, I. (2003). Gap detection in Mongolian gerbils (*Meriones unguiculatus*). *Hearing Research*, 176(1-2), 11–16.
Walton, J., Frisina, R., Ison, J., & O'Neill, W. (1997). Neural correlates of behavioral gap detection in the inferior colliculus of the young CBA mouse. *Journal of Comparative Physiology A: Sensory, Neural, and Behavioral Physiology*, 181(2), 161–176.
Wang, J., Van Wijhe, R., Chen, Z., & Yin, S. (2006). Is duration tuning a transient process in the inferior colliculus of guinea pigs? *Brain Research*, 1114(1), 63–74.
Wetzel, W., Ohl, F. W., Wagner, T., & Scheich, H. (1998a). Right auditory cortex lesion in Mongolian gerbils impairs discrimination of rising and falling frequency-modulated tones. *Neuroscience Letters*, 252(2), 115–118.
Wetzel, W., Wagner, T., Ohl, F. W., & Scheich, H. (1998b). Categorical discrimination of direction in frequency-modulated tones by Mongolian gerbils. *Behavioral Brain Research*, 91(1-2), 29–39.
Wood, C. C. (1976). Discriminability, response bias, and phoneme categories in discrimination of voice onset time. *The Journal of the Acoustical Society of America*, 60(6), 1381–1389.
Xu, L., Thompson, C. S., & Pfingst, B. E. (2005). Relative contributions of spectral and temporal cues for phoneme recognition. *The Journal of the Acoustical Society of America*, 117(5), 3255–3267.
Yao, J. D., Bremen, P., & Middlebrooks, J. C. (2015). Emergence of spatial stream segregation in the ascending auditory pathway. *The Journal of Neuroscience*, 35(49), 16199–16212.
Yost, W. A., & Shofner, W. P. (2009). Critical bands and critical ratios in animal psychoacoustics: An example using chinchilla data. *The Journal of the Acoustical Society of America*, 125(1), 315–323.
Zheng, Q. Y., Johnson, K. R., & Erway, L. C. (1999). Assessment of hearing in 80 inbred strains of mice by ABR threshold analyses. *Hearing Research*, 130(1-2), 94–107.

Chapter 5
Rodent Sound Localization and Spatial Hearing

Amanda M. Lauer, James H. Engel, and Katrina Schrode

Abstract Sound localization and directional hearing are fundamental for rodents to successfully navigate their environment, detect predators and prey, and locate conspecifics. In this chapter, the cues available to rodents for localizing and discriminating sounds in space are reviewed, including interaural level and timing differences and spectral cues. The neural circuits supporting sound localization are also introduced. Sound localization acuity and directional hearing behavior in azimuth and elevation that have been measured in terrestrial and subterranean rodent species are compared. Due to the limited scope of behavioral testing performed in most rodent species to date, binaural evoked potential studies that have been used to probe directional hearing in rodents are also considered as an alternative to labor-intensive behavioral measures. Finally, the trends observed across species are summarized and areas for future study are suggested.

Keywords Acoustic startle response · Binaural interaction component · Directional hearing · Interaural level differences · Interaural time differences · Minimum audible angles · Operant conditioning · Pinna cues · Precedence effect · Spatial masking release

5.1 Introduction

The ability to localize and separate sounds in space is essential for an animal's survival (Popper and Fay 2005). Animals must be able to locate conspecifics for mating and care of offspring. To perform these tasks accurately, the animal must not only identify the location of particular sounds in an environment but also avoid confusing them with unimportant sounds. This perceptual separation of sounds is nontrivial, because sounds in the environment (or the acoustic scene) tend to overlap in time.

A. M. Lauer (✉) · J. H. Engel · K. Schrode
Department of Otolaryngology, Center for Hearing and Balance, and David M. Rubenstein Center for Hearing Research, Johns Hopkins University School of Medicine, Baltimore, MD, USA
e-mail: alauer2@jhmi.edu; jengel13@jhmi.edu; kschrod1@jhmi.edu

M. L. Dent et al. (eds.), *Rodent Bioacoustics*, Springer Handbook of Auditory Research 67, https://doi.org/10.1007/978-3-319-92495-3_5

As a result, all of the sounds in an environment are received by the ear as a single composite waveform. Spatial cues can assist in parsing auditory objects from the scene. An animal's ability to exploit spatial cues for auditory scene analysis is likely correlated with the spatial resolution of its auditory system.

Rodents must be able to locate separated offspring and potential mates, processes that involve receiving vocal communication signals (see Schleich and Francescoli, Chap. 3). Prey species must be able to detect the approach of predators so that they can initiate escape responses, and predatory species presumably make use of spatial cues to aid prey capture. In this chapter, the acoustic cues available for localizing sounds, the basic circuits for processing these cues, and what is known about sound localization and directional hearing in rodents is reviewed.

5.2 Sound Localization Cues

In most animals, sound localization in the horizontal plane can be achieved using interaural time and intensity differences, depicted in Fig. 5.1. Lord Rayleigh first observed that because the ears are separated in space by the head, sounds arriving at the two ears have different levels (Rayleigh 1907). This is known as an acoustic shadowing effect, and it causes an interaural level difference (ILD) that serves as a cue for sound localization of high frequencies (>1 to 2 kHz). However, the cue is negligible at very low frequencies (<1 to 2 kHz) when the wavelength of the sound

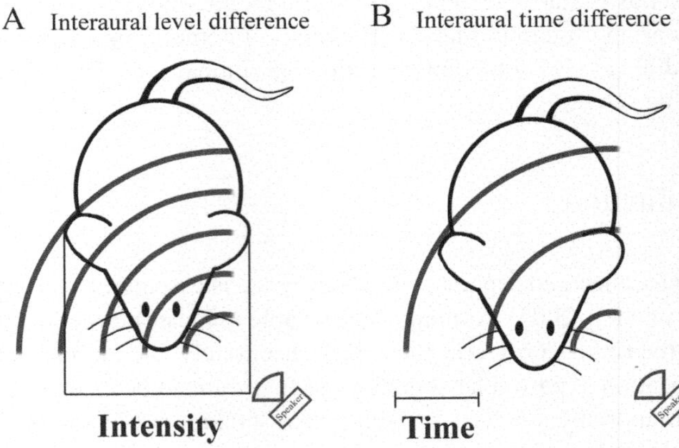

Fig. 5.1 Interaural cues available to a rodent for sound localization in the horizontal plane. (**a**) Interaural level differences (ILDs) occur when the sound arriving at one ear is louder than the sound arriving at the other ear due to head shadow effects. The head dampens the levels of sound with wavelengths that are not long enough to travel around the head. The high frequency sensitivity and small heads of many rodents render ILDs the most robust interaural cue. (**b**) Interaural time differences (ITDs) occur when a sound arrives at one ear earlier than at the other ear. The phase of the sound arriving at the two ears is different. The head size is too small and the range of audible frequencies is too high to enable meaningful ITDs in many rodents

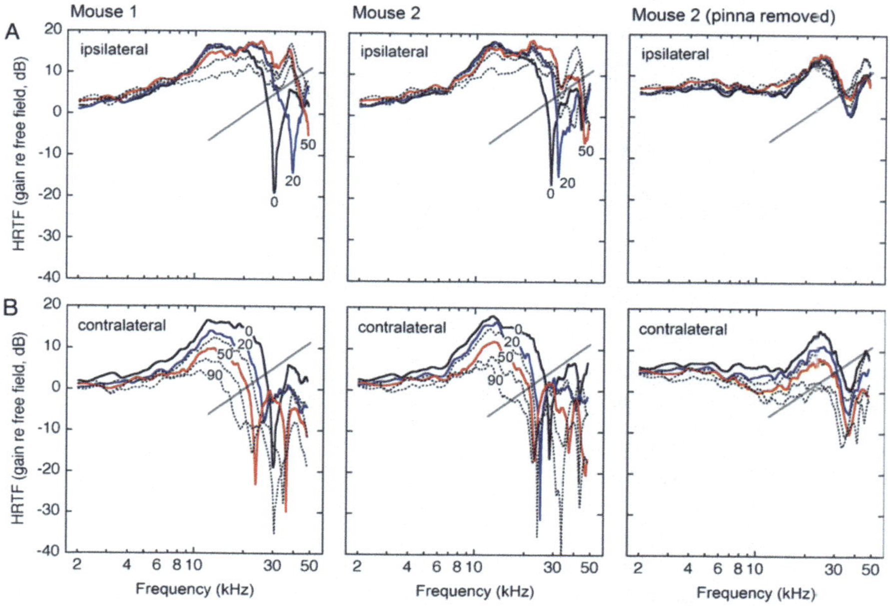

Fig. 5.2 Spectral cues for localization. Head-related transfer functions (*HRTF*) measured by placing miniature microphones in the ear canals of mice reveal a pattern of spectral peaks and notches that varies systematically with the position of a sound source, shown for the *ipsilateral* (**a**) and *contralateral* (**b**) interaural horizontal plane. *Numerical labels* indicate the azimuth of the sound source in degrees. Three locations are highlighted (*0° black, 20° blue, 50° red*) to highlight the location-dependent spectral shape differences. *Dashed lines* represent other speaker locations measured. *Gray lines* highlight the directional trends of the low frequency edges of the spectral notches. Removal of the pinna substantially reduces these notches (*right panels*). (Reprinted from Lauer et al. (2011) with permission)

is larger than the diameter of the head. At low frequencies there is a time difference in the sound reaching the two ears from a source directly in front of the receiver, creating an interaural time difference (ITD) that is readily detectable (Rayleigh 1907). This dichotomy of localization based on ILDs for high frequencies and ITDs for low frequencies is known as the duplex theory of sound localization. Most rodents must rely on high frequency ILD cues because of their small head size and range of hearing, although this reliance has only been tested in a small number of species, as discussed in Sect. 5.4. The reliance on high frequency ILD cues may be reduced for species with lower frequency hearing and larger heads, such as the chinchilla (*Chinchilla lanigera*), discussed later in this chapter.

High resolution head-related transfer functions, which characterize how an ear receives sound from a particular point in space, have not been measured for many rodent species, but spectral cues resulting from the interaction of the acoustic waveform with the pinnae and head shape also influence localization abilities in mice (Fig. 5.2). Classically, spectral cues have been implicated in localization in the median vertical plane (Hebrank and Wright 1974; Rice et al. 1992); however, there

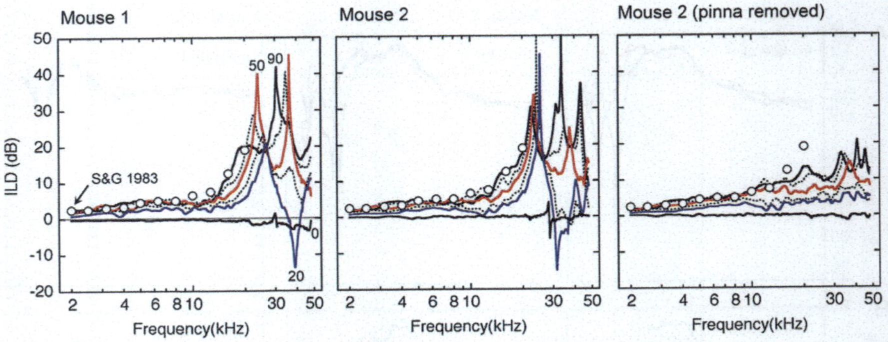

Fig. 5.3 Effects of azimuth on interaural level differences (ILDs). ILDs are enhanced by pinnae, especially for frequencies above 20 kHz. *Numerical labels* indicate the azimuth of the sound source in degrees. Three locations are indicated (*20° blue, 50° red, 90° black*) to highlight the location-dependent ILD differences. *Dashed lines* represent other speaker locations measured. For comparison, ILDs measured for pure tones at 90° (*open circles*) are replotted from Saunders and Garfinkle (1983). (Adapted from Lauer et al. (2011))

is evidence that laboratory mice can use these cues to boost ILDs in the horizontal plane (Fig. 5.3) (Lauer et al. 2011). For these cues to be effective, the sound must be relatively broadband and contain energy at high frequencies.

5.3 Sound Localization Circuits

The fundamentals of mammalian sound localization circuits and mechanisms have been reviewed in detail elsewhere (see Grothe et al. 2010; Heffner and Heffner 2016). Here, the major components of mammalian sound localization circuits are reviewed with a focus on rodents in particular. The binaural and monaural pathways are summarized in Fig. 5.4. The first nucleus in the brain to receive acoustic information from the ear is the ventral cochlear nucleus (VCN), a major subdivision of the cochlear nucleus. The absolute size of the VCN varies across rodent species (and varies across mammals in general) but is generally correlated with the size of the animal (Godfrey et al. 2016). The relative size of the VCN within the entire cochlear nucleus varies little between rodent species even when compared to other mammals (Godfrey et al. 2016).

The primary cells of the VCN that are involved in sound localization are bushy cells, which in rodents receive high fidelity input from the auditory nerve via specialized endbulb of Held synapses (Lauer et al. 2013). These cells project to the superior olivary complex (SOC), which computes binaural differences (Grothe et al. 2010). The SOC shows the most well-documented species diversity in rodents, particularly in the lateral superior olive (LSO) and the medial superior olive (MSO). Spherical bushy cells provide excitatory input to both the LSO and the MSO (Cant and Casseday 1986). Inhibitory inputs in the LSO and MSO arise from the medial

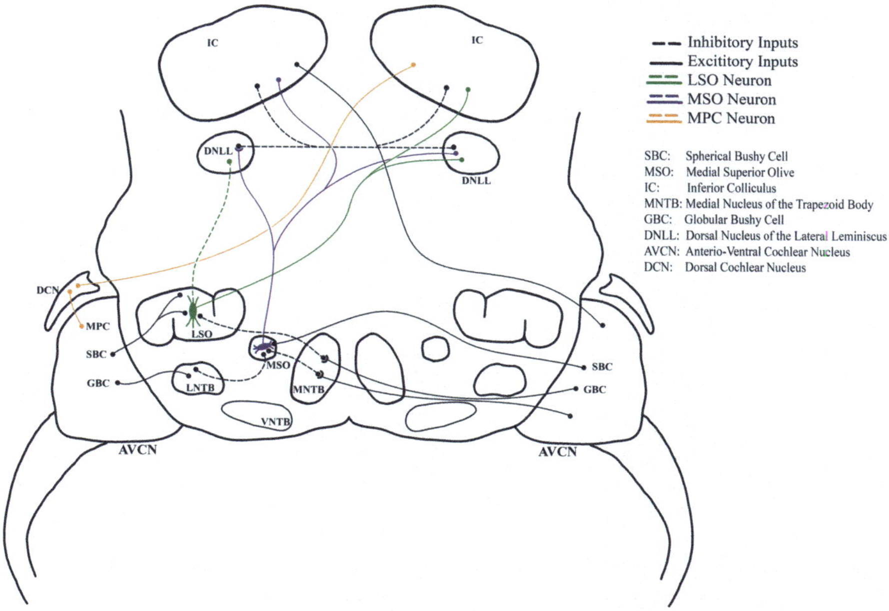

Fig. 5.4 Interaural level difference (ILD) and interaural time difference (ITD) encoding circuits. Excitatory inputs are represented with *solid lines*; inhibitory inputs are represented with *dashed lines*. **ILDs**: In the lateral superior olive (*LSO*), inhibitory projections come from within the ipsilateral medial nucleus of the trapezoid body (*MNTB*). MNTB in turn receives excitatory projections from contralateral anteroventral cochlear nucleus (*AVCN*) globular bushy cells (*GBC*s) (*solid black*). The inhibitory projections from MNTB to the LSO (*dashed black*) are combined with excitatory inputs projecting from spherical bushy cells (*SBC*s) (*solid black*) in the ipsilateral ACVN to create ILD sensitivity. In turn, the LSO sends inhibitory projections (*dashed green*) to the ipsilateral dorsal nucleus of the lateral lemniscus (*DNLL*) and inferior colliculus (*IC*) and excitatory projections (*green solid*) to the contralateral DNLL and IC. In the IC, the convergence of binaural inhibitory DNLL input (*dashed*) and monaural contralateral excitatory AVCN input give rise to ILD sensitivity. **ITDs**: Principal cells within the medial superior olive (*MSO*, *purple*) receive binaural inhibitory inputs from the MNTB and lateral nucleus of the trapezoid body (*LNTB*) as well as binaural excitatory inputs from SBCs in the contralateral and ipsilateral AVCN. The MSO neurons in turn send excitatory projections to the IC and DNLL. In addition to ILD and ITD, ventral cochlear nucleus (*VCN*) neurons (multipolar cells, *MPC*s) and dorsal cochlear nucleus (*DCN*) neurons encode spectral cues (*gold*). This information is also sent to the IC. (Adapted from Grothe et al. (2010))

and lateral nuclei of the trapezoid body (MNTB and LNTB, respectively) (Harrison and Warr 1962; Smith et al. 1991). The LSO is important for processing ILDs, and the MSO is important for ITDs, although there is evidence for ITD processing in the LSO (Joris and Yin 1995; Tollin and Yin 2005). In high-frequency hearing rodents, the LSO typically has far more cells than the MSO (Irving and Harrison 1967). The LSO and MSO of rodents specialized to hear low frequencies have approximately equal numbers of cells (Irving and Harrison 1967).

Table 5.1 Abbreviations

ABR	Auditory brainstem response
BIC	Binaural interaction component
DCN	Dorsal cochlear nucleus
DNLL	Dorsal nucleus of the lateral lemniscus
FFR	Frequency following response
IC	Inferior colliculus
ILD	Interaural level difference
ITD	Interaural time difference
LNTB	Lateral nucleus of the trapezoid body
LSO	Lateral superior olive
MAA	Minimum audible angle
MGN	Medial geniculate nucleus
MLR	Middle latency response
MNTB	Medial nucleus of the trapezoid body
MSO	Medial superior olive
VCN	Ventral cochlear nucleus

Both the MSO and the LSO send major projections to the dorsal nucleus of the lateral lemniscus (DNLL) (Glendenning et al. 1981). Neurons in the DNLL exhibit sensitivity to both ILDs and ITDs (Markovitz and Pollak 1994; Kelly et al. 1998). Projections from nearly all of the previously described nuclei converge on the inferior colliculus (IC). The IC receives input from the LSO, the MSO, and the DNLL (see Table 5.1 for all abbreviations) (Beyerl 1978). Different types of neurons in IC tend to be most sensitive to different types of localization cues (Ramachandran et al. 1999), but these neurons appear to integrate the various cues as well (Chase and Young 2005). In general, coding in the IC of rodents is similar to that seen in other mammals, but there are possible differences in the distributions of types of neurons (Davis et al. 2003).

The IC sends projections to the auditory thalamus or medial geniculate nucleus (MGN), which in turn sends projections to the auditory cortex (Huang and Winer 2000). Binaural neurons in auditory cortex encode both ILDs and ITDs (Kelly and Phillips 1991; Kyweriga et al. 2014). Performance on sound localization tasks is severely decreased following lesions of the auditory cortex in several mammalian species (e.g., Jenkins and Merzenich 1984; Thompson and Cortez 1983). However, rats do not show the same severe deficits in localization ability following auditory cortex lesions (Kelly and Glazier 1978; Kelly 1980). This result suggests that, at least in the rat, binaural processing does not strictly require the auditory cortex.

The use of spectral cues for sound localization is mediated by pathways originating in the dorsal cochlear nucleus (DCN) (Young and Davis 2002). The relative size of the DCN is generally consistent across rodents and other mammals (Godfrey et al. 2016). While the pathway for localizing spectral cues is typically considered monaural, many neurons in the DCN exhibit binaural responses (Mast 1969; Davis 2005), presumably due to the many bilateral descending projections. Projections from the DCN to the IC are mostly contralateral (Nordeen et al. 1983; Coleman and

Clerici 1987). The DCN also projects directly to the MGN in several species (Malmierca et al. 2002). Further processing of sounds in the vertical plane presumably occurs in the auditory cortex, although this has not been investigated in rodents.

5.4 Sound Localization

5.4.1 *Overview of Sound Localization Acuity*

Sound localization acuity is typically expressed as minimum audible angles (MAAs). These can be calculated using different criteria but typically a sensitivity level of d' = 1.0 or 1.5 is used. In the case of acoustic startle modification measurements, MAAs are reported based on statistically significant inhibition of the acoustic startle response. MAAs for all rodent species known to have been tested are summarized in Table 5.2. Species-specific data are discussed in more detail in the following sections.

5.4.2 *Mice*

Sound localization abilities have been measured in several species of mice. Laboratory mice (*Mus musculus*) are used in many physiological and anatomical studies of the development and function of the sound localization pathway. There are many strains of laboratory mice, but only a few strains have been tested. There are also many species of wild mice, but only a handful of wild species have been tested. Nevertheless, all mouse strains and species appear to be limited by small head sizes and high-frequency hearing ranges (see Dent, Screven, and Kobrina, Chap. 4). Laboratory strains may be further constrained by limited exposure to natural environmental sounds occurring in a large space, since they are typically housed in small shoebox-sized cages inside rooms dedicated to rodent housing. Many laboratory mice are also exposed to background noise from individually ventilated caging systems or biocontainment equipment (Lauer et al. 2009).

Location discrimination performance has been reported in three laboratory mouse strains tested with controlled head position: the C57BL/6J (Heffner et al. 2001; Behrens and Klump 2016), CBA/129SvEvTac (Lauer et al. 2011), and NMRI (Ehret and Dreyer 1984). Both C57BL6 and CBA/129SvEvTac strains show average MAAs around 30°. Minimum audible angles increase by 13° on average in the C57BL/6J strain as hereditary high-frequency hearing loss develops, though discrimination of large angular separations remains normal (Heffner et al. 2001).

Experiments in animals that were not confined to a fixed head location have reported smaller MAAs (about 7–15°) in the NMRI strain (open-loop stimulus presentation, Ehret and Dreyer 1984) and in the CBA/CaJ strain (acoustic startle

Table 5.2 Minimum audible angles (MAAs) in rodents

Species/strain	MAA (degrees)	References
Laboratory mouse (*Mus musculus*)	31–33°	Heffner et al. (2001) Lauer et al. (2011) Behrens and Klump (2016)
Grasshopper mouse (*Onychomys leucogaster*)	19°	Heffner and Heffner (1988b)
Spiny mouse (*Acomys cahirinus*)	19°	Heffner and Heffner (1992a) Heffner et al. (1994)
Laboratory rat (*Rattus norvegicus*)	8–14°	Kelly (1980) Kelly and Glazier (1978); Kavanagh and Kelly (1986)
Wild Norway rat (*Rattus norvegicus*)	19°	Heffner et al. (1994)
Kangaroo rat (*Dipodomys merriani*)	20°	Heffner and Masterton (1980)
Wood rat (species not identified)	19°	Heffner et al. (1994)
Chinchilla (*Chinchilla lanigera*)	16°	Heffner et al. (1994)
Mongolian gerbil (*Meriones unguiculatus*)	23–38°	Heffner and Heffner (1988b) Maier and Klump (2006) Maier et al. (2008) Lingner et al. (2012)
Blind mole rat (*Spalax ehrenbergi*)	29° (long duration stimuli)	Heffner and Heffner (1992b)
Naked mole rat (*Heterocephalus glaber*)	28° (long duration stimuli)	Heffner and Heffner (1993)
Pocket gopher (*Geomys bursarius*)	80° (long duration stimuli)	Heffner and Heffner (1990)
Fox squirrel (*Sciurus niger*)	15°	Heffner and Heffner (1992a)
Hamster (species not specified)	19°	Heffner and Heffner (1992a)
Groundhog (*Marmota monax*)	27°	Heffner and Heffner (1992a)
Chipmunk (species not specified)	35°	Heffner and Heffner (1992a)
Prairie dog (species not specified)	35°	Heffner and Heffner (1992a)

modification, Allen and Ison 2010). It is not surprising that the open-loop experiment reported improved discrimination because the freely moving animals could make use of loudness cues based on proximity to the speaker, and they were given sufficient time to correct any errors in their trajectories within a trial. The low sensitivity of the acoustic startle modification procedures renders any direct comparisons of performances across these studies problematic (Behrens and Klump 2016).

The poor hearing sensitivity below 4 kHz and small head size of mice limits the availability of ITD cues for horizontal sound localization. Thus, it has long been assumed that mouse species and other small-headed rodents must rely on ILD cues to localize sounds in the horizontal plane. A bias toward reliance on ILD cues is further reflected in the poorly developed medial superior olive (~210 neurons) compared to the lateral superior olive (~1190 neurons) and medial nucleus of the trapezoid body (~2270 neurons) in laboratory mice (Irving and Harrison 1967). Maximum ILDs can exceed 40 dB for frequencies above 20 kHz, whereas ILDs are around 10 dB and lower for frequencies below 20 kHz in laboratory mice (Fig. 5.3) (Saunders and Garfinkle 1983; Lauer et al. 2011). Maximum ITDs are only around 32 μs (Saunders and Garfinkle 1983).

Poor location discrimination in the median vertical plane might suggest that mice do not make use of monaural spectral cues. CBA/129 mice show an average MAA of 80° (Lauer et al. 2011). It should be noted that the mice used in those experiments were raised and housed in standard shoebox cages; therefore, their exposure to sounds in the vertical plane may have been limited. It is possible, but as yet untested, that mice raised in more complex environments might show improved vertical localization acuity. It is also possible that prey species of mice only need to perform crude front–back discriminations to initiate an escape response to evade predators (or locate conspecifics) or that aerial predators such as barn owls (*Tyto alba*) are not easily detected using acoustic cues.

Though the mouse pinna introduces direction-dependent patterns of spectral peaks and notches in both the horizontal and vertical planes, pinna-related spectral cues are most effective at enhancing ILD representation (Lauer et al. 2011). Further evidence that mice use spectral cues to aid localization comes from experiments demonstrating better localization for broadband stimuli than for narrowband stimuli (Lauer et al. 2011; Behrens and Klump 2016).

Experiments in laboratory mice have produced mixed results concerning the effective frequency range of stimuli that produce optimal location-discrimination performance. Young adult C57BL6 mice perform poorly when spectral energy is limited to frequencies below 20 kHz (Heffner et al. 2001). Experiments using acoustic startle reflex modification (Allen and Ison 2010) also showed reduced responsiveness to changes in sound location in azimuth for lower frequency sounds in the CBA/CaJ laboratory mouse strain. In contrast, operant experiments performed in CBA/129SvEvTac mice showed good discrimination of low pass noise with energy below 10 kHz (Lauer et al. 2011). ILDs are certainly available below 10 kHz (Fig. 5.3) (Saunders and Garfinkle 1983; Lauer et al. 2011). The reasons for the discrepancies remain unclear.

To date, sound localization has been tested only in two species of feral mice: the predatory grasshopper mouse (*Onychomys leucogaster*) and the spiny mouse (*Acomys cahirinus*). The grasshopper mouse is a nocturnal species that inhabits deserts and short grass prairies, consuming a diet of insects, including scorpions, and sometimes other rodents. Despite a presumed reliance on sound to localize its prey (Langley 1983), this species shows a fairly large average MAA of 19° for broadband noise bursts presented in the horizontal plane (Heffner and Heffner 1988a). Though its small interaural distance must limit interaural cues in this species, its performance is better than what is predicted based on maximal interaural distance and MAAs in other rodents tested using similar procedures (Heffner and Heffner 1988a). Heffner and Heffner (1988a) suggested that this species has an enhanced ability to make use of available interaural cues due to the selective pressures to localize prey. A similar MAA of 19° has been reported in the spiny mouse, a desert-dwelling omnivorous species, but experimental details are unavailable (Heffner and Heffner 1992a; Heffner et al. 1994). It remains to be determined if other feral mouse species show similar localization enhancements relative to the house mouse.

5.4.3 Rats

Sound localization has been studied in just a few of the numerous rat species. Domesticated laboratory rats (*Rattus norvegicus)* are commonly used in studies of the auditory system. They are usually subject to limited environmental experience, similar to laboratory mice and probably other common laboratory rodent species. Localization acuity in the horizontal plane has been measured in laboratory rats and in several feral rat species. Albino and pigmented laboratory rats show MAAs averaging around 8–14° for broadband sounds (Kelly and Glazier 1978; Kavanagh and Kelly 1986). Wild-caught Norway rats (i.e., *R. norvegicus)* show similar acuity, with MAAs of 10–12° (Heffner and Heffner 1985).

As in mice, laboratory rats rely primarily on ILD cues to discriminate sound location in the horizontal plane (Wesolek et al. 2010). Maximum ILDs are less than 10 dB for frequencies below 8 kHz, but can be as large as 20–40 dB for frequencies above 20 kHz. Maximum ITDs are 130–160 μs (Koka et al. 2008). Like mice, rats have small medial superior olivary nuclei compared to the lateral superior olive and medial nucleus of the trapezoid body (Irving and Harrison 1967). Even though laboratory rats have better hearing for frequencies below 4 kHz when compared to laboratory mice, they show poor localization of sounds with frequencies limited to below 2 kHz (Wesolek et al. 2010). This has been cited as evidence that laboratory rats do not rely on ITDs for localization in the horizontal plane. However, earlier studies showed better performance on lower frequency stimuli (Masterton et al. 1975; Kelly and Kavanagh 1986). The discrepancies are attributed to the confounding presence of overtones in the study by Masterton and colleagues and to the transient ILD cues associated with brief rapid-onset stimuli by Kelly and Kavanagh (1986).

Directional transfer functions measured in laboratory rats reveal spectral dips, or notches, at frequencies between 16 kHz and 30 kHz (Koka et al. 2008). That study also showed that pinna removal reduced spectral notches, ILDs, and ITDs. Localization in the median vertical plane has not been reported. It is possible that, as in laboratory mice, pinna cues primarily benefit localization in the horizontal plane in laboratory rats.

Sound localization data are also available for several species of feral rats. Kangaroo rats (*Dipodomys merriani*), which move bipedally, tend to live in dry, hot regions and subsist mainly on seeds. The best MAA in this species is about 20° (Heffner and Masterton 1980). These animals have enlarged bullae and hypertrophied medial superior olivary nuclei compared to mammals of similar size (Heffner and Masterton 1980). These anatomical specializations accompany a broad range of hearing with good sensitivity from about 0.125 kHz to 32 kHz (Heffner and Masterton 1980). Kangaroo rats can localize tones between 0.25 and 32 kHz with a dip in performance at 4 kHz, presumably due to ambiguous ITD and ILD cues (Heffner and Masterton 1980). Wood rats (species not identified), more commonly known as pack rats, have similar horizontal sound localization acuity with a reported MAA of 19° (Heffner et al. 1994).

5.4.4 Chinchillas

The chinchilla is native to the Andes Mountains of South American and they live in herds. Domesticated chinchillas are popular in auditory research because of their broad hearing range (see Dent, Screven, and Kobina, Chap. 4). Relative to other rodents, chinchillas are good at sound localization with average MAAs of about 16° for broadband sounds in left–right discriminations in the horizontal plane (Heffner et al. 1994). In contrast to most other rodent species tested, chinchillas can localize both low and high frequency noise and pure tones, indicating that they use both ITDs and ILDs (Heffner et al. 1994). Maximum ILDs fall below 10 dB for frequencies below 5 kHz and range from 10 dB to 30 dB above 5 kHz (Koka et al. 2011). Maximum ITDs range from 236 μs to 336 μs (Koka et al. 2011).

A minimum audible angle of 23° in the median vertical plane has been determined for chinchillas (Heffner et al. 1995). Pinna removal experiments resulted in small performance deficits in left–right discrimination of low-pass filtered noise in the horizontal plane (but not broadband noise), small impairments in front–back localization, and large impairments in the median vertical plane (Heffner et al. 1996). Directional transfer functions show prominent spectral notches between 6 kHz and 18 kHz, and pinna removal reduces spectral notches, ILDs, and ITDs as observed in rats (Koka et al. 2011).

5.4.5 Gerbils

Mongolian gerbils (*Meriones unguiculatus*) are a semidesert-dwelling species that has adapted to living in a harsh climate. Considered a diurnal species, they spend the hottest and coldest parts of their days in their underground burrows. Much of the winter is also spent underground. Domesticated gerbils are popular subjects for auditory research due to their good low frequency hearing and well-developed medial superior olive.

Gerbils are able to localize broadband sounds about as well as other small-headed rodents, with a best MAA of 23° (Heffner and Heffner 1988b). Location discrimination of both low and high frequency tones and narrowband noise indicates that gerbils use both ITDs and ILDs for sound localization (Heffner and Heffner 1988b; Maier and Klump 2006), presumably due to their broad hearing range and well-developed medial superior olive (Portfors and Von Gersdorff 2013). Localization performance decreases with age in gerbils (Maier et al. 2008). The maximum ITD is 120 μs (Maki and Furukawa 2005). Maximum ILDs are less than 10 dB for frequencies below 10 kHz and up to 30 dB for frequencies above 10 kHz (Maki and Furukawa 2005).

Head-related and directional transfer functions reveal prominent spectral notches above 10 kHz that vary slightly with changing posture in gerbils (Maki and Furukawa 2005). These spectral cues reportedly vary across individual animals, but they appear to confer the strongest cues in azimuth or in combined azimuth and elevation. Localization in the vertical plane has not been reported in the literature.

5.4.6 Subterranean Species

While many rodent species spend at least some time sheltering inside underground burrows, some species live almost exclusively underground. Rodents living in subterranean environments must navigate through a system of tunnels and burrows in which airborne sound propagation is limited primarily to low frequencies, interaural cues are not prominent, and visual stimulation is minimal (see Schleich and Francescoli, Chap. 3). Subterranean species are either eusocial or solitary, and they primarily consume vegetarian diets. Sound localization has been studied in three subterranean species. The blind mole rat (*Spalax ehrenbergi*), the naked mole rat (*Heterocephalus glaber*), and the pocket gopher (*Geomys bursarius*) hear primarily in the low-frequency range and show best thresholds that are 20–35 dB higher than in most other mammals (Dent, Screven, and Kobrina, Chap. 4).

None of the three species tested is able to localize brief sounds well. Blind mole rats show chance levels of performance at 180° when localizing sounds of short duration (100 ms), but can produce MAAs of 180° for long-duration noise bursts (2.5 s), and MAAs as small as 29° for long-duration trains of stimuli (Heffner and Heffner 1992b). Naked mole rats can show MAAs as small as 63° for single long-duration noise bursts and as small as 28° for long-duration stimuli presented in trains, as in the blind mole rats (Heffner and Heffner 1993). Pocket gophers also require long-duration stimuli for location discrimination, and even then the MAA is an unimpressive 80° (Heffner and Heffner 1990). Performance is at chance for all three species when tested using more conventional stimuli.

The poor sound localization abilities in these three species led to the hypothesis that their auditory brainstem pathways might be underdeveloped compared to other mammals. Surprisingly, the auditory brainstem pathways are reported to be relatively normal in size (Heffner and Heffner 1990, 1993). A recent study noted a lack of superior paraolivary nuclei and potassium/sodium HCN1 (hyperpolarization-activated cyclic nucleotide-gated channel 1) channels in most brainstem nuclei of the naked mole rat brainstem (Gessele et al. 2016). Because HCN channels are required for fast integration of binaural inputs (Koch et al. 2004; Khurana et al. 2012), the absence of these channels may explain the need for long-duration stimuli in order to integrate enough information from binaural inputs to perform localization.

Localization in the median vertical plane has not been tested in subterranean rodents, though Heffner and Heffner (1992b) noted difficulties obtaining performances above chance levels on front-back discriminations for short-duration stimuli. Presumably, these animals would show poor localization acuity in the vertical plane because they do not have prominent pinnae, which are required for producing monaural spectral cues. The extremely poor localization in subterranean rodents is somewhat surprising given that they communicate via low frequency vocalizations that propagate well in their tunnels and burrows (Lange et al. 2007; Schleich and Francescoli, Chap. 3). It is possible that subterranean rodents actually depend more on the detection of substrate vibrations than sound in air.

5.4.7 Other Species

Aspects of directional hearing have been evaluated in a few other rodent species, though experimental details are unavailable. Reported MAAs are as follows: 15° for fox squirrels (*Sciurus niger*); 19° for hamster (species not specified); 27° for groundhog (*Marmota monax*); 35° for chipmunks (species not specified); and 35° for prairie dogs (*Cynomus*, species not specified) (Heffner and Heffner 1992a; Heffner et al. 1994; Heffner and Heffner 2016). Though localization acuity has not been reported, acoustic measurements performed in guinea pigs show that high-frequency, pinna-based spectral cues are present and may aid localization in the horizontal and vertical planes (Carlile and Pettigrew 1987; Hartung and Sterbing 1997). Maximum ILDs range from 10 dB or less below 10 kHz to over 20 dB above 10 kHz (Carlile and Pettigrew 1987).

5.5 Behavioral Studies in Complex Acoustic Environments

Several studies have investigated other aspects of binaural hearing in rodents. In addition to being able to locate an acoustic target, rodents must be able to detect and attend to sounds in the presence of other simultaneously occurring sounds in space. The ability to focus on one sound in the presence of competing sounds is commonly known as the *cocktail party effect*. In nature, competing sounds might come from running water, animal vocalizations, weather events, etc. Two behavioral studies performed in the free field shed light on how this process unfolds in rodents. Ison and Agrawal (1998) measured the effects of low and high frequency tones masked by spatially congruent noise or contralateral noise on the acoustic startle reflex of young adult and old CBAxC57 hybrid mice. This strain maintains excellent hearing sensitivity through late adulthood (Frisina et al. 2011). The mice were lightly sedated in order to reduce head movements. In young adult mice, spatial separation of the signal and masker produced an unmasking effect for high frequency signals, such that inhibition of the startle was greater for spatially separated than for spatially coincident stimuli. A similar benefit of spatial separation was not observed for low frequency stimuli, indicating that ILD cues support spatial release from masking in laboratory mice. Old mice also demonstrated a benefit from spatially separated sounds, indicating that these processes remain intact in mice with good high-frequency hearing. A similar increase in acoustic startle response suppression for spatially separated signal and masker combinations was observed in socially housed rats (but not isolated rats) after fear conditioning to the prepulse stimulus but not before fear conditioning or after extinction (Du et al. 2010), indicating that the emotional salience of the prepulse stimulus is a factor.

In another study, sound localization acuity in the presence of low-pass filtered noise presented from a six-speaker horizontal array was measured in gerbils using operant conditioning procedures and in humans using a similar stimulus presentation paradigm (Lingner et al. 2012). Localization performance was reduced when detecting low frequency targets, demonstrating the interfering effect of competing signals on sound location discrimination. Interestingly, playing correlated masking noise resulted in more masking compared to uncorrelated noise, which is an effect that is opposite to what has been reported in humans. Additionally, gerbils require a much higher signal-to-noise ratio to localize low-frequency signals in noise when compared to humans. The species difference is attributed to the smaller head size and broader peripheral auditory filters in the gerbil.

Rodents must also maintain their ability to localize sounds in the presence of echoes when sounds occur in reverberant environments. Echo suppression, localization dominance, and summing localization are aspects of the phenomenon known as *the precedence effect*, during which pairs of sounds are localized as coming from the leading location or both locations depending on the time delay between the lead and lag sounds (Litovsky et al. 1999; Brown et al. 2015). Kelly (1974) first examined localization of spatially separate click pairs at a range of lead–lag delays in albino laboratory rats of the Wistar strain that were trained on a conditioned licking suppression paradigm. Click pairs were presented in the free field from speakers located 90° to the left and right of the midline, and animals were trained to detect when a left–right click pair sequence was switched to right–left. Good discrimination performance occurred for click delays between 0.25 ms and 16.0 ms, with a lower limit for discrimination of about 31–62 μs. Hoeffding and Harrison (1979) demonstrated that discrimination of left–right click pairs depends on both relative time and intensity differences in albino Sprague-Dawley rats trained on an operant conditioning task. Finally, spatially separated and delayed stimuli produce fear-conditioned prepulse inhibition of the acoustic startle response in socially housed Sprague-Dawley rats, indicating that aspects of the precedence effect may affect sensorimotor gating (Du et al. 2009a).

Gerbils have also been tested on behavioral tasks to measure the precedence effect. Subjects were trained to approach tone pips emitted by one of two speakers and were tested with varied lead–lag delays (Wolf et al. 2010). Localization dominance was achieved for delays of about 0.1 ms to 3.2 ms for frequencies below 12 kHz. Performance at 24 kHz dropped off at 0.8 ms. Only half of the animals were able to learn the approach task and perform to criterion, suggesting that alternative behavioral paradigms might be more appropriate for investigating the precedence effect in rodents. In aggregate, these studies suggest that the precedence effect in rodents operates similarly to humans, although the exact effects of time, intensity, and frequency spectrum may differ somewhat across species.

5.6 Binaural Evoked Potential Studies

Laboratory rodents have been used in numerous studies of the physiological basis for binaural hearing, often at the level of single neurons. These studies are beyond the scope of this chapter, and the reader is referred to Grothe and colleagues (2010) for an overview of the major findings. Here, studies using evoked potentials to probe aspects of binaural hearing in rodents are reviewed in order to encourage investigators to adopt similar functional assays when behavioral testing is not feasible (e.g., wild species that cannot be kept in the lab for extended periods) or practical (lack of requisite equipment and animal behavior expertise). Evoked potentials, like psychoacoustic tests, can be performed in both humans and animals, facilitating comparisons across species.

Several studies have used auditory brainstem response (ABR) or middle latency response (MLR) measurements to investigate aspects of binaural hearing. These experiments are usually performed in anesthetized animals to eliminate head movements, but they have occasionally been performed in awake rodent preparations. The binaural interaction component (BIC) of the ABR is indicated by calculating the amplitude difference in stimulating both ears simultaneously versus each ear separately and then adding the two monaural responses. Typically, later ABR waves (4 and/or 5, depending on the species and recording conditions) and MLRs are smaller in amplitude for binaural stimulation than for the summed left and right

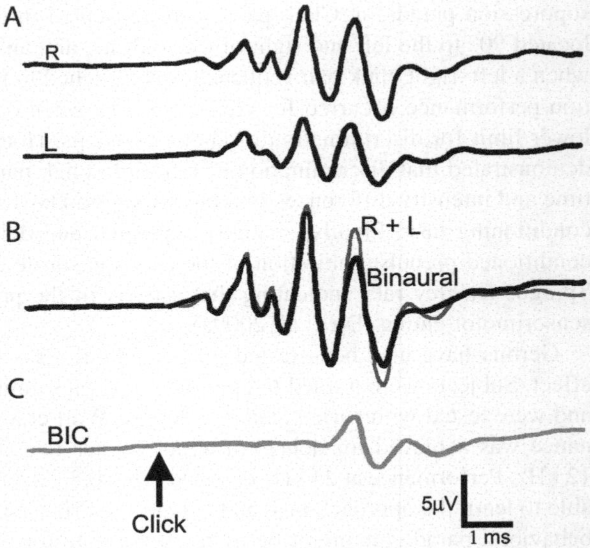

Fig. 5.5 Binaural interaction component (*BIC*). Auditory brainstem responses are measured separately for the left (*L*) and right (*R*) ears (**a**) and for both ears stimulated simultaneously (**b**). The sum of the potentials measured from the right and left ears is larger than the binaurally evoked potential in (**b**). The difference in the summed potential and the binaural potential is calculated and shown in (**c**). (Adapted from responses measured in guinea pigs by Wada and Starr (1989))

monaural stimulation as shown in Fig. 5.5. This effect is attributed to binaural interactions in the lateral lemniscus, most likely reflecting inhibitory mechanisms in the LSO and possibly MSO (Wada and Starr 1989; Laumen et al. 2016b). Sizable BICs have been demonstrated in mice (Henry 1982), guinea pigs (Dobie and Berlin 1979; Wada and Starr 1989), and gerbils (Laumen et al. 2016b). The amplitudes and latencies of the BICs are observed at a range of ITDs and ILDs, though the effects of ITD tend to be larger than the effects of ILDs (Goksoy et al. 2005; Laumen et al. 2016b). The amplitude of the BIC correlates with ILDs and to some extent with ITDs, indicating that BICs may provide a useful measure of binaural processing in the horizontal plane in species that cannot readily be tested using behavioral methods (Laumen et al. 2016a).

Several studies have investigated spatial masking release of frequency following responses (FFRs) elicited by rat pain vocalizations masked by broadband noise. In humans, FFRs measured in the presence of background noise can be unmasked by binaural difference cues (Wilson and Krishnan 2005). In Sprague-Dawley rats, a chatter vocalization with a fundamental frequency of 2.1 kHz or harmonic complexes designed to mimic the vocalizations (2.1 kHz, 4.2 kHz, 6.3 kHz components) elicit FFRs in IC (Du et al. 2009b, c) and amygdala (Du et al. 2012). Unmasking occurs when the signal is presented in the presence of binaurally correlated noise and with disparate ITDs. The amygdala effects are further amplified when the auditory association cortex is blocked bilaterally or when the harmonic complex is paired with a fear-conditioning stimulus (Du et al. 2009c, 2012). The IC effects are mediated by blocking inhibitory input from the contralateral DNLL and excitatory input from the contralateral IC (Du et al. 2009b). These studies indicate that subcortical structures show evidence of binaural unmasking in rats, and that the spatial effects may be facilitated based on the emotional or biological meaning of the stimulus. A more recent study from Li and colleagues has shown that a break in interaural correlation or an interaurally correlated amplitude gap can reduce the FFR elicited by steady-state narrowband noise and recorded in the IC of rats (Wang and Li 2015). Interaural correlation is known to affect sound localization, spatial unmasking, and auditory scene analysis. This study provides some of the first evidence of how disruptions to interaural correlation are represented in the auditory brain. Similar spatially unmasked FFRs could potentially be recorded using scalp electrodes in rodents to investigate binaural processes. Binaural unmasking of FFRs has not been studied in other rodent species.

5.7 Summary

Of the approximately 2000 extant species of rodents, only a small proportion has been tested for spatial hearing abilities. Due to the heavy time commitments, the particular expertise required for performing behavioral testing, and the lack of funding to support psychophysical assessment of hearing in nonlaboratory species, these kinds of detailed assessments are becoming increasingly rare. This is unfortunate because using a diverse array of species can yield exciting discoveries about diversity in the nervous system, reveal unexpected findings that can be applied to human health and disease, aid in conserving wild species and supporting biodiversity, and help overcome the limitations of a given model species, such as genetic homogeneity (Brenowitz and Zakon 2015).

Based on the data reviewed in this chapter, there are several general conclusions that are likely to apply to most untested rodent species. First, rodents do not show particularly acute resolution of differences in angular separation of sounds in azimuth, with best MAAs around 10–30°. This limitation is most likely to be related to the small head sizes, the range of audible frequencies, and the availability of interaural cues required for fine localization. It would be interesting to test a rodent species with a large head, such as the capybara (*Hydrochoerus hydrochaeris*), to determine if small MAAs are possible in rodents with large heads.

Second, limited low-frequency hearing results in a reliance on ILDs in many surface-dwelling rodents, since available ITD cues are smaller than physiologically feasible limits. Those species with low frequency sensitivity (e.g., gerbils) are still constrained by head size, but they may be able to use ITDs to some extent.

Third, subterranean rodents with low-frequency hearing are very poor at localizing sounds, and this appears to be due to the paucity of fast ion channels in brainstem circuits.

Fourth, species with pinnae may be using spectral cues primarily to aid localization in the horizontal plane rather than the median vertical plane, but more testing is needed to confirm this observation.

Several papers have proposed that sound localization evolved to guide the eyes toward a sound source, particularly prey (Heffner and Heffner 1992a). This vision hypothesis is based on a correlation between width of field of best vision and sound localization acuity, independent of head size. The poor localization evident in rodents would seem to support the vision hypothesis. Rodents generally have poor vision and are most active in the dark; therefore, acute sound localization is not needed to direct gaze. Rodents with extremely poor vision, such as mole rats, show almost no localization acuity for brief stimuli, lending further support for the vision hypothesis. However, more recent work argues that the basic neural mechanisms for coding binaural cues is very consistent across mammals (with some exceptions to the rule) and that evolutionary pressure is primarily on the animal's ability to localize sounds based on the most informative subset of interaural cues available (Phillips et al. 2012).

With data available from so few of the approximately 2000 rodent species, it is difficult to determine if either hypothesis can explain the evolution of rodent sound localization. Based on the available data, it appears that rodents with higher best frequency sensitivity have smaller MAAs (Fig. 5.6), perhaps due to stronger ILD or spectral cues above 10 kHz. Similarly, species with higher frequency-hearing limits show better MAAs (Fig. 5.7). In contrast, rodents with lower frequency hearing, and possibly stronger ITD encoding, do not show better MAAs (Fig. 5.8). This is likely related to the small head size and limited available ITDs.

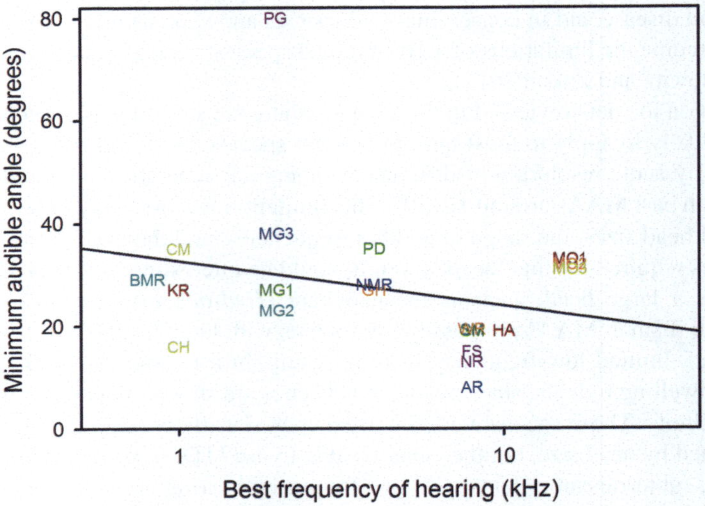

Fig. 5.6 Best minimum audible angle (MAA) plotted for each species tested as a function of that species' *best frequency of hearing* (see text for MAA references). Where MAA data are available from multiple studies for a species, the results from each study are plotted and numbered. *Line* represents best fit regression. *Abbreviations*: AR, albino rat; BMR, blind mole rat; CH, chinchilla; CM, chipmunk; FS, fox squirrel; GH, ground hog; GM, grasshopper mouse; HM, hamster; KR, kangaroo rat; MG, Mongolian gerbil; MO, house mouse; NMR, naked mole rat; PD, prairie dog; PG, pocket gopher; RA, norway rat; SM, spiny mouse; WR, wood rat

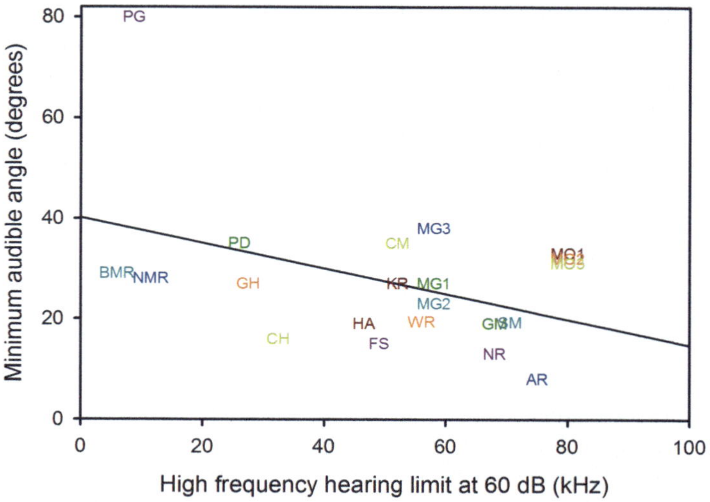

Fig. 5.7 Best minimum audible angle (MAA) plotted for each species as a function of that species' *high-frequency hearing limit* defined as the highest frequency audible at 60 dB SPL. *Line* represents best fit regression. *Abbreviations*: AR, albino rat; BMR, blind mole rat; CH, chinchilla; CM, chipmunk; FS, fox squirrel; GH, ground hog; GM, grasshopper mouse; HM, hamster; KR, kangaroo rat; MG, Mongolian gerbil; MO, house mouse; NMR, naked mole rat; PD, prairie dog; PG, pocket gopher; RA, norway rat; SM, spiny mouse; WR, wood rat

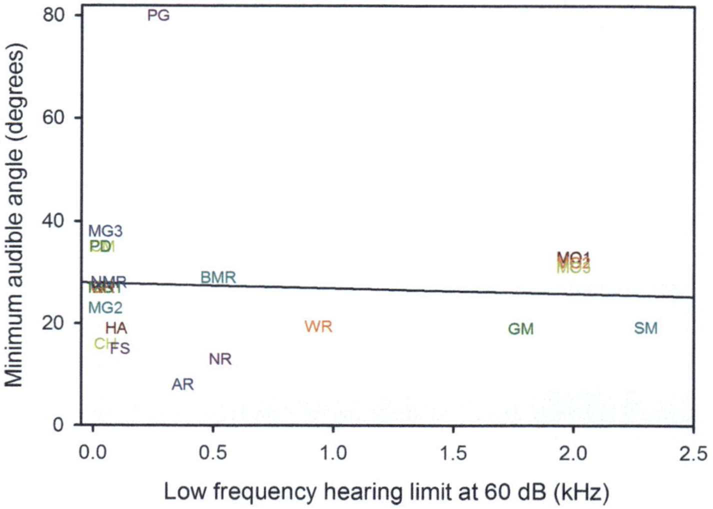

Fig. 5.8 Best minimum audible angle (MAA) plotted for each species as a function of that species' *low frequency hearing limit*, defined as the lowest frequency audible at 60 dB SPL. *Line* represents best fit regression. *Abbreviations*: AR, albino rat; BMR, blind mole rat; CH, chinchilla; CM, chipmunk; FS, fox squirrel; GH, ground hog; GM, grasshopper mouse; HM, hamster; KR, kangaroo rat; MG, Mongolian gerbil; MO, house mouse; NMR, naked mole rat; PD, prairie dog; PG, pocket gopher; RA, norway rat; SM, spiny mouse; WR, wood rat

Though it is clear that rodents are not specialized for highly acute sound localization, there are many areas for further study. For instance, a recent study in the Vietnamese pygmy dormouse (*Typhlomys chaplensis*), which is a blind species that climbs among tree branches, suggests that they may use ultrasonic echolocation to navigate in space (Panyutina et al. 2017). The neurophysiological basis for this is unknown in rodents. Spatial sound perception and the role of vocalizations in spatial navigation need to be studied in order to recognize potential phenotypic diversity and better understand how various species have evolved to interact with their acoustic environments. More experiments on complex hearing phenomena, such as spatial release from masking and the precedence effect, would increase our understanding of how rodents may be affected by competing sounds and background noise, and how these processes are similar or different in model species compared to humans. Virtually nothing is known about how spatial hearing interacts with other sensory systems to aid navigation and orientation in rodents. The role of top–down modulation of spatial hearing is also not understood in rodents. Finally, the increasing accessibility of genetic tools, such as optogenetics, CRISPR, and other technologies, opens up new possibilities for understanding species-specific circuit specializations and the role of specific neural populations in affecting sound localization behavior.

Compliance with Ethics Requirements A. Lauer, J. H. Engel, and K. Schrode declare that they have no conflicts of interest. Amanda Lauer has received grants from the National Institutes of Health, Action on Hearing Loss, the American Hearing Research Foundation, the National Organization for Hearing Research, the Tinnitus Research Consortium, the Capita Foundation, Johns Hopkins University, and the David M. Rubenstein Fund for Hearing Research. Katrina Schrode has received funding from the National Science Foundation and the National Institutes of Health.

References

Allen, P. D., & Ison, J. R. (2010). Sensitivity of the mouse to changes in azimuthal sound location: Angular separation, spectral composition, and sound level. *Behavioral Neuroscience*, 124(2), 265–277.

Behrens, D., & Klump, G. M. (2016). Comparison of mouse minimum audible angle determined in prepulse inhibition and operant conditioning procedures. *Hearing Research*, 333, 167–178.

Beyerl, B. D. (1978). Afferent projections to the central nucleus of the inferior colliculus in the rat. *Brain Research*, 145, 209–223.

Brenowitz, E. A., & Zakon, H. H. (2015). Emerging from the bottleneck: Benefits of the comparative approach to modern neuroscience. *Trends in Neurosciences*, 38(5), 273–278.

Brown, A. D., Stecker, G. C., & Tollin, D. J. (2015). The precedence effect in sound localization. *Journal of the Association for Research in Otolaryngology*, 16(1), 1–28.

Cant, N. B., & Casseday, J. H. (1986). Projections from the anteroventral cochlear nucleus to the lateral and medial superior olivary nuclei. *The Journal of Comparative Neurology*, 247, 457–476.

Carlile, S., & Pettigrew, A. G. (1987). Directional properties of the auditory periphery in the guinea pig. *Hearing Research*, 31(2), 111–122.

Chase, S. M., & Young, E. D. (2005). Limited segregation of different types of sound localization information among classes of units in the inferior colliculus. *The Journal of Neuroscience*, 25, 7575–7585.

Coleman, J. R., & Clerici, W. J. (1987). Sources of projections to subdivisions of the inferior colliculus in the rat. *The Journal of Comparative Neurology*, 262, 215–226.

Davis, K. A. (2005). Contralateral effects and binaural interactions in dorsal cochlear nucleus. *Journal of the Association for Research in Otolaryngology*, 6, 280–296.

Davis, K. A., Ramachandran, R., & May, B. J. (2003). Aauditory processing of spectral cues for sound localization in the inferior colliculus. *Journal of the Association for Research in Otolaryngology*, 4, 148–163.

Dobie, R. A., & Berlin, C. I. (1979). Binaural interaction in brainstem-evoked responses. *Archives of Otolaryngology*, 105(7), 391–398.

Du, Y., Li, J., Wu, X., & Li, L. (2009a). Precedence-effect-induced enhancement of prepulse inhibition in socially reared but not isolation-reared rats. *Cognitive, Affective, & Behavioral Neuroscience*, 9(1), 44–58.

Du, Y., Ma, T., Wang, Q., Wu, X., & Li, L. (2009b). Two crossed axonal projections contribute to binaural unmasking of frequencyfollowing responses in rat inferior colliculus. *European Journal of Neuroscience*, 30(9), 1779–1789.

Du, Y., Huang, Q., Wu, X., Galbraith, G. C., & Li, L. (2009c). Binaural unmasking of frequency-following responses in rat amygdala. *Journal of Neurophysiology*, 101(3), 1647–1659.

Du, Y., Wu, X., & Li, L. (2010). Emotional learning enhances stimulus-specific top-down modulation of sensorimotor gating in socially reared rats but not isolation-reared rats. *Behavioural Brain Research*, 206(2), 192–201.

Du, Y., Wang, Q., Zhang, Y., Wu, X., & Li, L. (2012). Perceived targetmasker separation unmasks responses of lateral amygdala to the emotionally conditioned target sounds in awake rats. *Neuroscience*, 225, 249–257.

Ehret, G., & Dreyer, A. (1984). Sound localization in the horizontal plane by the house mouse (*Mus musculus*). In D. Varju & H. U. Schnitzler (Eds.), *Localization and orientation in biology and engineering* (pp. 60–62). Berlin: Springer-Verlag.

Frisina, R. D., Singh, A., Bak, M., Bozorg, S., et al. (2011). F1 (CBA× C57) mice show superior hearing in old age relative to their parental strains: Hybrid vigor or a new animal model for "golden ears"? *Neurobiology of Aging*, 32(9), 1716–1724.

Gessele, N., Garcia-Pino, E., Omerbašić, D., Park, T. J., & Koch, U. (2016). Structural changes and lack of HCN1 channels in the binaural auditory brainstem of the naked mole-rat (*Heterocephalus glaber*). *PLoS One*, 11(1), e0146428.

Glendenning, K. K., Brusno-Bechtold, J. K., Thompson, G. C., & Masterton, R. B. (1981). Ascending auditory afferents to the nuclei of the lateral leminscus. *The Journal of Comparative Neurology*, 197, 673–703.

Godfrey, D. A., Lee, A. C., Hamilton, W. D., Benjamin III, L. C., et al. (2016). Volumes of cochlear nucleus regions in rodents. *Hearing Research*, 339, 161–174.

Goksoy, C., Demirtas, S., Yagcioglu, S., & Ungan, P. (2005). Interaural delay-dependent changes in the binaural interaction component of the guinea pig brainstem responses. *Brain Research*, 1054(2), 183–191.

Grothe, B., Pecka, M., & McAlpine, D. (2010). Mechanisms of sound localization in mammals. *Physiological Reviews*, 90(3), 983–1012.

Harrison, J. M., & Warr, W. B. (1962). A study of the cochlear nuclei and ascending auditory pathways of the medulla. *The Journal of Comparative Neurology*, 119, 341–379.

Hartung, K., & Sterbing, S. J. (1997). Generation of virtual sound sources for electrophysiological characterization of auditory spatial tuning in the guinea pig. In J. Syka (Ed.), *Acoustical signal processing in the central auditory system* (pp. 407–412). New York: Springer-Verlag.

Hebrank, J., & Wright, D. (1974). Spectral cues used in the localization of sound sources on the median plane. *The Journal of the Acoustical Society of America*, 56(6), 1829–1834.

Heffner, H., & Masterton, B. (1980). Hearing in glires: Domestic rabbit, cotton rat, feral house mouse, and kangaroo rat. *The Journal of the Acoustical Society of America*, 68(6), 1584–1599.
Heffner, H. E., & Heffner, R. S. (1985). Sound localization in wild Norway rats (*Rattus norvegicus*). *Hearing Research*, 19(2), 151–155.
Heffner, H. E., & Heffner, R. S. (2016). The evolution of mammalian sound localization. *Acoustics Today*, 12(1), 20–35.
Heffner, R. S., & Heffner, H. E. (1988a). Sound localization in a predatory rodent, the northern grasshopper mouse (*Onychomys leucogaster*). *Journal of Comparative Psychology*, 102(1), 66–71.
Heffner, R. S., & Heffner, H. E. (1988b). Sound localization and use of binaural cues by the gerbil (*Meriones unguiculatus*). *Behavioral Neuroscience*, 102(3), 422–428.
Heffner, R. S., & Heffner, H. E. (1990). Vestigial hearing in a fossorial mammal, the pocket gopher (*Geomys bursarius*). *Hearing Research*, 46(3), 239–252.
Heffner, R. S., & Heffner, H. E. (1992a). Visual factors in sound localization in mammals. *The Journal of Comparative Neurology*, 317(3), 219–232.
Heffner, R. S., & Heffner, H. E. (1992b). Hearing and sound localization in blind mole rats (*Spalax ehrenbergi*). *Hearing Research*, 62(2), 206–216.
Heffner, R. S., & Heffner, H. E. (1993). Degenerate hearing and sound localization in naked mole rats (*Heterocephalus glaber*), with an overview of central auditory structures. *The Journal of Comparative Neurology*, 331(3), 418–433.
Heffner, R. S., Heffner, H. E., Kearns, D., Vogel, J., & Koay, G. (1994). Sound localization in chinchillas. I: Left/right discriminations. *Hearing Research*, 80(2), 247–257.
Heffner, R. S., Heffner, H. E., & Koay, G. (1995). Sound localization in chinchillas. II. Front/back and vertical localization. *Hearing Research*, 88(1), 190–198.
Heffner, R. S., Koay, G., & Heffner, H. E. (1996). Sound localization in chinchillas III: Effect of pinna removal. *Hearing Research*, 99(1), 13–21.
Heffner, R. S., Koay, G., & Heffner, H. E. (2001). Sound-localization acuity changes with age in C57BL/6J mice. In J. F. Willott (Ed.), *Handbook of mouse auditory research: From behavior to molecular biology* (pp. 31–35). Boca Raton, FL: CRC Press.
Henry, K. R. (1982). Genetic influences on binaural summation and recovery rate of the brainstem auditory evoked response. *Acta Otolaryngologica*, 93(1–6), 1–7.
Hoeffding, V., & Harrison, J. M. (1979). Auditory discrimination: Role of time and intensity in the precedence effect. *Journal of the Experimental Analysis of Behavior*, 32(2), 157–166.
Huang, C. L., & Winer, J. A. 2000. Auditory thalamocortical projections in the cat: Laminar and areal patterns of input. *The Journal of Comparative Neurology*, 427(2), 302–331.
Irving, R., & Harrison, J. M. (1967). The superior olivary complex and audition: A comparative study. *The Journal of Comparative Neurology*, 130(1), 77–86.
Ison, J. R., & Agrawal, P. (1998). The effect of spatial separation of signal and noise on masking in the free field as a function of signal frequency and age in the mouse. *The Journal of the Acoustical Society of America*, 104(3), 1689–1695.
Jenkins, W. M., & Merzenich, M. M. (1984). Role of cat primary auditory cortex for sound-localization behavior. *Journal of Neurophysiology*, 52, 819–847.
Joris, P. X., &Yin, T. C. (1995). Envelope coding in the lateral superior olive. I. Sensitivity to interaural time differences. *Journal of Neurophysiology*, 73, 1043–1062.
Kavanagh, G. L., & Kelly, J. B. (1986). Midline and lateral field sound localization in the albino rat (*Rattus norvegicus*). *Behavioral. Neuroscience*, 100(2), 200–205.
Kelly, J. B. (1974). Localization of paired sound sources in the rat: Small time differences. *The Journal of the Acoustical Society of America*, 55(6), 1277–1284.
Kelly, J. B. (1980). Effects of auditory cortical lesions on sound localization by the rat. *Journal of Neurophysiology*, 44(6), 1161–1174.
Kelly, J. B., & Glazier, S. J. (1978). Auditory cortex lesions and discrimination of spatial location by the rat. *Brain Research*, 145(2), 315–321.
Kelly, J. B., & Kavanagh, G. L. (1986). Effects of auditory cortical lesions on pure-tone sound localization by the albino rat. *Behavioral Neuroscience*, 100(4), 569–575.

Kelly, J. B., & Phillips, D. P. (1991). Coding of interaural time differences of transients in auditory cortex of *Rattus norvegicus*: Implications for the evolution of mammalian sound localization. *Hearing Research*, 55, 39–44.

Kelly, J. B., Buckthought, A. D., & Kidd, S. A. (1998). Monaural and binaural response properties of single neurons in the rat's dorsal nucleus of the lateral lemniscus. *Hearing Research*, 122, 25–40.

Khurana, S., Liu, Z., Lewis, A. S., Rosa, K., et al. (2012). An essential role for modulation of hyperpolarization-activated current in the development of binaural temporal precision. *The Journal of Neuroscience*, 32(8), 2814–2823.

Koch, U., Braun, M., Kapfer, C., & Grothe, B. (2004). Distribution of HCN1 and HCN2 in rat auditory brainstem nuclei. *European Journal of Neuroscience*, 20(1), 79–91.

Koka, K., Read, H. L., & Tollin, D. J. (2008). The acoustical cues to sound location in the rat: Measurements of directional transfer functions. *The Journal of the Acoustical Society of America*, 123(6), 4297–4309.

Koka, K., Jones, H. G., Thornton, J. L., Lupo, J. E., & Tollin, D. J. (2011). Sound pressure transformations by the head and pinnae of the adult chinchilla (*Chinchilla lanigera*). *Hearing Research*, 272(1), 135–147.

Kyweriga, M., Stewart, W., Cahill, C., & Wehr, M. (2014). Synaptic mechanisms underlying interaural level difference selectivity in rat auditory cortex. *Journal of Neurophysiology*, 112, 2561–2571.

Lange, S., Burda, H., Wegner, R. E., Dammann, P., et al. (2007). Living in a "stethoscope": Burrow acoustics promote auditory specializations in subterranean rodents. *Naturwissenschaften*, 94(2), 134–138.

Langley, W. M. (1983). Relative importance of the distance senses in grasshopper mouse predatory behaviour. *Animal Behaviour*, 31(1), 199–205.

Lauer, A. M., Slee, S. J., & May, B. J. (2011). Acoustic basis of directional acuity in laboratory mice. *Journal of the Association for Research in Otolaryngology*, 12(5), 63–645.

Lauer, A. M., May, B. J., Hao, Z. J., & Watson, J. (2009). Analysis of environmental sound levels in modern rodent housing rooms. *Lab Animal*, 38(5), 154–160.

Lauer, A. M., Connelly, C. J., Graham, H., & Ryugo, D. K. (2013). Morphological characterization of bushy cells and their inputs in the laboratory mouse (*Mus musculus*) anteroventral cochlear nucleus. *PLoS One*, 8, e73308.

Laumen, G., Tollin, D. J., Beutelmann, R., & Klump, G. M. (2016a). Aging effects on the binaural interaction component of the auditory brainstem response in the Mongolian gerbil: Effects of interaural time and level differences. *Hearing Research*, 337, 46–58.

Laumen, G., Ferber, A. T., Klump, G. M., & Tollin, D. J. (2016b). The physiological basis and clinical use of the binaural interaction component of the auditory brainstem response. *Ear and Hearing*, 37(5), 276–290.

Lingner, A., Wiegrebe, L., & Grothe, B. (2012). Sound localization in noise by gerbils and humans. *Journal of the Association for Research in Otolaryngology*, 13(2), 237–248.

Litovsky, R. Y., Colburn, H. S., Yost, W. A., & Guzman, S. J. (1999). The precedence effect. *The Journal of the Acoustical Society of America*, 106(4), 1633–1654.

Maier, J. K., & Klump, G. M. (2006). Resolution in azimuth sound localization in the Mongolian gerbil (*Meriones unguiculatus*). *The Journal of the Acoustical Society of America*, 119(2), 1029–1036.

Maier, J. K., Kindermann, T., Grothe, B., & Klump, G. M. (2008). Effects of omni-directional noise exposure during hearing onset and age on auditory spatial resolution in the Mongolian gerbil (*Meriones unguiculatus*)—a behavioral approach. *Brain Research*, 1220, 47–57.

Maki, K., & Furukawa, S. (2005). Reducing individual differences in the external-ear transfer functions of the Mongolian gerbil. *The Journal of the Acoustical Society of America*, 118(4), 2392–2404.

Malmierca, M. S., Merchán, M. A., Henkel, C. K. & Oliver, D. L. (2002). Direct projections from cochlear nuclear complex to auditory thalamus in the rat. *The Journal of Neuroscience*, 22, 10891–10897.

Markovitz, N. S. & Pollak, G. D. (1994). Binaural processing in the dorsal nucleus of the lateral lemniscus. *Hearing Research*, 73, 121–140.
Mast, T. E. (1969). Binaural interaction and contralateral inhibition in dorsal cochlear nucleus of the chinchilla. *Journal of Neurophysiology*, 33, 108–115.
Masterton, B., Thompson, G. C., Bechtold, J. K., & RoBards, M. J. (1975). Neuroanatomical basis of binaural phase-difference analysis for sound localization: A comparative study. *Journal of Comparative and Physiological Psychology*, 89(5), 379–386.
Nordeen, K. W., Killackey, H. P. & Kitzes, L. M. (1983). Ascending auditory projections to the inferior colliculus in the adult gerbil, *Meriones unguiculatus*. *The Journal of Comparative Neurology*, 214, 131–143.
Phillips, D. P., Quinlan, C. K., & Dingle, R. N. (2012). Stability of central binaural sound localization mechanisms in mammals and the Heffner hypothesis. *Neuroscience & Biobehavioral Reviews*, *36*(2), 889–900.
Panyutina, A. A., Kuznetsov, A. N., Volodin, I. A., Abramov, A. V., & Soldatova, I. B. (2017). A blind climber: The first evidence of ultrasonic echolocation in arboreal mammals. *Integrative Zoology*, 12(2), 172–184.
Popper, A. N., & Fay, R. R. (2005). *Sound source localization*. New York: Springer Science+Business Media.
Portfors, C. V., & Von Gersdorff, H. (2013). Macrocircuits for sound localization use leaky coincidence detectors and specialized synapses. *Neuron*, 78(5), 755–757.
Ramachandran, R., Davis, K. A., & May, B. J. (1999). Single-unit responses in the inferior colliculus of decerebrate cats. I. Classification based on frequency response maps. *Journal of Neurophysiology*, 82, 152–163.
Rayleigh, L. (1907). XII. On our perception of sound direction. *The London, Edinburgh, and Dublin Philosophical Magazine and Journal of Science*, Series 6, 13(74), 214–232.
Rice, J. J., May, B. J., Spirou, G. A., & Young, E. D. (1992). Pinna-based spectral cues for sound localization in cat. *Hearing Research*, 58(2), 132–152.
Saunders, J. C., & Garfinkle, T. J. (1983). Peripheral anatomy and physiology. In J. F. Willott (Ed.), *The auditory psychobiology of the mouse* (pp. 131–168). Springfield, IL: Charles C. Thomas.
Smith, P. H., Joris, P. X., Carney, L. H., & Yin, T. C. T. (1991). Projections of physiologically characterized globular bushy cell axons from the cochlear nucleus of the cat. *The Journal of Comparative Neurology*, 304, 387–407.
Thompson, G. C., & Cortez, A. M. (1983). The inability of squirrel monkeys to localize sound after unilateral ablation of auditory cortex. *Behavioral Brain Research*, 8, 211–216.
Tollin, D. J., & Yin, T. C. T. (2005). Interaural phase and level difference sensitivity in low-frequency neurons in the lateral superior olive. *The Journal of Neuroscience*, 25, 10648–10657.
Wada, S. I., & Starr, A. (1989). Anatomical bases of binaural interaction in auditory brain-stem responses from guinea pig. *Electroencephalography and Clinical Neurophysiology*, 72(6), 535–544.
Wang, Q., & Li, L. (2015). Auditory midbrain representation of a break in interaural correlation. *Journal of Neurophysiology*, 114(4), 2258–2264.
Wesolek, C. M., Koay, G., Heffner, R. S., & Heffner, H. E. (2010). Laboratory rats (*Rattus norvegicus*) do not use binaural phase differences to localize sound. *Hearing Research*, 265(1), 54–62.
Wilson, J. R., & Krishnan, A. (2005). Human frequency-following responses to binaural masking level difference stimuli. *Journal of the American Academy of Audiology*, 16(3), 184–195.
Wolf, M., Schuchmann, M., & Wiegrebe, L. (2010). Localization dominance and the effect of frequency in the Mongolian gerbil, *Meriones unguiculatus*. *Journal of Comparative Physiology A: Neuroethology, Sensory, Neural, and Behavioral Physiology*, 196(7), 463–470.
Young, E. D., & Davis, K. A. (2002). Circuitry and function of the dorsal cochlear nucleus. In D. Oertel, R. R. Fay, & A. N. Popper (Eds.), *Integrative functions in the mammalian auditory pathway* (pp. 160–206). New York: Springer Science+Business Media.

Chapter 6
Anatomy of Vocal Communication and Hearing in Rodents

M. Fabiana Kubke and J. Martin Wild

Abstract Many animals produce sounds to communicate different types of information. More often than not, such sounds are vocal in nature and elicit a predictable behavioral response from the listener. While much of the literature on vocal communication derives from classic neuroethological studies on a number of vertebrates, rodents are fast becoming the group of choice to study vocalizations for a variety of reasons, not the least of which is the advantage they offer for genetic manipulation. Central to the study of vocal communication is the need to understand how the nervous system mediates vocal production and how the auditory system accesses the information within a communication signal that leads to an appropriate behavioral response. A key goal is to determine the essential features of communication signals, what information they transmit, how they are categorized, and in combination with information derived from other sensory modalities, how they are interpreted and linked to a context-appropriate motor response. There is a substantial body of literature on the anatomy and physiology of the neural pathways that mediate vocalizations in rodents, but exciting new research lines are investigating the role of learning in vocal communication and how the rodent nervous system processes complex vocal communication signals.

Keywords Auditory processing · Auditory system · Vocal control · Vocal learning

6.1 Introduction

Sounds in nature are a rich source of information, and animals can use what they hear, for example, to identify and escape from an approaching predator or identify and capture moving prey (Bradbury et al. 1998). Animals also actively produce

M. F. Kubke (✉) · J. M. Wild
Department of Anatomy and Medical Imaging, and Eisdell Moore Centre, University of Auckland, Auckland, New Zealand
e-mail: f.kubke@auckland.ac.nz; jm.wild@auckland.ac.nz

M. L. Dent et al. (eds.), *Rodent Bioacoustics*, Springer Handbook of Auditory Research 67, https://doi.org/10.1007/978-3-319-92495-3_6

sounds—frequently vocal sounds—to communicate different types of information (Suthers et al. 2016).

Communication signals often elicit a predictable response by the receiver, so the study of vocal communication cannot be divorced from the role it plays in natural behavior. Thus it comes as no surprise that a large proportion of those working on vocal communication systems identify themselves as neuroethologists. Tinbergen (1963) identified four fundamental problems that ethologists seek to solve: those related to causation, survival, evolution, and ontogeny. In the context of vocal communication, these four problems can be reframed as the following questions:

(1) What are the specific mechanisms that mediate the production of vocal communication signals?
(2) What roles do these signals play in survival and reproduction?
(3) What are the ancestral features that characterize vocal communications and how have they evolved in different lineages?
(4) How are these vocal signals acquired and/or modified during development?

Throughout the neuroethological literature, studies of vocal communication can address any of these four questions, and this framework is used in this chapter.

Vocal signals are used to communicate a wide range of information, such as alerting others to the presence of a predator, communicating distress, resolving conflict, initiating courtship, mating, and defending a territory (Bradbury et al. 1998). Vocalizations are initiated in a variety of contexts, and their role is to produce a predictable response from the listener (Griffiths and Warren 2004; Bennur et al. 2013). Acoustic signals must convey the right kind of information, and the listener, in turn, must be able to detect, discriminate, and categorize the information contained in the signal in order to produce an appropriate response. Most vertebrates have vocal repertoires in the audible range, and many vertebrates also produce vocal signals with frequencies in the ultrasound (over 20 kHz). In general, hearing sensitivities are such that they match the spectral range of the animal's vocal signals (Wilczynski and Ryan 2010).

This chapter describes what is known about how vocal signals are produced, detected, and categorized by rodents. Sections 6.2 and 6.3 describe what is known about how communications are produced and the neural pathways that form part of the vocal production and auditory systems. How the nervous system is able to process information contained in vocal communication signals is considered in Sect. 6.4, and how these principles apply to specific examples of rodent communication is covered in Sect. 6.5. The role of learning in the shaping of communication signals and the evidence for learning in rodents are discussed in Sect. 6.6 and, finally, Sect. 6.7 considers how vocal signals are integrated into an appropriate behavioral response. All abbreviations are defined in Table 6.1.

Table 6.1 Abbreviations

AAF	Anterior auditory field
AC	Auditory cortex in text; arytenoid cartilage in Fig. 6.1
A1	Primary auditory cortex
A2	Secondary auditory cortex
AI	Primary auditory field
AII	Secondary auditory field
AM	Amplitude modulation
AN	Auditory nerve
AVCN	Anterior ventral cochlear nucleus
CIC	Central nucleus of the inferior colliculus
CN	Cochlear nucleus complex
DCN	Dorsal cochlear nucleus
DNLL	Dorsal nucleus of the lateral lemniscus
DM	Dorsomedial field
DP	Dorsoposterior field
FM	Frequency modulation
IC	Inferior colliculus
IHC	Inner hair cell
LSO	Lateral superior olive
MGB	Medial geniculate body
MNTB	Medial nucleus of the trapezoid body
MSO	Medial superior olive
NLL	Nuclei of the lateral lemniscus
NRA	Nucleus retroambiguus
OHC	Outer hair cell
PAG	Periaqueductal gray
PIL	Posterior intralaminar nucleus
PP	Peripendicular nucleus
PVCN	Posterior ventral cochlear nucleus
RA	Robustus archopallialis
SOC	Superior olivary complex
SPON	Superior paraolivary nucleus
UF	Ultrasound field
USV	Ultrasound vocalization
VCN	Ventral cochlear nucleus
VNLL	Ventral nucleus of the lateral lemniscus
VNTB	Ventral nucleus of the trapezoid body

6.2 Vocal Production Mechanisms

In any discussion of vocalizations, it is helpful first to make a distinction between *phonation* and *vocalization*. Although the verb *phonate* is defined (by the Concise Oxford Dictionary) as "to utter a vocal sound," more is frequently implied by *vocalization* than just the production of a vocal sound. A cough is a vocal sound, but usually would not be called a vocalization. The basic requirements of vocal sound production (i.e., phonation) in terrestrial mammals are a larynx and an externally directed air stream. When the latter passes through the former with sufficient speed and/or resistance, audible sounds are produced, even in excised larynges. In life, when terrestrial mammals vocalize, modulation of sounds takes place to a greater or lesser degree through the combined action of expiration (usually), partial or full closure of the laryngeal glottis (by adduction of the vocal folds in mammals), and the use of a supralaryngeal component of the vocal tract to ensure a filtering and articulation of sounds produced by the vocal organ (Simonyan and Horwitz 2011). All these actions entail the coordinated neural control of a great many muscles (Jürgens 2009), and lack of this coordination or control in humans is reflected in a wide range of vocal impairments (or impediments) from stuttering to the total inability to vocalize, such as can result from cerebral vascular accidents (CVA or stroke) or brain lesions. The ability to phonate (make sounds) in these cases, however, is usually retained, unless the lesion includes the midbrain periaqueductal gray (see Sect. 6.2.1).

In addition, it is evident that vocal production can also be impaired due to a variety of defects. These may be of a genetic (e.g., *FoxP2*; Fisher and Scharff 2009), signaling pathway (Hedgehog; Tabler et al. 2017) or molecular (cadherin-6; Fischer and Hammerschmidt 2011; Nakagawa et al. 2012) nature or due to problems affecting the larynx itself. Laryngeal issues range from nodules on the vocal folds to laryngeal carcinoma. There are also a variety of developmental ciliathapies that recently have been highlighted in a thorough examination of the developmental biology of the larynx (Tabler et al. 2017) in which mice were used as model experimental subjects.

Comparative aspects of the mammalian larynx are considered in Harrison (1995), who notes that rodents, including rats, mice, gerbils (Family: Muridae), and hamsters (Family Cricetinae), possess a larynx that does not differ in any significant way from that of other mammals (Fig. 6.1), even though, unlike most other mammals, they produce both a variety of ultrasonic sounds (> 20 kHz) and sounds (cries) audible to the human ear in different behavioral and social contexts. This apparent lack of laryngeal specialization in rodents has puzzled researchers for many years (Roberts 1975; Riede 2013). A recent analysis in mice has shown how ultrasounds can be produced by an air jet passing through an inter-arytenoid glottis, which then impinges on the planar inner wall of the upper thyroid (and perhaps the epiglottis). Feedback may be generated in this way, but exactly how is not yet clear (Mahrt et al. 2016). This analysis in no way denies the contributions of the laryngeal nerves, anatomy, and neural control involved in the respiratory-vocal mechanisms that mediate ultrasonic vocalizations or audible cries (Herbst 2016).

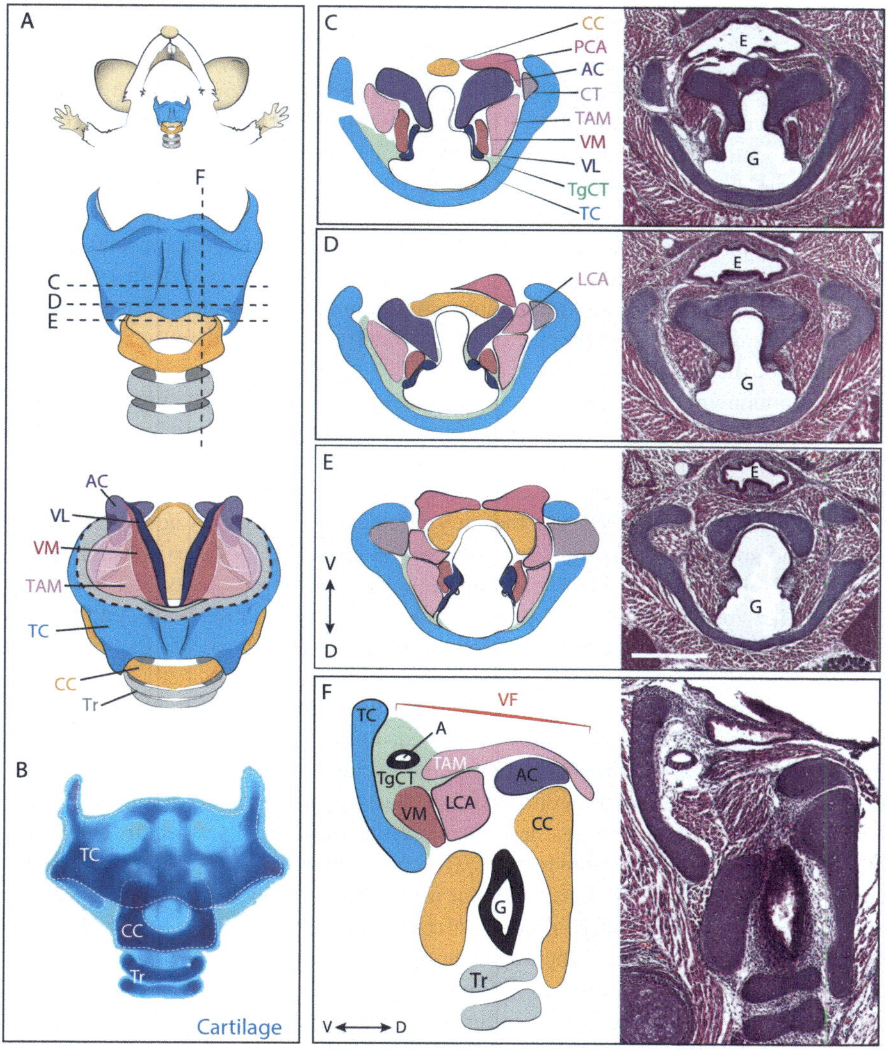

Fig. 6.1 Anatomy of the mouse larynx. (**a**) Diagram representing ventral view of mouse laryngeal anatomy. *Dashed lines* indicate sectional plane represented in panels **c–f**. (**b**) Ventral view of an excised adult larynx stained with alcian blue, marking cartilage. (**c–e**) H&E staining of horizontal sections of E18.5 mouse larynx. Sectional plane is indicated in **a**. Diagrams indicate anatomy observed in sections. (**f**) H&E staining of sagittal section of E18.5 mouse larynx. Diagram indicates anatomy represented in section. *Scale bar* indicates 500 μm. *V* and *D* indicate dorso-ventral axes. *Abbreviations*: A, artery; AC, arytenoid cartilage; CC, cricoid cartilage; CT, cricothyroid muscle; E, esophagus; G, glottis; LCA, lateral cricoarytenoid muscle; PCA, posterior cricoarytenoid muscle; TAM, thyroarytenoid muscle; TC, thryoid cartilage; TgCT, thyroglottal connective tissue; Tr, trachea; VL, vocal ligament; VM, vocalis muscle; VF, vocal fold. (Image and legend from Tabler et al. (2017), distributed under the terms of the Creative Commons Attribution License. doi: https://doi.org/10.7554/eLife.19153.004)

6.2.1 Neural Control

Forty muscles, their nerves, and the brain nuclei involved in mammalian vocalization are listed by Jürgens (2009). The cricothyroid muscle, which is involved in changes in pitch, is innervated by the external branch of the superior laryngeal nerve, while all other laryngeal muscles are innervated by the recurrent branch of the inferior laryngeal nerve. In humans, at least, this simple textbook description does not do justice to the variability of innervation of laryngeal muscles by the laryngeal nerves or the degree to which the muscles are innervated from both sides. The motoneurons projecting to both nerves lie in nucleus ambiguus (the nucleus of the vagal cranial nerve X), which extends as a narrow column deep in the reticular formation of the medulla.

This fundamental brainstem circuitry is headed by a midbrain center, which in mammals comprises parts of the periaqueductal gray (PAG) (Jürgens 2002). The PAG neurons project their axons downstream to various nuclei involved in respiratory-vocal control, the principal one being nucleus retroambiguus (NRA) in the caudal medulla. The NRA is the only nucleus to have direct access to all the cranial and spinal motoneurons involved in this control (Holstege 1989; Holstege et al. 1997). Electrical or chemical stimulation of appropriate parts of PAG evoke realistic-like vocalizations (Phillips and Peek 1975; Zhang et al. 1994), and lesions of PAG result in mutism in animals, including humans (Brown 1965; Esposito et al. 1999).

This last observation is central to the idea that PAG and its limbic associated (anterior cingulate) afferent cortex (Newman et al. 1989; Dujardin and Jürgens 2005) control unlearned, innate, or emotional vocalizations (e.g., cries and laughter in humans), whereas the precentral motor (laryngeal) cortex is required for the learning and production of learned vocalizations, specifically, speech in humans (Jürgens 2009; Simonyan and Horwitz 2011; Simonyan 2014). Songbirds, likewise, possess a telencephalic nucleus (robustus arcopallialis, RA) that is essential for the learning and production of learned vocalizations ("song", Nottebohm et al. 1976) but the homologous relationship of this nucleus to any part of the cortex is unclear. Nevertheless, laryngeal motor cortex in humans and RA in songbirds both project directly upon vocal motoneurons in nucleus ambiguus in humans (Kuypers 1958; Iwatsubo et al. 1990) and in nucleus tracheosyringealis in songbirds (Nottebohm et al. 1976). Those projections are not found in nonhuman primates or in birds that do not produce song, none of which learn their vocalizations (Wild et al. 1997; Simonyan and Jürgens 2003). It occasioned some surprise, therefore, when the laryngeal motor cortex was shown in mice to have a direct, albeit very sparse, projection upon laryngeal motoneurons (Arriaga et al. 2012), supporting the idea that ultrasonic vocalizations (USVs) are learned by mice or have a learned component (Arriaga and Jarvis 2013). However, whether auditory feedback is necessary for the learning of USVs in mice is controversial (Hammerschmidt et al. 2012; Arriaga and Jarvis 2013), but mice lacking a cerebral cortex appear to develop normal USVs (Hammerschmidt et al. 2015). In addition, innate emotional vocalizations in certain strains of very young *FoxP2* mutant mice are not affected by the experimental knockdown (Gaub et al. 2010), suggesting that the midbrain vocal control pathway remains intact in these mice.

6.3 Organization of the Auditory System

What is known about the organization of the auditory system in vertebrates has come from a variety of disciplines and a large number of species. The description of the organization of the auditory system below follows that of Malmierca (2003) for the rat.

6.3.1 The Middle Ear

The presence of three ossicles in the middle ear is one of the distinguishing characteristics of mammals, although substantial variation and evolutionary novelties are seen in the organization of mammalian ears (Fritzsch et al. 2013; Mason 2013). For example, in many rodents the ear is semi-isolated from the skull, having a bony shell that connects the ear to the skull by small bridges and cartilages (Fleischer 1978). As a rule of thumb, animals whose middle ears are smaller and stiffer and have smaller tympanic membrane and footplate areas are better at hearing high frequencies, whereas those that have larger middle ears that are less stiff and have larger tympanic and foot plate areas are better at hearing lower frequencies (Rosowski 1994).

The size of the middle ear generally scales to body size, although there are some exceptions. The *microtype ear* is of particular interest (Fleischer 1978). Found only in species with small body size, these middle ears tend to be smaller than what would be expected for their body size, the malleus tends to be bigger than expected, and the area of the stapedial footplate tends to be smaller. Mason suggested that these are specializations for high frequency hearing (Rosowski 1994; Mason 2013). Not all small mammals, however, have a microtype ear. For example, voles, hamsters, gerbils, and dormice have freely mobile ossicles (where the malleus complex is anchored to the tympanic membrane through a ligament) or transitional ossicles, and no microtype ears are found in the squirrel-related clade (Fleischer 1978; Mason 2013). In the mouse-related clade, specializations are found that probably evolved from the microtype ear (Lavocat and Parent 1985; Mason 2013).

Guinea pigs (*Cavia sp.*), chinchillas (*Chinchilla sp.*), and their relatives (Ctenohystrica) have an unusual middle ear related to, but somewhat distinct from, the typical freely mobile ear (Mason 2013). The malleus has a distinctive morphology with a small anterior process or the anterior process is missing altogether, such as in the guinea pig ear. An unusual stapedious muscle is found in guinea pig and chinchilla, although it has been lost in many members of Ctenohystrica. In species that lack the stapedius muscle, the malleus and incus are fused (Rosowski 1994; Mason 2013).

Subterranean mammals show middle ear specializations, including the loss of middle ear muscles, and adaptations in the arrangement of the ear ossicles that lead to less overall middle ear stiffness, which may improve low-frequency hearing. Burda and colleagues (1992) described a flattening of the articulation at the incudal-mallear joint, an enlarged incus, varied stapes morphologies, and a reduction in middle ear muscles associated with subterranean life. The reduction in middle ear muscles is said to be an adaptation for low frequency hearing, as increased stiffness due to contraction of the stapedius muscle reduces low frequency transmission. It should be noted, however, that small mammals that show low-frequency hearing, such as sciurids and Mongolian gerbils (*Meriones unguiculatus*), do not have reduced middle ear muscles.

Fossorial species that have limited high frequency hearing also show some middle ear adaptations, including an enlargement of the floorplate of the stapes; the size of the stapes and the mass of the malleus and incus are not different from those of nonfossorial species (Lavocat and Parent 1985; Mason 2001). As indicated here, describing the mammalian middle ear simply as a three-ossicle ear does not do justice to the amount of variation present.

6.3.2 The Inner Ear

Mammals typically have an elongate basilar membrane with "inner" and "outer" sensory hair cells and supporting cells in the organ of Corti (Fig. 6.2). The inner hair cells (IHC) are involved in the transduction of vibrations in the fluids of the inner ear (caused by sounds) into a neural signal that is transmitted to the brain via the auditory nerve (AN), and the outer hair cells (OHC) receive efferent projections from the brainstem. The size and degree of coiling of the cochlea, the bony structure that houses the basilar membrane, varies among mammals, and the organization of hair cells along the basilar membrane exhibit profound differences when compared to other vertebrates (Fritzsch et al. 2013; Manley 2017).

The coiled cochlea of therian mammals formed after the appearance of the three-ossicle middle ear between 220 and 150 Ma, before marsupials and placentals split into two separate lineages (Manley 2017). One characteristic of the inner ear of therian mammals is that bone integrates into the organ of Corti (Fig. 6.2), encapsulates the acoustic ganglion, and forms longitudinal ridges on the side of the basilar membrane, which may provide the basilar membrane with greater stiffness to support higher frequency hearing.

In a study of the inner ears of rodent species, West (1985) found a correlation between the number of spiral coils and the octave range of hearing. While the number of turns was not correlated to the length of the basilar membrane, the basilar membrane length appeared to be related to the upper and lower limits of the hearing range. Heffner and colleagues (2001) showed that rodents have higher variability in the upper limit of their hearing range than other mammals, and animals with smaller

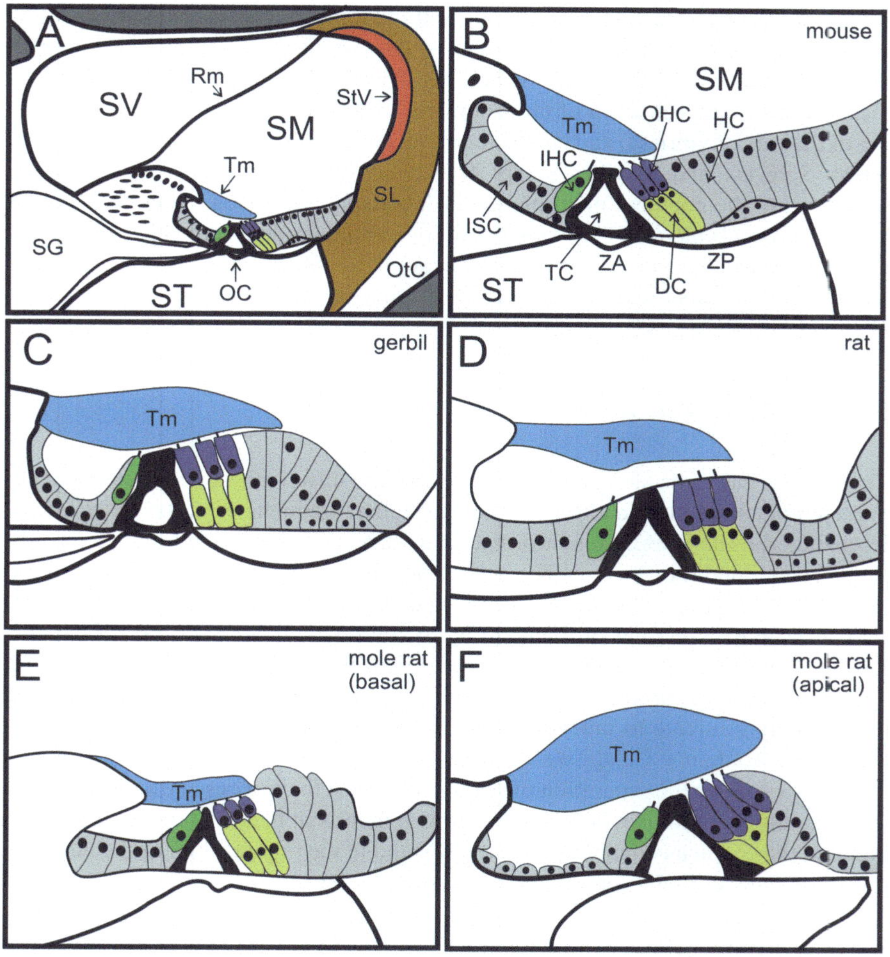

Fig. 6.2 Schematic organization of the inner ear of mouse (**a**) and organ of Corti in the cochlear duct of mouse (**b**), gerbil (**c**), rat (**d**), and mole rat (**e**: basal end, **f**: apical end). *Abbreviations*: DC, Deiters' cells; HC, Hensen's cells; IHC, inner hair cells; ISC, inner sulcus cells; OC, organ of Corti; OHC, outer hair cells, OtC, otic capsule; Rm, Reissner's membrane; SG, spiral ganglion; SL, spiral ligament; SM, scala media; ST, scala timpani; StV, stria vascularis; SV, scala vestibuli; TC, tunnel of Corti; Tm, tectorial membrane; ZA, arcuate zone of basilar membrane; ZP, pectinate zone of basilar membrane. (©2018 Kubke and Vlajkovic, distributed under the terms of the Creative Commons Attribution License. doi: https://doi.org/10.17608/k6.auckland.5926732.v1)

heads tended to have higher frequency hearing limits. High frequency hearing may be important for sound localization in animals with small heads, with the exception of subterranean mammals in which sound localization is not expected to play an important role. Subterranean rodents have a higher hair cell number in the inner ear than rats, and there are indications that the regions that map best frequency may be over-represented (Lange et al. 2007).

A distinguishing feature of modern small mammals is their sensitivity to ultrasound frequencies, although a number of small rodents (e.g., chipmunks, *Tamias striatuts*; hamsters) show some of the lowest frequency hearing (below 100 Hz) (Manley 2017). Presentation of ultrasound up to 50 kHz can elicit reflexes (twitching of the ears or vibrissa) and can be used successfully in conditioning experiments (Dent, Screven, and Kobrina, Chap. 4). Sensitivity to ultrasound also has been shown using the cochlear microphonic in laboratory rats, Mongolian gerbils, and kangaroo rats (*Dupodomis merriami*). Although some rodents are able to hear and process ultrasounds, no obvious specialization for ultrasound hearing has been found in rodent ears (Sales and Pye 1974).

6.3.3 The Spiral Ganglion

The sensory hair cells along the cochlea respond to different frequency components that are contained in an auditory stimulus (highest frequencies at the base; lowest frequencies at the apex). This information is then carried as spike trains in the AN to the central auditory system. The cell bodies of the neurons that form the AN lie in the spiral ganglion, which is housed in the modiolus of the bony cochlea (Fig. 6.2).

The peripheral endings of the majority of spiral ganglion cells (Type I) synapse on IHCs and a smaller population of peripheral fibers (Type II) synapses on the OHCs. The central processes of spiral ganglion neurons enter the brainstem and immediately bifurcate to innervate different regions of the cochlear nucleus complex (CN) (Ramón y Cajal 1904; Lorente de No 1933). This early bifurcation is thought to support the separation of parallel channels that process different features of the auditory stimulus. The tonotopic arrangement that begins at the basilar membrane in the cochlea is maintained in the AN and in the organization of the auditory processing nuclei in the central nervous system to the cortex.

6.3.4 The Cochlear Nuclear Complex

In mammals, afferents in the AN travel from the cochlea to the CN (Fig. 6.3a, black lines), which can be divided into three regions: dorsal cochlear nucleus (DCN), posterior and anterior ventral cochlear nuclei (PVCN and AVCN, respectively). The descending branch of the AN synapses with cells in the PVCN, and axons continue beyond the PVCN to innervate the dorsal cochlear nucleus (DCN). The ascending branch of the AN synapses on neurons of the AVCN (Fig. 6.3a). Type I afferents from the AN make large calyceal synapses on cells of the AVCN; the calyceal synapses maintain the temporal features of the auditory stimulus. In the PVCN, the AN synapses form boutons on the dendrites of octopus cells. The basic patterns and the parallel pathways of the auditory system emerge in the CN (Winer and Schreiner 2005).

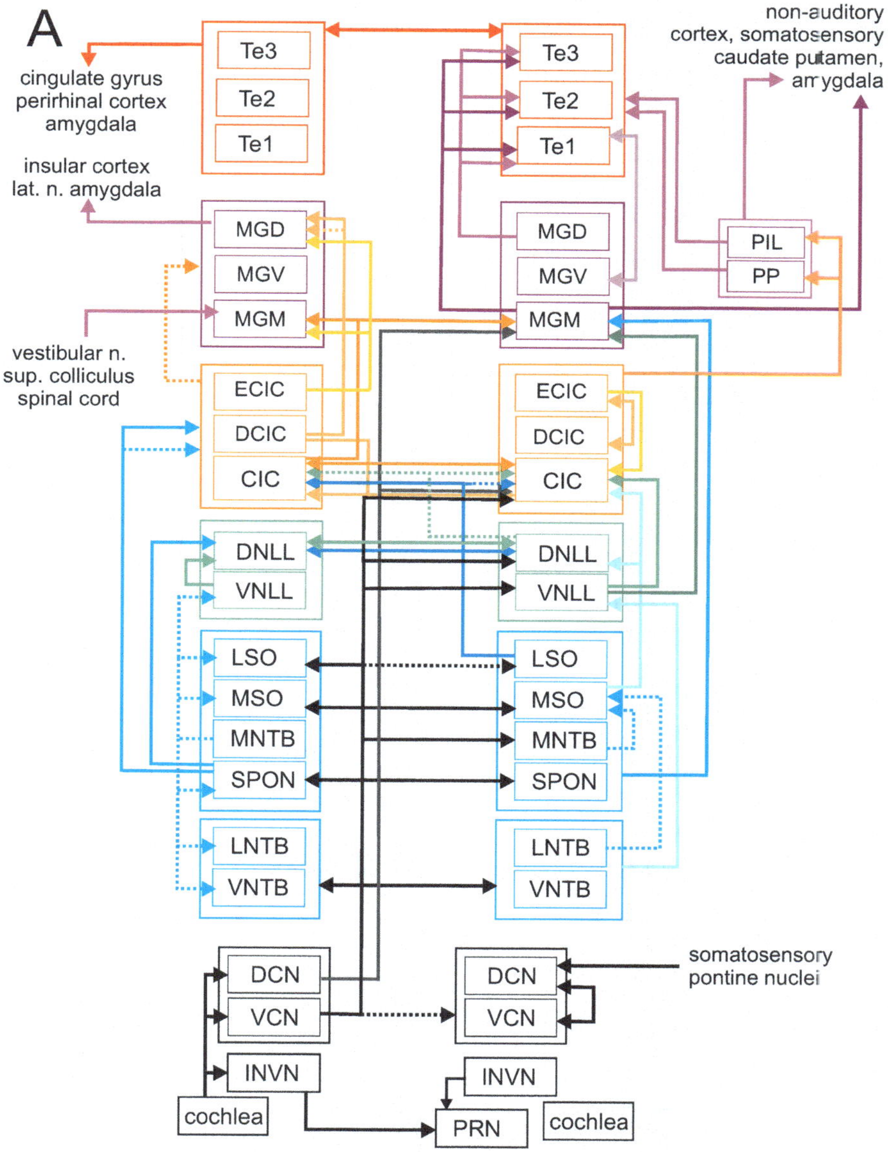

Fig. 6.3 Schematic of the auditory pathway showing ascending (**a**) and descending (**b**) projections. Projections from the ear and cochlear nucleus complex are shown in *black*; superior olivary complex in *blue*; nuclei of the lateral lemniscus in *green tones*; inferior colliculus in *orange tones*, auditory thalamus in *purple tones* and auditory cortex in *red tones*. *Dotted lines* represent known inhibitory projections. *Abbreviations*: CIC, central nucleus of the inferior colliculus; DCIC, dorsal cortex of the inferior colliculus; DCN, dorsal cochlear nucleus; DNLL, dorsal nucleus of the lateral lemniscus; ECIC, external cortex of the inferior colliculus; INVN, interstitial nucleus of the vestibulocochlear nerve; LNTB, lateral nucleus of the trapezoid body; LSO, lateral superior olive; MGD, dorsal subdivision of the medial geniculate body; MGM, medial subdivision of the medial geniculate body;

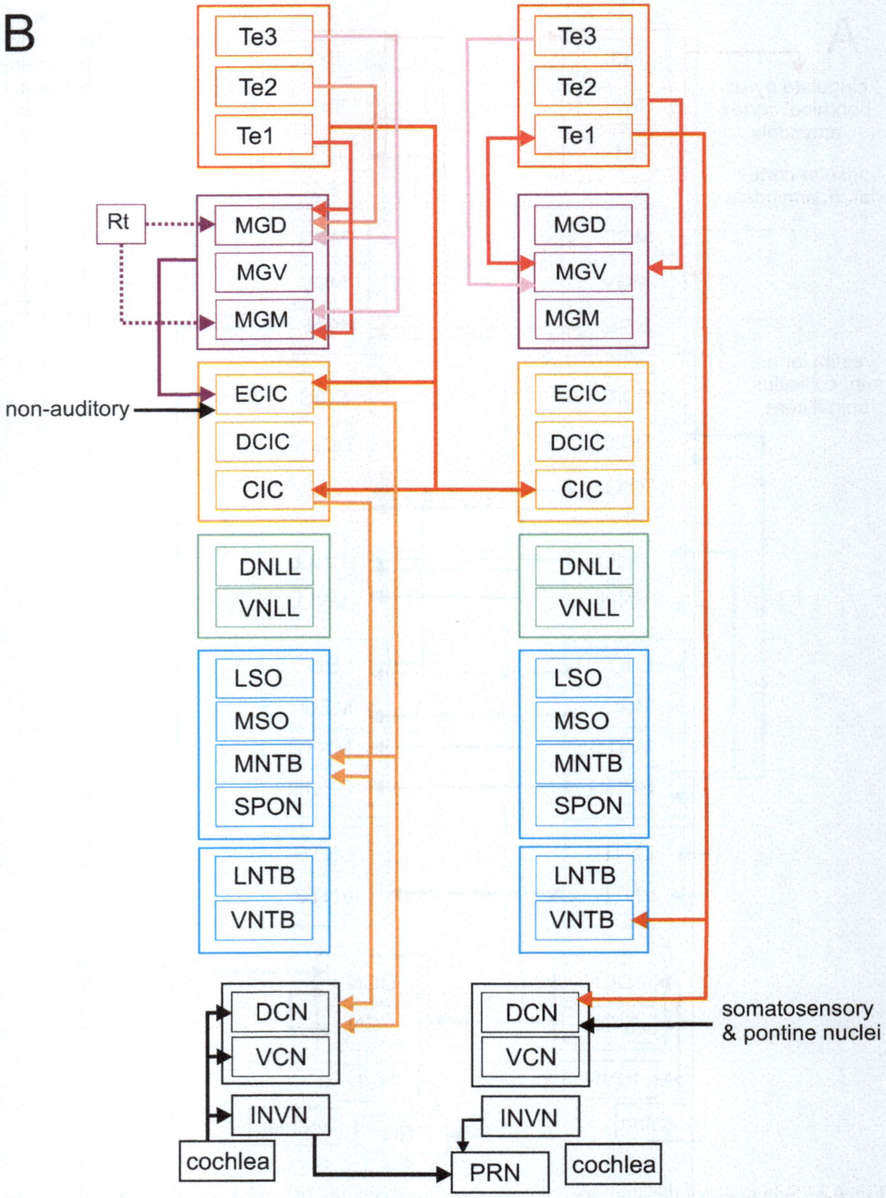

Fig. 6.3 (continued) MGV, ventral subdivision of the medial geniculate body; MNTB, medial nucleus of the trapezoid body; MSO, medial superior olive; PIL, posterior intralaminar nucleus; PP, peripendicular nucleus; PRN, pontine reticular nucleus; Rt, reticular thalamic nucleus; SPON, superior paraolivary nucleus; Te1, primary auditory cortex; Te2, caudal auditory field; Te3, rostroventral auditory field; VCN, ventral cochlear nucleus; VNLL, ventral nucleus of the lateral lemniscus; VNTB, ventral nucleus of the trapezoid body. (© 2017, Kubke and Wild, distributed under the terms of the Creative Commons Attribution License. doi: https://doi.org/10.17608/k6.auckland.4969529.v1)

The DCN is laminated in rodents, as it is in most mammals, and consists of three outer layers and a central core, the latter being poorly developed in rats (Malmierca 2003). Willard and Ryugo (1983) describe the mouse DCN as consisting of two superficial layers that surround a core below which is a fourth basal layer that contains the fibers that will join the dorsal acoustic stria. The dendrites and cell bodies of the pyramidal (fusiform) cells, which are the main projection cells of the DCN, are found in the outer layers. The DCN is interesting in that it receives inputs not just from the auditory nerve and higher auditory areas, but also from nonauditory regions of the brain, and it is possible that more complex transformations of the auditory input may take place there (Montgomery and Bodznick 1994; Wigderson et al. 2016), including the cancellation of self-generated sounds (Singla et al. 2017).

In rats and some other rodents, a group of cells forms the interstitial nucleus of the vestibulocochlear nerve or auditory nerve nucleus, located lateral to VCN (between the Schwann glia cell border and the zone of bifurcation of the auditory nerve). These cells, which receive collaterals from primary afferents and project mainly to the contralateral pontine nucleus, may be involved in the acoustic reflex (Harrison and Warr 1962). The axons of these cells have a variety of targets but do not appear to project to other nuclei of the auditory system (López et al. 1999; Nodal and López 2003).

Since acoustic stimuli can be deconstructed relatively easily into simpler acoustic parameters, such as frequency and amplitude modulation (FM, AM) and duration, it comes as no surprise that a large body of literature describes the properties of auditory neurons in the context of stimulation with simple tones or reduced representations of time-varying artificial stimuli. Each auditory nerve fiber carries information over a narrow frequency range and may also carry some AM information, which is transferred to neurons in the CN (Heil and Peterson 2015). Some neurons in the CN encode AM and amplify the depth of the AM envelope with high fidelity (Frisina 2001; Joris et al. 2004), but it is perhaps in the DCN where more complex transformations of spectral and temporal aspects of the auditory input occur (Winer and Schreiner 2005).

The branching of the auditory nerve to synapse on cells in different parts of the CN suggests that this is where the mapping of different features of the stimulus begins to emerge and diverge. Neurons in the ventral cochlear nucleus extract and enhance timing and frequency inputs from the auditory nerve, and their projections establish two broad parallel pathways. Neurons in the AVCN project to the superior olivary complex (SOC) where interaural time and intensity differences are used for sound localization (the sound localization pathway). Some neurons in the PVCN bypass the SOC and project to the nuclei of the lateral lemniscus (NLL) and the central nucleus of the inferior colliculus (CIC), conveying spectrotemporal representations (the sound identification pathway) (Eggermont 2001). Both the sound localization and sound identification pathways converge at the level of the inferior colliculus (IC).

6.3.5 The Superior Olivary Complex

The series of nuclei in the SOC are found in the ventral tegmentum; their numbers and sizes vary between species (Fig. 6.3, blue tones) (Malmierca 2003). The lateral superior olive (LSO), medial superior olive (MSO), and the medial nucleus of the trapezoid body (MNTB) can be consistently identified in mammals. In rats, the ventral nucleus of the trapezoid body (VNTB) is conspicuous, as is a superior paraolivary nucleus (SPON) that projects to the ipsilateral IC (Malmierca 2003). Most of the inputs to these nuclei originate in the AVCN, and some neurons of the SOC project topographically and bilaterally back to the CN or to the cochlea, where they modulate dynamic aspects of the basilar membrane. Bilateral inputs to neurons in the SOC form the basis of the interaural comparisons between intensity and temporal information that are used for sound localization (Winer and Schreiner 2005).

6.3.6 The Lateral Lemniscus and Nuclei of the Lateral Lemniscus

The lateral lemniscus is a large tract that ascends through the ventrolateral brainstem to terminate in the IC. Embedded within the lateral lemniscus are several groups of cells collectively known as the nuclei of the lateral lemniscus (NLL) (Fig. 6.3, green tones). The NLL vary in size in different species: they are described as modest in rats, well developed in cats, and hypertrophied in some echolocating bats (Winer and Schreiner 2005). Many axons of the lateral lemniscus make collaterals that synapse on NLL neurons, while other axons continue to the IC without branching. The lateral lemniscus carries axons from the contralateral VCN and DCN and axons from the SOC that terminate in the IC. The axons of the octopus cells of the PVCN terminate in the ventral nucleus of the lateral lemniscus (VNLL) (Fig. 6.3a). The majority of axons from NLL neurons join the lateral lemniscus and terminate in the ipsilateral CIC; some neurons from the dorsal NLL (DNLL) cross the midline to synapse in the contralateral IC. The functional division of the NLL into a ventral mainly monaural component and a dorsal binaural component reflects parallel processing along the auditory pathway (Malmierca 2003). Although not mentioned by Malmierca, an intermediate division of the NLL that projects bilaterally to the IC has been described in mice (Willard and Ryugo 1983). Excitatory and inhibitory inputs converge in DNLL where the cells respond to most signals that contain energy within their excitatory response region. A strong inhibitory projection originating in DNLL targets neurons in the IC (Pollak 2013).

6.3.7 The Inferior Colliculus

The auditory midbrain (the IC in mammals) is considered the main point of convergence of auditory input in the brainstem. The IC receives tonotopically organized inputs from the ascending axons of the lateral lemniscus as well as axons entering the IC via the commissure of Probst, which connects the ICs on the right and left side of the brain. The IC also receives descending inputs from cortical areas and sends descending projections to the lower auditory nuclei (Fig. 6.3b). Aside from its auditory inputs, the IC receives projections from other regions of the brain, thereby allowing multisensory integration (Fig. 6.3a, b; orange tones).

The largest region of the IC is formed by the CIC, which is surrounded by a smaller lateral nucleus and a dorsal cortex (Malmierca 2003). Each can be further subdivided into smaller regions based on cellular architecture and connectivity. In the mouse, Willard and Ryugo (1983) describe four IC regions: a central nucleus, a dorsomedial nucleus, an external cortex (which is divided into two layers with different inputs), and a dorsal cortex. Different regions of the IC receive different combinations of inputs (Fig. 6.3a, b). The CIC is primarily auditory and retains the tonotopic organization of the ascending fibers; the lateral nucleus is multisensory and the dorsal cortex receives inputs from cerebral cortex. In addition to a tonotopic organization, neurons in the IC also appear to be organized on the basis of other internal representations (e.g., onset latency, thresholds, receptive fields), probably resulting from differences in ascending inputs (Cant and Benson 2006).

The lateral nucleus of the IC receives ascending inputs from the dorsal column nuclei and the trigeminal system, which carry somatosensory information from the body, and descending cortical inputs. This region of IC receives little, if any, input from the lateral lemniscus, and auditory information is brought in from commissural projections that have no apparent tonotopic mapping.

The dorsal cortex of the IC contains broadly tuned neurons that respond more robustly to vocal stimuli than to noise (Aitkin et al. 1994). Although cells appear to be organized in a laminated structure, these do not seem to be related to tonotopy. The dorsal cortex also receives commissural axons from contralateral IC and descending axons from cortical areas. Cortical inactivation (through cooling of the auditory cortex, AC) of female rats had effects on neuronal responses in IC, particularly the dorsal and external cortices (Popelář et al. 2016). Ascending axons from the IC bilaterally innervate the medial geniculate body (MGB) in the thalamus (Winer and Schreiner 2005). In addition, efferents from the dorsal cortex of the IC project to midbrain tegmental areas that are implicated in the production of vocalizations.

The IC integrates inhibitory and excitatory inputs, and the neurons are tuned to temporal features of sound (e.g., sound duration, frequency modulation) (Eggermont 2001). About 30–50% of neurons in the IC show combination sensitivity that can be facilitatory or inhibitory (Woolley and Portfors 2013). The response patterns of neurons in the IC to tones and white noise are not too different from the responses found at lower levels of the auditory pathway, but since these change when inhibition is blocked locally, it is likely that the IC response patterns are at least partly generated locally rather than inherited from the afferent fibers (Pollak 2013).

Responses in IC neurons could be shaped by a balance of the magnitude of the incoming excitatory and inhibitory inputs or, alternatively, by the timing of the two inputs. The timing hypothesis is particularly attractive since it can explain neuronal responses to FM (Kuo and Wu 2012); however, the timing hypothesis assumes a linear transformation of inputs, and many IC neurons do not appear to linearly transform their inputs (Pollak 2013). The auditory system also tracks the duration of stimuli quite reliably, and neurons sensitive to the duration of the stimulus have been found in the IC of frogs, bats, chinchilla, rats, mice, and guinea pigs (Xia et al. 2000; Sayegh et al. 2011). In mice, Xia and colleagues (2000) described duration sensitive IC neurons that respond within the range of natural mouse vocalizations (3–300 ms). The duration-sensitive neurons could be grouped into four categories: duration selective, short-duration selective, bandpass, and all pass.

From this level on, the auditory system is often described as being organized in *core* (lemniscal) tonotopic regions and *belt* (nonlemniscal) regions where there is less tonotopic organization and often a higher degree of influence from nonauditory inputs. Lemniscal and nonlemniscal inputs from the IC project to different regions of the auditory thalamus.

6.3.8 The Medial Geniculate Body

The auditory thalamus of mammals (medial geniculate body, MGB) is generally divided into ventral, dorsal, and medial (magnocellular) subdivisions (Fig. 6.3, purple tones). The ventral MGB receives lemniscal inputs from the CIC, whereas nonlemniscal projections terminate in the dorsal and medial subdivisions of the MGB (He 2003; Lee 2013). Like the IC, the MGB receives cortical descending inputs; however, the inputs to MGB are more robust than those terminating in the IC (Anderson et al. 2007; Anderson and Linden 2011).

As in the CIC, the ventral MGB is tonotopically organized. Most axons originating from the IC terminate in this subdivision, and the response properties of neurons in this region resemble the properties seen in CIC (Malmierca 2003; Hackett 2011). The ventral MGB of Mongolian gerbils has been further divided into a pars lateralis, a pars ovoidea, and a rostral pole (Saldeitis et al. 2014). Neurons from the ventral MGB project to primary auditory areas in cerebral cortex. Ventral MGB neurons receive descending inputs from primary AC (A1) and, to a lesser extent, from association cortices that show tonotopic organization (Winer 1992).

The dorsal subdivision of the MGB has a more complex organization. It, too, receives inputs from the CIC but also from the peripheral portions of IC, from the reticular thalamic nucleus, and from ventral MGB and other thalamic regions. In the Mongolian gerbil, the dorsal MGB can be subdivided further into a deep dorsal nucleus and a dorsal MGB proper (Saldeitis et al. 2014). Dorsal MGB projects mainly to association auditory regions of cortex and has been implicated in maintaining auditory attention (Kraus et al. 1994).

The medial portion of MGB receives afferents from the lateral nucleus of the IC, axonal terminations from vestibular nuclei, spinal cord nuclei, nuclei of the SOC and lateral lemniscus, and from the midbrain superior collicullus. Neurons in medial MGB project to all auditory cortices but also to nonauditory cortices (e.g., somatosensory and prefrontal cortex), putamen, and amygdala. All cortical auditory areas and some nonauditory areas project back to this medial division, suggesting that the medial MGB might play a role as a multisensory arousal system. One interesting feature is that the different portions of the MGB do not appear to connect with each other, suggesting that the different regions represent true separate pathways with different processing tasks (Winer 1992; Hu 2003; Malmierca 2003).

Each thalamic nucleus receives a unique combination of inputs from different portions of the IC and provides a distinct set of inputs to different areas of cortex (Anderson and Linden 2011; Hackett 2011). A direct projection from lower auditory brainstem that bypasses the IC provides the MGB with short-latency inputs that may prepare the thalamus for the auditory information that will arrive from IC (Pannese et al. 2015). Descending inputs from cortex, in turn, modulate the responses in MGB (He 2003). A proposed role for thalamus stems from the observation that thalamic neurons have two modes of firing: bursting and tonic. Burst firing is commonly seen in anaesthetized animals. In awake animals, thalamic neurons primarily show tonic firing patterns but switch to bursting in association with a salient stimulus. It is then suggested that the switch from tonic to bursting may act as a wake-up call alerting the cortex to the presence of a biologically relevant stimulus (Sherman 2001; He 2003). Thus, the auditory thalamus might modify the responsiveness of AC in the context of attention. Inputs from the amygdala put the auditory thalamus in a position to evaluate the emotional valence of auditory stimuli (Anderson and Linden 2011).

The posterior intralaminar nucleus (PIL) and the peripeduncular nucleus (PP) of the posterior parathalamic nuclei receive inputs from IC and, in turn, project to AC. Malmierca (2003) thus considers them part of the rat thalamocortical pathway.

6.3.9 Auditory Regions of the Cerebral Cortex

The auditory cortex receives inputs from a variety of regions in the forebrain, including the contralateral AC and other auditory-related areas, and subcortical inputs originating from the auditory thalamus and other sensory thalamic nuclei (Hackett 2011). Using anatomical criteria, regions of the cortex receiving input from the auditory thalamus are defined as auditory (Hackett 2011). As is the case in IC and MGB, the AC can be separated into a primary region that is tonotopically organized and primarily concerned with specific auditory processing (core) and other regions involved with more complex, cross-modality processing (belt).

The AC has been divided into different regions in different species. As many as fourteen cortical fields have been identified in the cat, and twelve cortical fields were identified in the moustached bat (He 2003; Shamma and Fritz 2009). In rats, the AC consists of a primary AC (Te1, area 41) and nonprimary auditory fields: a caudally positioned Te2 and a rostroventral Te3 (Fig. 6.3, red tones) (Winer 1992). In mice, Willard and Ryugo (1983) divide the AC into three areas: primary or koniocortex (area 41), area 22 (dorsal to area 41), and area 36 (ventral to area 41). Stiebler and colleagues (1997) divided the mouse AC into five fields: two that are tonotopically organized (primary auditory field or AI and anterior auditory field AAF), an ultrasound field (UF) that responds to frequencies above 40 kHz, a secondary auditory field (AII), and a dorsoposterior field (DP). Tsukano and colleagues (2017) considered a different organization in the mouse: AI, a dorsomedial field (DM), A2 and AFF, all tonotopically organized; and two nontonotopic fields, dorsoanterior field (DA) and dorsoposterior field (DP). Ultrasound responses are found not in a separate field but within A1, AAF, A2, and DM. In the Mongolian gerbil, Budinger and colleagues (2000) identify eight fields: two koniocortical fields (AI and AAF) with a dorsal subfield of AI, two smaller tonotopic fields (DP and a ventroposterior field), a dorsal field, and ventrally the anteroventral, ventral, and ventromedial fields.

Outputs from AC target a number of cortical and subcortical structures, including not only a large majority of the auditory nuclei of the ascending auditory pathway (Schofield and Coomes 2006; Hackett 2011) but also the basal ganglia and other premotor and brainstem nuclei, the amygdala, and others (Fig. 6.3b, red tones) (Winer 2006). Descending inputs to the CN are usually described as targeting the DCN. In guinea pigs, A1 and the dorsocaudal auditory field (but not secondary cortex) project to all subdivisions of the ipsilateral CN in addition to some contralateral projections that originate in low frequency regions (Jacomme et al. 2003).

Responses to features of acoustic stimuli, such as FM and AM, are seen in all cortical areas (Harris et al. 2011; Honma et al. 2013). In guinea pigs, responses in A1 are nonhomogenous: the dorsal belt of AC responds more strongly to broadband noise than to pure tones, whereas the ventral belt responds better to pure tones (Suta et al. 2008). Wang and colleagues (2016) examined the responses to stimulus duration in the A1 of mice and found that about half of the neurons were sensitive to duration, a proportion similar to that seen in the IC. However, after blocking the neurotransmitter GABA (gamma-aminobutyric acid), 60% of those neurons lost their duration sensitivity, underscoring how similar representations of acoustic features may arise at different levels of the auditory system.

The AC shows a collection of maps and selective filters, and cortical neurons appear to respond to higher order features of sounds (Nelken 2008). The responses observed in AC are influenced by anaesthesia. Responses to sustained stimuli and nonsynchronized responses to long repetitive stimuli are almost absent under anesthesia (Wang 2007). Temporal responses to sinusoidal AM stimuli in Mongolian gerbil AC are altered under anaesthesia, although the effects of anaesthesia in IC are not as pronounced (Ter-Mikaelian et al. 2007). These state-dependent responses in the cortex suggest that cortical neurons engage with acoustic stimuli such that they are actively transformed in a context-dependent way (Wang 2007; Harris et al. 2011).

Although cortical neurons are good at tracking fast temporal fluctuations of auditory stimuli, there is a general decrease in the temporal following rates from periphery to central areas (Escabí and Read 2003; Gaucher et al. 2013b). Spike trains appear to carry less information in A1 and MGB than in IC, although there is increasing redundancy of information moving from IC to MGB to A1 (Huetz et al. 2011). It seems now that both a rate and a temporal code may operate along the entire auditory axis. Temporal codes, rather than rate codes, may underlie discrimination of vocalizations in cortex, in which the computing unit is probably a neuronal population (Eggermont 2001; Ter-Mikaelian et al. 2013).

Traditionally, ascending projections of the auditory system are referred to as either lemniscal or nonlemniscal. The lemniscal pathway carries tonotopic projections from core regions of the auditory nuclei and runs in parallel with projections from the nonlemniscal pathway, which carries information under the influence of nonauditory centers. Recordings in mice (Anderson and Linden 2011) suggest that it is more appropriate to categorize the ascending inputs to cortex into three rather than two pathways: a tonotopic, a nontonotopic or diffuse, and a polysensory pathway. In this view, the tonotopic pathway would be made up of projections of the central IC to ventral MGB, which then projects to laminae III/IV of A1. The nontonotopic or diffuse pathway would be made up of IC inputs from the external and dorsal cortex of IC to the dorsal MGB, which then projects to layers I, I, III, IV, and VI of secondary AC. The polysensory pathway would include the medial MGB that receives projections from all IC regions plus ventral NLL, DCN, and nonauditory areas, and projects to all cortices (layers I, III, IV, and VI) (Eggermont 2001; Anderson and Linden 2011).

6.4 Processing of Vocal Signals

Vocal signals are rhythmic sequences of units with characteristic temporal and spectral structures, and they can be described based on agreed classification criteria that recognize the variation in the structure and the sequence of different units (Espmark et al. 2000). Information is contained in the signal's inherent properties (e.g., repetition, ordering, overlapping, and timing of elements) (Kershenbaum et al. 2016). Vocal output may also carry elements that are not used as part of the information conferred by the signal, as when ultrasound frequencies are present as a by-product of the vocal production but may be outside the hearing range of the listener. Different parts of the sequence may also carry different meanings (Okanoya and Screven, Chap. 2). The initial segment of a sequence, for example, may only operate to get the listener's attention, and the segment that follows carries the information needed to release a behavioral response. There is no a priori reason, then, to expect a 1:1 matching between the vocal signal (influenced by production mechanisms) and the perceptual construct (influenced by sensory and perceptual mechanisms) (Dent, Screven, and Kobrina, Chap. 4). Those parts of the vocalization that do carry biologically relevant information must have enough distinctiveness so that they can be grouped in a perceptual construct that the animal can use to adequately respond, especially when additional cues (such as vision or olfaction) are not available to supplement the categorization (Marler 1957).

How the nervous system accesses the information contained within a communication signal to produce a response is central to the study of vocal communication. The process begins with the transformation of the input from a time-varying acoustic waveform to a perceptual abstract representation or *auditory object* (Bizley and Cohen 2013). Species-specific vocalizations, like other natural sounds, are quite stereotypical (to the extent that they can be described with a certain degree of accuracy) but with embedded variation that depends on the type of call and the individual caller. The nervous system thus needs to account for this inherent variability to produce an efficient neural representation (Eggermont 2001; Theunissen and Elie 2014).

Researchers need to understand how acoustic structures are transformed into neural responses, how neuronal response patterns are used to identify and categorize a signal and, ultimately, how this information is linked to the motor decision and control regions that mediate the response. The challenge is to establish what constitutes the neural representation of an auditory categorical percept, keeping in mind that there is not necessarily a 1:1 map with a behavioral response and acknowledging the influence of experience, attention, and hormonal states (King et al. 2015; Hurley and Kalcounis-Rueppell, Chap. 8). The more that is known about the relationship between stimulus and behavior, the easier it is to identify the underlying neural code (Gentner and Margoliash 2003; Sanes and Woolley 2011).

Artificial stimuli, such as those used to establish the best frequency of a neuron, do not vary much in the time domain, and the response of the neurons can often be described in terms of the overall number of spikes produced. Variations in spike probabilities related to the onset and offset of the stimulus have also been used to describe cell types, especially in the CN and IC. The neural responses in the auditory system are often shaped by interactions of excitatory and inhibitory inputs to create specific neuronal responses that can be influenced by development and experience (Pollak 2013; Theunissen and Elie 2014). Many auditory neurons respond preferentially to a subset of categories of vocal stimuli (selectivity) or to specific acoustic features (specificity) (Gentner and Margoliash 2003).

Where responses are neatly synchronized to the temporal properties of the stimulus (isomorphic) it is possible that information is being relayed rather than processed. Feature extraction of the stimulus is more likely to occur in nonisomorphic representations. Auditory neurons may carry stimulus information in their average firing rate (rate code) and/or in the fine grain temporal pattern of the spike train (temporal code) (Huetz et al. 2011; Woolley and Portfors 2013). A temporal code is often thought to be in play when the stimulus can be discriminated based on the temporal pattern of spikes produced (but not on the basis of total spike count alone) or when the pattern in which the spikes are produced varies in accordance with the rate of change of the stimulus (Huetz et al. 2011). When neuronal responses to natural stimuli are very reliable and reproducible across presentation trials, yet with spike patterns that are different for different stimuli, then it is likely that a temporal code is being used (Huetz et al. 2011; Gaucher et al. 2013b). For nonvarying artificial stimuli, a rate code approach may be sufficient to describe the neuronal responses, but a temporal code seems to more appropriately describe the neural representations of vocal signals and natural stimuli that show a high degree of spectrotemporal variation (Gaucher et al. 2013b).

A second consideration when examining neural responses to natural stimuli is to determine whether the representation of a combination of features can be achieved by a single neuron (feature detectors) or whether it requires a coordinated set of neurons (Huetz et al. 2011; Schneidman 2016). The *feature detector hypothesis* posits that a neuron (or a small set of neurons) can respond to particular features or combinations of features within the acoustic stimulus. The existence of feature detectors requires that stimuli be processed hierarchically in parallel pathways that converge on feature detector neurons that respond to specific combinations of features (Gentner and Margoliash 2003). Alternatively, *the population hypothesis* states that perception of a signal emerges through distributed neuronal assemblies in which each neuron provides some coarse representation of a particular stimulus feature which, when put together, creates the representation of the stimulus as a whole (Gentner and Margoliash 2003; Schneidman 2016). Since the construction of an auditory object requires combining multiple sets of information, its representation is likely to be distributed and to span multiple cortical networks (Eggermont 2001; He 2003).

While natural stimuli can be described in terms of their individual features, the ability to categorize natural sounds depends on more than just the sum of the parts (Geissler and Ehret 2004). Neuronal responses obtained by presenting isolated features of vocal signals contrast with those obtained when the entire vocalization is used as a stimulus. Neurons may respond strongly to natural and behaviorally significant sounds, but not to their simpler components (Woolley and Portfors 2013). In most neurons of the MGB of the guinea pig, for example, a neuron's response to communication signals could not be predicted based on its frequency tuning curve (Tanaka and Taniguchi 1991). Similar results are seen in the IC of other mammals and the auditory midbrain of birds (Woolley and Portfors 2013). Thus, responses to natural stimuli involve nonlinear transformations that produce a representation of the higher order statistics of natural sounds (Theunissen and Elie 2014).

The search for a neural code ultimately seeks to identify the set of rules that relate neural activity to a stimulus or a behavior (Eggermont 2001). The code may take different forms at different levels of processing and for different types of stimuli. Whatever the pattern of the neural responses may be, it can be considered to be part of a neural code in as much as it occurs under natural conditions and is elicited by natural stimuli. As Eggermont (2001) pointed out, knowing that the information necessary to represent a stimulus property is present in the firing of a neuron is different from knowing whether the nervous system uses part or all of the information.

6.5 Vocal Communication in Rodents

Rodents produce calls that have energy in the audible and/or ultrasound range (Okanoya and Screven, Chap. 2). Special attention has been focused on the production of ultrasonic vocalizations that include the infant calls that elicit parental responses, or calls between adults associated with mating and agonistic behaviors. The auditory system shows robust and preferential response to conspecific vocalizations over other natural sounds (including heterospecific vocalizations; Dent, Screven, and Kobrina, Chap. 4). The main question that needs to be answered is whether there is something special about the neural substrates through which vocal communication signals are processed or whether preferential responses to biologically significant sounds merely emerge through repeated exposure (Poremba et al. 2013).

Selectivity for communication sounds first emerges in the IC (Pollak 2013). In DNLL, responses to vocalizations are nonspecific, but in IC, neurons are reported to selectively respond to some natural vocalizations but not others (Pollak 2013). Blocking GABA and glycine inhibition results in an expansion of the tuning curves of IC neurons and reduces the selectivity to vocalizations (Woolley and Portfors 2013). In the auditory midbrain of birds, neurons are also tuned to specific spectrotemporal modulations of the conspecific song (Woolley 2012). Responses to vocalizations in rodents are seen in the IC (Suta et al. 2003; Portfors et al. 2009), and selectivity and specificity can arise through rate codes or temporal codes. The neural responses to natural vocalizations in mice and guinea pigs are described in the next sections.

6.5.1 *Adult Mouse Vocalizations*

Males of laboratory mice strains produce at least two types of USVs when they encounter a female or her urine, each related to a different phase of copulation (Portfors 2007; Nyby 2010). Male mice USVs seem to be regulated by androgens and pheromones, such as those found in female urine, and they are probably processed redundantly by the vomeronasal and olfactory systems (Nyby 2010). Females responding to male songs appear to maintain close contact with the vocalizing males, and they will approach a speaker playing USV sounds (Portfors 2007; Nyby 2010).

Neurons in IC are selective to conspecific vocalizations (Portfors et al. 2011; Pollak 2013). In female mice, Garcia-Lazaro and colleagues (2015) reported that while only 9% of neurons in the CN respond to vocalizations, responses are seen in 59% of neurons in the IC. Robust responses to USVs in the IC are shaped by inhibition and are sensitive to perturbations of the acoustic features of the USVs (Portfors and Felix 2005; Egnor and Seagraves 2016). Responses in IC are modulated by dopamine and may also be modulated by social interactions through the influence of serotoninergic inputs (Gittelman et al. 2013; Egnor and Seagraves 2016). The IC of female mice shows heterogeneous responses, increased encoding efficiency, and a distinct neural representation for different vocalizations (Holmstrom et al. 2010).

In general, hearing sensitivities match the spectral range of the vocal signals (Wilczynski and Ryan 2010; Dent, Screven, and Kobrina, Chap. 4); however, ultrasound frequencies are not highly represented in the IC tonotopic map despite their behavioral relevance (Portfors et al. 2009). Instead, IC neurons respond to conspecific vocalizations with energies outside the neuron's tuning curve, perhaps by exploiting cochlear distortions created by combinations of high frequencies in the USV (Holmstrom et al. 2010; Woolley and Portfors 2013).

A population code is proposed to lead to the representation of each vocalization (Portfors et al. 2009). In A1, neuronal responses to USVs seem to be correlated with the neuron's frequency tuning curve. Neurons responded more strongly to natural rather than time-reversed versions of USVs, and responses were also less robust when the USVs had been temporally distorted (Carruthers et al. 2013).

6.5.2 Guinea Pig Vocalizations

Guinea pigs produce at least eleven types of vocalizations composed of multiple acoustical attributes (Berryman 1976). Neurons within individual nuclei in the auditory pathway may be responsible for extracting particular features or combinations of features (Suta et al. 2008).

Responses in the IC of guinea pigs are less selective than in other rodents and a lower proportion of neurons show call selectivity (Suta et al. 2003; Syka 2010). Neuronal responses are higher for vocalizations than for tones, noise, or time-reversed presentations of the vocalization but show little preference for one call over another (Suta et al. 2003; Pollak 2013). Vocalizations are encoded spatially across the tonotopic map, matching the spectral content of the vocalization. A small number of neurons appeared to be sufficient to encode a representation of the vocalizations, although increasing the number of neurons involved increased discrimination (Lyzwa et al. 2015).

The responses in the thalamic MGB show less selectivity than in IC (Syka 2010). Neurons in the MGB in guinea pigs respond both to tones and to species-specific vocalizations, although these appear to be nonselective (Tanaka and Taniguchi 1991; Philibert et al. 2005). The representation of spectral features appears to be preserved in calls with wider frequency spectra but to be less precise for lower frequency calls (e.g., chutter and purr). In ventral and medial MGB, Suta and colleagues (2008) showed that neurons that phase locked to the fundamental frequency of the call also showed responses to the call that were strongly dependent on its spectral composition. In contrast, Tanaka and Taniguchi (1991) recorded from MGB and concluded that the majority of neurons showed discharge patterns to vocalizations that could not be predicted by the neuron's frequency tuning curves.

Neurons in MGB and A1 in awake and anaesthetized animals showed similar firing rates for vocalizations with only a few producing spike trains that carried a significant amount of information (Gaucher et al. 2013a). A significant amount of information was contained in the discharge patterns of neurons in MGB and AC in awake animals as compared to anaesthetized animals (Huetz et al. 2009). The reliability of spike timing was similar for both the natural and time-reversed version of the vocalization and was higher in nonanaesthetized animals, although the responses were not mirror images of each other, suggesting they were sensitive to a specific sequence of signal presentation (Huetz et al. 2009).

The information in the temporal spike patterns also increases between MGB and AC. Severe loss of spectral information did not seem to prevent AI neurons from correctly classifying individual calls, but the representation of natural calls became degraded following temporal disturbances of the vocal stimulus (Ter-Mikaelian et al. 2013). Wang and colleagues (1995) suggested that the coding in cortex is different for sounds that have a behavioral relevance versus sounds that do not. Ultimately, the representation of acoustic vocal communication signals must be integrated with other regions of the brain where the combination of context and internal state can elicit an appropriate behavior.

6.6 Learning in the Context of Vocal Production

Manipulating the behavioral significance of a particular acoustic feature may lead to a corresponding representation of that feature in the brain. This has been shown in rat cortex using conditioned stimuli for the representation of tones, sound intensity, and temporal information (de Villers-Sidani et al. 2007). For example, rat pups exposed to different sequences of tone pips differentially represent those frequencies in cortex based on the pups' developmental experience (Köver et al. 2013). Thus, early experience may help to shape the representation of auditory categories in the cortex (Han et al. 2007).

Contextual learning, whereby a signal becomes associated with a new context as a result of experience (Janik and Slater 1997, 2000), is common in mammals, including rodents such as rats and guinea pigs, and may underlie the association of conspecific vocalizations with a biological meaning. It appears that the 22 kHz group of rat calls may not be innately recognized as an alarm call but acquires that meaning as a consequence of associative learning (Wöhr et al. 2010). The USVs produced by mouse pups carry no behavioral relevance for nonmaternal females, so it has been suggested that the relevance emerges as the result of experience raising pups, which leads to the recognition of the pup call (Shepard et al. 2015). Thus, learning can influence the central mapping and meaning of acoustic structures (Hurley and Kalcounis-Rueppell, Chap. 8).

The possibility of vocal learning in the courtship USVs of mice was recently examined (Arriaga et al. 2012). There may be a vocal control pathway shared by all vertebrates that mediates the production of innate vocalizations. This *innate pathway* involves the PAG that acts on vocal motoneurons via the lower brainstem respiratory/vocal nuclei (or equivalent nuclei in nonmammalian species) (see Sect. 6.2.1). Learning acquisition involves a cortical pathway (present in humans and absent in nonhuman primates) that has direct descending projections from cortical regions into primary laryngeal motoneurons (Fitch et al. 2010). An equivalent cortico-bulbar pathway is present in songbirds that also show learning acquisition (Doupe and Kuhl 1999).

Arriaga and colleagues (2012) showed activation of immediate early gene *erg1* in singing mice in a cortical region near the anterior commissure that included some parts of motor cortices and anterodorsal striatum, but those increases in expression were not seen in mice that were hearing but not singing. After injecting a trans-synaptic label into the laryngeal muscles, retrogradely labeled neurons were observed in the same region of the motor cortex and thereby established a connection between these cortical structures and laryngeal motoneurons. Chemical lesions in this area did not prevent the mice from singing; their songs had a similar syllable composition, although some changes in the spectral composition were seen within days. Mice that were deafened also produced courtship USVs with similar syllable composition and some deteriorated spectral composition, but the deterioration was seen only after a period of eight months. These results are in contrast with those of other studies. For instance, the vocalizations of congenitally deaf transgenic mice or mice that were deafened before the onset of hearing appear to be no different from the vocalizations produced by hearing animals (Hammerschmidt et al. 2012; Heckman et al. 2016). When separated from their mothers, mice pups with a *FoxP2* knockout (a gene that has been implicated in vocal learning, see Sect. 6.2.1) vocalize less in the sonic range and utter fewer ultrasound whistles than do the heterozygous and wild-type counterparts (Scharff and Haesler 2005), although it should be noted that *FoxP2* knockout pups also exhibit a number of general brain abnormalities. Furthermore, mice lacking a cerebral cortex appear to develop normal USVs (Hammerschmidt et al. 2015). Arriaga and colleagues (2012) acknowledged that the projection from cortex to the medulla in mice is not robust and that further studies are needed to establish the possibility of learning acquisition in mice.

6.7 Behavioral Responses to Vocal Communication Signals

Vocal signals in nature do not happen in isolation. Short range signals are usually accompanied by stimulation of other sensory modalities (e.g., olfaction, vision) that contribute to the elicitation of a behavioral response. Grasshopper mice (*Onychomys leucogaster*) produce their adult type IV vocalization along with clear postural displays: standing on their hind legs, preferentially from an elevated position, suggesting that both acoustic and visual information play a part in the signal (Hafner and Hafner 1979). Guinea pigs that were presented with wideband noise paired with a light stimulus showed that the visual input had a suppressive effect on the auditory responses in AC (Kubota et al. 2017). The auditory system has substantial connections with nonauditory centers in the brain, especially through the nonlemniscal pathways, which may provide the contextual information to incorporate attention and affective qualities in stimulus processing (Pannese et al. 2015).

Holstege and Subramanian (2016) proposed that vocal behavior in humans was controlled by an"emotional motor pathway" that connects the amygdala, hypothalamus, and other areas with the PAG, brainstem, and spinal cord, and a voluntary pathway originating in motor cortex. Nonlemniscal regions of the thalamus send axons to a number of limbic structures (e.g., the amygdala, association cortex, and striatum) (Hu 2003). The amygdala is implicated in the processing of emotion and affective value of stimuli, and it shows early activation when socially relevant stimuli, such as conspecific calls, are presented. Playback of artificial 20 kHz sine waves to rats increases the expression of the immediate early gene *cFos* in PAG, amygdala, hypothalamus, and thalamus (Wöhr and Schwarting 2010). Lesions of the amygdala reduce both male mouse courtship behavior and female directed USVs (Matsumoto et al. 2012). Olfactory stimuli are relayed to a number of brain regions, including amygdala and lateral entorhinal cortex (Hashikawa et al. 2016). The amygdala may play a role in mediating approach or evasive behaviors toward odors, and neurons there may be under the regulation of sex hormones (Hashikawa et al. 2016). The hypothalamus, a target for amygdalar inputs and somatosensory inputs from genital areas, also may be critical for the expression of innate social behaviors such as sexual behavior (Hashikawa et al. 2016). The hypothalamus influences the striatum in relation to motivation and reward, mediated by dopaminergic cells in the ventral tegmental area. The striatum is thought to participate in reward-related decisions and action selection (Lee 2015). The dorsal raphe nucleus, a brain region that is activated by arousal, acoustic stimuli, and social interactions, provides serotonergic inputs to all auditory nuclei, particularly the IC, where levels of serotonin increase during social interactions in mice (Pollak 2013; Pannese et al. 2015). Therefore, understanding how the auditory system interacts with nonauditory regions to elicit appropriate responses to vocal signals is important.

A central role in production and perception of vocal behavior is attributed to the PAG, which is an intermediary between hypothalamus, brainstem, and motoneurons in the spinal cord (Pannese et al. 2015; Hashikawa et al. 2016). Lesions of PAG accelerate male mounting behavior and trigger aggressive behaviors in rats. The PAG has been suggested to have a behavior-gating function, helping to select and initiate behaviors that are associated with a social dimension. Thus socially relevant vocal communication signals involve the processing of features along the auditory pathway plus integration with other brain centers that mediate the appropriate behavioral response. The study of vocal communication signals is not an auditory/motor problem, but a systems level problem.

6.8 Conclusions

According to Theunissen and Elie (2014), studies of vocal communication have followed two main approaches that now may be merging. On the one hand, auditory physiologists have sought to explain auditory processing by studying neuronal responses to isolated features of natural sounds. A neuroethological approach has, on the other hand, sought to take a more integrated approach that considers the natural context in which the social signals are used. A key goal in the analysis of vocal communication signals is to determine the essential features of these signals, what information they transmit, how they are categorized, and equally important, how they act in combination with information derived from other sensory modalities, are interpreted, and are acted upon by the receiver.

As Pollak and colleagues (2003) say, it is important to understand not just "how acoustic information is progressively transformed" but also "the functional consequences of those transformations." Auditory neurons clearly respond to vocal communication signals and, especially in the IC and higher levels, show nonlinearities that emphasize that the responses to vocal communication signals cannot be explained by the isolated features of the calls. Further, it is important to consider the social role that vocal communication signals play, along with other environmental and social variables that contribute to the processing of vocal signals, even (or especially) when they may not be easily reproduced in the laboratory (Beecher 1996). Mice, in particular, increasingly have become a species of choice to study hearing function and disease, primarily from a genetic perspective (Ohlemiller et al. 2016; Ohlemiller, Chap. 7), but other rodent models, for which the communication signals and responses they elicit in natural settings can be readily described, may significantly contribute to our understanding of vocal communication (Hauser 1999; Bennur et al. 2013).

Acknowledgments The authors are grateful to Micheal Dent, Art Popper, and Peggy Walton for useful improvements to the chapter and to Srdjan Vlajkovic for help with the interpretation of inner ear histological material.

Compliance with Ethics Statement M. Fabiana Kubke declares that she has no conflicts of interest.

J. Martin Wild declares that he has no conflicts of interest.

References

Aitkin, L., Tran, L., & Syka, J. (1994). The responses of neurons in subdivisions of the inferior colliculus of cats to tonal, noise and vocal stimuli. *Experimental Brain Research*, 98(1), 53–64.

Anderson, L. A., & Linden, J. F. (2011). Physiological differences between histologically defined subdivisions in the mouse auditory thalamus. *Hearing Research*, 274(1-2), 48–60.

Anderson, L. A., Wallace, M. N., & Palmer, A. R. (2007). Identification of subdivisions in the medial geniculate body of the guinea pig. *Hearing Research*, 228(1-2), 156–167.

Arriaga, G., & Jarvis, E. D. (2013). Mouse vocal communication system: Are ultrasounds learned or innate? *Brain and Language*, 124(1), 96–116.

Arriaga, G., Zhou, E. P., & Jarvis, E. D. (2012). Of mice, birds, and men: The mouse ultrasonic song system has some features similar to humans and song-learning birds. *PLoS One*, 7(10), e46610.

Beecher, M. D. (1996). Birdsong learning in the laboratory and field. In D. E. Kroodsma & E. H. Miller (Eds.), *Ecology and evolution of acoustic communication in birds* (pp. 61–78). Ithaca, New York: Comstock Publishing.

Bennur, S., Tsunada, J., Cohen, Y. E., & Liu, R. C. (2013). Understanding the neurophysiological basis of auditory abilities for social communication: A perspective on the value of ethological paradigms. *Hearing Research*, 305, 3–9.

Berryman, J. C. (1976). Guinea-pig vocalizations: Their structure, causation and function. *Zeitschrift für Tierpsychologie*, 41(1), 80–106.

Bizley, J. K., & Cohen, Y. E. (2013). The what, where and how of auditory-object perception. *Nature Reviews, Neuroscience*, 14(10), 693–707.

Bradbury, J. W., Vehrencamp, S., & Bradbury, J. W. (1998). *Principles of animal communication* (1st ed.). Sunderland, MA: Sinauer Associates.

Brown, J. L. (1965). Loss of vocalization caused by lesions in the nucleus mesencephalicus lateralis of the redwinged blackbird. *American Zoologist*, 5, 693.

Budinger, E., Heil, P., & Scheich, H. (2000). Functional organization of auditory cortex in the Mongolian gerbil (*Meriones unguiculatus*). III. Anatomical subdivisions and corticocortical connections. *European Journal of Neuroscience*, 12(7), 2425–2451.

Burda, H., Bruns, V., & Hickman, G. C. (1992). The ear in subterranean Insectivora and Rodentia in comparison with ground-dwelling representatives. I. Sound conducting system of the middle ear. *Journal of Morphology*, 214(1), 49–61.

Cant, N. B., & Benson, C. G. (2006). Organization of the inferior colliculus of the gerbil (*Meriones unguiculatus*): Differences in distribution of projections from the cochlear nuclei and the superior olivary complex. *The Journal of Comparative Neurology*, 495(5), 511–528.

Carruthers, I. M., Natan, R. G., & Geffen, M. N. (2013). Encoding of ultrasonic vocalizations in the auditory cortex. *Journal of Neurophysiology*, 109(7), 1912–1927.

Doupe, A. J., & Kuhl, P. K. (1999). Birdsong and human speech: Common themes and mechanisms. *Annual Review of Neuroscience*, 22(1), 567–631.

Dujardin, E., & Jürgens, U. (2005). Afferents of vocalization-controlling periaqueductal regions in the squirrel monkey. *Brain Research*, 1034(1-2), 114–131.

Eggermont, J. (2001). Between sound and perception: Reviewing the search for a neural code. *Hearing Research*, 157, 1–42.

Egnor, S. R., & Seagraves, K. M. (2016). The contribution of ultrasonic vocalizations to mouse courtship. *Current Opinion in Neurobiology*, 38, 1–5.

Escabí, M. A., & Read, H. L. (2003). Representation of spectrotemporal sound information in the ascending auditory pathway. *Biological Cybernetics*, 89, 350–362.

Espmark, Y., Amundsen, T., & Rosenqvist, G. (2000). *Animal signals: Signalling and signal design in animal communication*. Trondheim (Norway): Tapir Academic Press.

Esposito, A., Demeurisse, G., Alberti, B., & Fabbro, F. (1999). Complete mutism after midbrain periaqueductal gray lesion. *Neuroreport*, 10(4), 681–685.

Fischer, J., & Hammerschmidt, K. (2011). Ultrasonic vocalizations in mouse models for speech and socio-cognitive disorders: Insights into the evolution of vocal communication. *Genes, Brain, and Behavior*, 10(1), 17–27.

Fisher, S. E., & Scharff, C. (2009). FOXP2 as a molecular window into speech and language. *Trends in Genetics*, 25(4), 166–177.

Fitch, W. T., Huber, L., & Bugnyar, T. (2010). Social cognition and the evolution of language: Constructing cognitive phylogenies. *Neuron*, 65(6), 795–814.

Fleischer, G. (1978). *Evolutionary principles of the mammalian middle ear*. Berlin: Springer-Verlag.

Frisina, R. D. (2001). Subcortical neural coding mechanisms for auditory temporal processing. *Hearing Research*, 158(1-2), 1–27.

Fritzsch, B., Pan, N., Jahan, I., Duncan, J. S., et al. (2013). Evolution and development of the tetrapod auditory system: An organ of Corti-centric perspective. *Evolution & Development*, 15(1), 63–79.

Garcia-Lazaro, J. A., Shepard, K. N., Miranda, J. A., Liu, R. C., & Lesica, N. A. (2015). An overrepresentation of high frequencies in the mouse inferior colliculus supports the processing of ultrasonic vocalizations.*PLoS One*, 10(8), e0133251.

Gaub, S., Groszer, M., Fisher, S. E., & Ehret, G. (2010). The structure of innate vocalizations in *Foxp2*-deficient mouse pups. *Genes, Brain and Behavior*, 9(4), 390–401.

Gaucher, Q., Huetz, C., Gourévitch, B., & Edeline, J.-M. (2013a). Cortical inhibition reduces information redundancy at presentation of communication sounds in the primary auditory cortex. *The Journal of Neuroscience*, 33(26), 10713–10728.
Gaucher, Q., Huetz, C., Gourévitch, B., Laudanski, J., et al. (2013b). How do auditory cortex neurons represent communication sounds? *Hearing Research*, 305, 102–112.
Geissler, D. B., & Ehret, G. (2004). Auditory perception vs. recognition: Representation of complex communication sounds in the mouse auditory cortical fields. *European Journal of Neuroscience*, 19(4), 1027–1040.
Gentner, T. Q., & Margoliash, D. (2003). The neuroethology of vocal communication: Perception and cognition. In A. M. Simmons, R. R. Fay & A. N. Popper (Eds.), *Acoustic communication* (pp. 324–386). New York: Springer Science+Business Media.
Gittelman, J. X., Perkel, D. J., & Portfors, C. V. (2013). Dopamine modulates auditory responses in the inferior colliculus in a heterogeneous manner. *Journal of the Association for Research in Otolaryngology*, 14(5), 719–729.
Griffiths, T. D., & Warren, J. D. (2004). What is an auditory object? *Nature Reviews, Neuroscience*, 5(11), 887–892.
Hackett, T. A. (2011). Information flow in the auditory cortical network. *Hearing Research*, 271(1-2), 133–146.
Hafner, M. S., & Hafner, D. J. (1979). Vocalizations of grasshopper mice (Genus *Onychomys*). *Journal of Mammalogy*, 60(1), 85–94.
Hammerschmidt, K., Whelan, G., Eichele, G., & Fischer, J. (2015). Mice lacking the cerebral cortex develop normal song: Insights into the foundations of vocal learning. *Scientific Reports*, 5, 8808.
Hammerschmidt, K., Reisinger, E., Westekemper, K., Ehrenreich, L., et al. (2012). Mice do not require auditory input for the normal development of their ultrasonic vocalizations. *BMC Neuroscience*, 13(1), 40.
Han, Y. K., Köver, H., Insanally, M. N., Semerdjian, J. H., & Bao, S. (2007). Early experience impairs perceptual discrimination. *Nature Neuroscience*, 10(9), 1191–1197.
Harris, K. D., Bartho, P., Chadderton, P., Curto, C., et al. (2011). How do neurons work together? Lessons from auditory cortex. *Hearing Research*, 271(1-2), 37–53.
Harrison, D. F. N. (1995). *The anatomy and physiology of the mammalian larynx*. Cambridge, UK: Cambridge University Press.
Harrison, J. M., & Warr, W. B. (1962). A study of the cochlear nuclei and ascending auditory pathways of the medulla. *The Journal of Comparative Neurology*, 119(3), 341–379.
Hashikawa, K., Hashikawa, Y., Falkner, A., & Lin, D. (2016). The neural circuits of mating and fighting in male mice. *Current Opinion in Neurobiology*, 38, 27–37.
Hauser, M. D. (1999). *The design of animal communication*. Cambridge, MA: MIT Press.
He, J. (2003). Corticofugal modulation of the auditory thalamus. *Experimental Brain Research*, 153(4), 579–590.
Heckman, J., McGuinness, B., Celikel, T., & Englitz, B. (2016). Determinants of the mouse ultrasonic vocal structure and repertoire. *Neuroscience & Biobehavioral Reviews*, 65, 313–325.
Heffner, R. S., Koay, G., & Heffner, H. E. (2001). Audiograms of five species of rodents: Implications for the evolution of hearing and the perception of pitch. *Hearing Research*, 157(1–2), 138–152.
Heil, P., & Peterson, A. J. (2015). Basic response properties of auditory nerve fibers: A review. *Cell and Tissue Research*, 361(1), 129–158.
Herbst, C. T. (2016). Biophysics of vocal production in mammals. In R. A. Suthers, W. T. Fitch, R. R. Fay, & A. N. Popper (Eds.), *Vertebrate sound production and acoustic communication* (pp. 159–189). New York: Springer International Publishing.
Holmstrom, L. A., Eeuwes, L. B. M., Roberts, P. D., & Portfors, C. V. (2010). Efficient encoding of vocalizations in the auditory midbrain. *The Journal of Neuroscience*, 30(3), 802–819.
Holstege, G. (1989). Anatomical study of the final common pathway for vocalization in the cat. *The Journal of Comparative Neurology*, 284(2), 242–252.

Holstege, G., & Subramanian, H. H. (2016). Two different motor systems are needed to generate human speech. *The Journal of Comparative Neurology*, 524(8), 1558–1577.

Holstege, G., Kerstens, L., Moes, M. C., & VanderHorst, V. G. J. M. (1997). Evidence for a periaqueductal gray–nucleus retroambiguus–spinal cord pathway in the rat. *Neuroscience*, 80(2), 587–598.

Honma, Y., Tsukano, H., Horie, M., Ohshima, S., et al. (2013). Auditory cortical areas ativated by slow frequency-modulated sounds in mice.*PLoS One*, 8(7), e68113.

Hu, B. (2003). Functional organization of lemniscal and nonlemniscal auditory thalamus. *Experimental Brain Research*, 153(4), 543–549.

Huetz, C., Philibert, B., & Edeline, J.-M. (2009). A spike-timing code for discriminating conspecific vocalizations in the thalamocortical system of anesthetized and awake guinea pigs. *The Journal of Neuroscience*, 29(2), 334–350.

Huetz, C., Gourévitch, B., & Edeline, J.-M. (2011). Neural codes in the thalamocortical auditory system: From artificial stimuli to communication sounds. *Hearing Research*, 271(1-2), 147–158.

Iwatsubo, T., Kuzuhara, S., Kanemitsu, A., Shimada, H., & Toyokura, Y. (1990). Corticofugal projections to the motor nuclei of the brainstem and spinal cord in humans. *Neurology*, 40(2), 309.

Jacomme, A. V., Nodal, F. R., Bajo, V. M., Manunta, Y., et al. (2003). The projection from auditory cortex to cochlear nucleus in guinea pigs: An in vivo anatomical and in vitro electrophysiological study. *Experimental Brain Research*, 153(4), 467–476.

Janik, V. M., & Slater, P. J. B. (1997). Vocal learning in mammals. In P. J.B. Slater, J. S. Rosenblatt, C. T. Snowdon, & M. Milinski (Eds.), *Advances in the study of behavior* (pp. 59–99). New York: Academic Press.

Janik, V. M., & Slater, P. J. B. (2000). The different roles of social learning in vocal communication. *Animal Behaviour*, 60(1), 1–11.

Joris, P. X., Schreiner, C. E., & Rees, A. (2004). Neural processing of amplitude-modulated sounds. *Physiological Reviews*, 84(2), 541–577.

Jürgens, U. (2002). Neural pathways underlying vocal control. *Neuroscience & Biobehavioral Reviews*, 26(2), 235–258.

Jürgens, U. (2009). The neural control of vocalization in mammals: A review. *Journal of Voice*, 23(1), 1–10.

Kershenbaum, A., Blumstein, D. T., Roch, M. A., Akçay, Ç., et al. (2016). Acoustic sequences in non-human animals: A tutorial review and prospectus. *Biological Reviews of the Cambridge Philosophical Society*, 91(1), 13–52.

King, J., Insanally, M., Jin, M., Martins, A. R. O., et al. (2015). Rodent auditory perception: Critical band limitations and plasticity. *Neuroscience*, 296, 55–65.

Köver, H., Gill, K., Tseng, Y.-T. L., & Bao, S. (2013). Perceptual and neuronal boundary learned from higher-order stimulus probabilities. *The Journal of Neuroscience*, 33(8), 3699–3705.

Kraus, N., McGee, T., Littman, T., Nicol, T., & King, C. (1994). Nonprimary auditory thalamic representation of acoustic change. *Journal of Neurophysiology*, 72(3), 1270–1277.

Kubota, M., Sugimoto, S., Hosokawa, Y., Ojima, H., & Horikawa, J. (2017). Auditory-visual integration in fields of the auditory cortex. *Hearing Research*, 346, 25–33.

Kuo, R. I., & Wu, G. K. (2012). The generation of direction selectivity in the auditory system. *Neuron*, 73(5), 1016–1027.

Kuypers, H. G. J. M. (1958). Some projections from the peri-central cortex to the pons and lower brain stem in monkey and chimpanzee. *The Journal of Comparative Neurology*, 110(2), 221–255.

Lange, S., Burda, H., Wegner, R. E., Dammann, P., et al. (2007). Living in a "stethoscope": Burrow acoustics promote auditory specializations in subterranean rodents. *Die Naturwissenschaften*, 94(2), 134–138.

Lavocat, R., & Parent, J.-P. (1985). Phylogenetic analysis of middle ear features in fossil and living rodents. In *Evolutionary relationships among rodents* (pp. 333–354). NATO Advanced Science Institutes Series A: Life Sciences. Boston: Springer.

Lee, C. C. (2013). Thalamic and cortical pathways supporting auditory processing. *Brain and Language*, 126(1), 22–28.

Lee, C. C. (2015). Exploring functions for the non-lemniscal auditory thalamus. *Frontiers in Neural Circuits*, 9, 69.

López, D. E., Saldaña, E., Nodal, F. R., Merchán, M. A., & Warr, W. B. (1999). Projections of cochlear root neurons, sentinels of the rat auditory pathway. *The Journal of Comparative Neurology*, 415(2), 160–174.

Lorente de No, R. (1933). Anatomy of the eighth nerve. III. General plan of stucture of the primary cochlear nuclei. *Laryngoscope*, 43, 327–350.

Lyzwa, D., Herrmann, J. M., & Wörgötter, F. (2015). Natural vocalizations in the mammalian inferior colliculus are broadly encoded by a small number of independent multi-units. *Frontiers in Neural Circuits*, 9, 91.

Mahrt, E., Agarwal, A., Perkel, D., Portfors, C., & Elemans, C. P. H. (2016). Mice produce ultrasonic vocalizations by intra-laryngeal planar impinging jets. *Current Biology*, 26(19), R880–R881.

Malmierca, M. S. (2003). The structure and physiology of the rat auditory system: An overview. *International Review of Neurobiology*, 56, 147–211.

Manley, G. A. (2017). Comparative auditory neuroscience: Understanding the evolution and function of ears. *Journal of the Association for Research in Otolaryngology*, 18(1), 1–24.

Marler, P. (1957). Specific distinctiveness in the communication signals of birds. *Behaviour*, 11(1), 13–38.

Mason, M. J. (2001). Middle ear structures in fossorial mammals: A comparison with non-fossorial species. *Journal of Zoology*, 255(4), 467–486.

Mason, M. J. (2013). Of mice, moles and guinea pigs: Functional morphology of the middle ear in living mammals. *Hearing Research*, 301, 4–18.

Matsumoto, Y. K., Okanoya, K., & Seki, Y. (2012). Effects of amygdala lesions on male mouse ultrasonic vocalizations and copulatory behaviour. *Neuroreport*, 23(11), 676–680.

Montgomery, J. C., & Bodznick, D. (1994). An adaptive filter that cancels self-induced noise in the electrosensory and lateral line mechanosensory systems of fish. *Neuroscience Letters*, 174(2), 145–148.

Nakagawa, R., Matsunaga, E., & Okanoya, K. (2012). Defects in ultrasonic vocalization of Cadherin-6 knockout mice. *PLoS One*, 7(11), e49233.

Nelken, I. (2008). Processing of complex sounds in the auditory system. *Current Opinion in Neurobiology*, 18(4), 413–417.

Newman, D. B., Hilleary, S. K., & Ginsberg, C. Y. (1989). Nuclear terminations of corticoreticular fiber systems in rats. *Brain, Behavior and Evolution*, 34(4), 253–264.

Nodal, F. R., & López, D. E. (2003). Direct input from cochlear root neurons to pontine reticulospinal neurons in albino rat. *The Journal of Comparative Neurology*, 460(1), 80–93.

Nottebohm, F., Stokes, T. M., & Leonard, C. M. (1976). Central control of song in the canary, *Serinus canarius*. *The Journal of Comparative Neurology*, 165(4), 457–486.

Nyby, J. G. (2010). Adult house mouse (*Mus musculus)* ultrasonic calls: Hormonal and pheromonal regulation. In S. M. Brudzynski (Ed.), *Handbook of behavioral neuroscience* (pp. 303–310). New York: Elsevier.

Ohlemiller, K. K., Jones, S. M., & Johnson, K. R. (2016). Application of mouse models to research in hearing and balance. *Journal of the Association for Research in Otolaryngology*, 17(6), 493–523.

Pannese, A., Grandjean, D., & Frühholz, S. (2015). Subcortical processing in auditory communication. *Hearing Research*, 328, 67–77.

Philibert, B., Laudanski, J., & Edeline, J. M. (2005). Auditory thalamus responses to guinea-pig vocalizations: A comparison between rat and guinea-pig. *Hearing Research*, 209(1-2), 97–103.

Phillips, R. E., & Peek, F. W. (1975). Brain organization and neuromuscular control of vocalization in birds. In P. Wright, P. Caryl, & D. Vowles (Eds.), *Hormones and behavior in vertebrates* (pp. 243–274). Amsterdam: Elsevier.

Pollak, G. D. (2013). The dominant role of inhibition in creating response selectivities for communication calls in the brainstem auditory system. *Hearing Research*, 305, 86–101.
Pollak, G. D., Burger, R. M., & Klug, A. (2003). Dissecting the circuitry of the auditory system. *Trends in Neurosciences*, 26(1), 33–39.
Popelář, J., Šuta, D., Lindovský, J., Bureš, Z., et al. (2016). Cooling of the auditory cortex modifies neuronal activity in the inferior colliculus in rats. *Hearing Research*, 332, 7–16.
Poremba, A., Bigelow, J., & Rossi, B. (2013). Processing of communication sounds: Contributions of learning, memory, and experience. *Hearing Research*, 305, 31–44.
Portfors, C. V. (2007). Types and functions of ultrasonic vocalizations in laboratory rats and mice. *Journal of the American Association for Laboratory Animal Science*, 46(1), 28–34.
Portfors, C. V., & Felix, R. A. (2005). Spectral integration in the inferior colliculus of the CBA/CaJ mouse. *Neuroscience*, 136(4), 1159–1170.
Portfors, C. V., Roberts, P. D., & Jonson, K. (2009). Over-representation of species-specific vocalizations in the awake mouse inferior colliculus. *Neuroscience*, 162(2), 486–500.
Portfors, C. V., Mayko, Z. M., Jonson, K., Cha, G. F., & Roberts, P. D. (2011). Spatial organization of receptive fields in the auditory midbrain of awake mouse. *Neuroscience*, 193, 429–439.
Ramón y Cajal, S. (1904). *Textura del sistema nervioso del hombre y de los vertebrados* (Edicion Facsimil 1992, Volume II). Alicante, Spain: Graficas Vidal Leuka.
Riede, T. (2013). Stereotypic laryngeal and respiratory motor patterns generate different call types in rat ultrasound vocalization. *Journal of Experimental Zoology Part A: Ecological Genetics and Physiology*, 319(4), 213–224.
Roberts, L. H. (1975). The functional anatomy of the rodent larynx in relation to audible and ultrasonic cry production. *Zoological Journal of the Linnean Society*, 56(3), 255–264.
Rosowski, J. J. (1994). Outer and middle ears. In R. R. Fay & A. N. Popper (Eds.), *Comparative hearing: Mammals* (pp. 172–247). New York: Springer-Verlag.
Saldeitis, K., Happel, M. F. K., Ohl, F. W., Scheich, H., & Budinger, E. (2014). Anatomy of the auditory thalamocortical system in the mongolian gerbil: Nuclear origins and cortical field-, layer-, and frequency-specificities. *The Journal of Comparative Neurology*, 522(10), 2397–2430.
Sales, G., & Pye, D. (1974). *Ultrasonic communication by animals*. New York: Springer-Verlag.
Sanes, D. H., & Woolley, S. M. N. (2011). A behavioral framework to guide research on central auditory development and plasticity. *Neuron*, 72(6), 912–929.
Sayegh, R., Aubie, B., & Faure, P. A. (2011). Duration tuning in the auditory midbrain of echolocating and non-echolocating vertebrates. *Journal of Comparative Physiology A: Neuroethology, Sensory, Neural, and Behavioral Physiology*, 197(5), 571–583.
Scharff, C., & Haesler, S. (2005). An evolutionary perspective on FoxP2: Strictly for the birds? *Current Opinion in Neurobiology*, 15(6), 694–703.
Schneidman, E. (2016). Towards the design principles of neural population codes. *Current Opinion in Neurobiology*, 37, 133–140.
Schofield, B. R., & Coomes, D. L. (2006). Pathways from auditory cortex to the cochlear nucleus in guinea pigs. *Hearing Research*, 216, 81–89.
Shamma, S. A., & Fritz, J. B. (2009). Auditory cortex: Models. In L. R. Squire (Ed.), *Encyclopedia of neuroscience* (pp. 709–714). Oxford, UK: Academic Press.
Shepard, K. N., Lin, F. G., Zhao, C. L., Chong, K. K., & Liu, R. C. (2015). Behavioral relevance helps untangle natural vocal categories in a specific subset of core auditory cortical pyramidal neurons. *The Journal of Neuroscience*, 35(6), 2636–2645.
Sherman, S. M. (2001). Tonic and burst firing: Dual modes of thalamocortical relay. *Trends in Neurosciences*, 24(2), 122–126.
Simonyan, K. (2014). The laryngeal motor cortex: Its organization and connectivity. *Current Opinion in Neurobiology*, 28, 15–21.
Simonyan, K., & Jürgens, U. (2003). Efferent subcortical projections of the laryngeal motorcortex in the rhesus monkey. *Brain Research*, 974(1-2), 43–59.
Simonyan, K., & Horwitz, B. (2011). Laryngeal motor cortex and control of speech in humans. *The Neuroscientist*, 17(2), 197–208.

Singla, S., Dempsey, C., Warren, R., Enikolopov, A. G., & Sawtell, N. B. (2017). A cerebellum-like circuit in the auditory system cancels responses to self-generated sounds. *Nature Neuroscience*, 20(7), 943–950.

Stiebler, I., Neulist, R., Fichtel, I., & Ehret, G. (1997). The auditory cortex of the house mouse: Left-right differences, tonotopic organization and quantitative analysis of frequency representation. *Journal of Comparative Physiology A: Sensory, Neural, and Behavioral Physiology*, 181(6), 559–571.

Suta, D., Kvasnák, E., Popelár, J., & Syka, J. (2003). Representation of species-specific vocalizations in the inferior colliculus of the guinea pig. *Journal of Neurophysiology*, 90(6), 3794–3808.

Suta, D., Popelár, J., & Syka, J. (2008). Coding of communication calls in the subcortical and cortical structures of the auditory system. *Physiological Research/Academia Scientiarum Bohemoslovaca*, 57(Suppl 3), 149–159.

Suthers, R. A., Fitch, W. T., Fay, R. R., & Popper, A. N. (Eds.). (2016). *Vertebrate sound production and acoustic communication* (Vol. 53). New York: Springer International Publishing.

Syka, J. (2010). Subcortical responses to species-specific vocalizations. In S. M. Brudzynski (Ed.), *Handbook of behavioral neuroscience* (pp. 99–112). Oxford, UK: Elsevier.

Tabler, J. M., Rigney, M. M., Berman, G. J., Gopalakrishnan, S., et al. (2017). Cilia-mediated hedgehog signaling controls form and function in the mammalian larynx. *eLife*, 6.

Tanaka, H., & Taniguchi, I. (1991). Responses of medial geniculate neurons to species-specific vocalized sounds in the guinea pig. *Japanese Journal of Physiology*, 41(6), 817–829.

Ter-Mikaelian, M., Sanes, D. H., & Semple, M. N. (2007). Transformation of temporal properties between auditory midbrain and cortex in the awake Mongolian gerbil. *The Journal of Neuroscience*, 27(23), 6091–6102.

Ter-Mikaelian, M., Semple, M. N., & Sanes, D. H. (2013). Effects of spectral and temporal disruption on cortical encoding of gerbil vocalizations. *Journal of Neurophysiology*, 110(5), 1190–1204.

Theunissen, F. E., & Elie, J. E. (2014). Neural processing of natural sounds. *Nature Reviews, Neuroscience*, 15(6), 355–366.

Tinbergen, N. (1963). On aims and methods of Ethology. *Zeitschrift für Tierpsychologie*, 20(4), 410–433.

Tsukano, H., Horie, M., Ohga, S., Takahashi, K., et al. (2017). Reconsidering tonotopic maps in the auditory cortex and lemniscal auditory thalamus in mice. *Frontiers in Neural Circuits*, 11, 1–8.

de Villers-Sidani, E., Chang, E. F., Bao, S., & Merzenich, M. M. (2007). Critical period window for spectral tuning defined in the primary auditory cortex (A1) in the rat. *The Journal of Neuroscience*, 27(1), 180–189.

Wang, X. (2007). Neural coding strategies in auditory cortex. *Hearing Research*, 229(1-2), 81–93.

Wang, X., Merzenich, M. M., Beitel, R., & Schreiner, C. E. (1995). Representation of a species-specific vocalization in the primary auditory cortex of the common marmoset: Temporal and spectral characteristics. *Journal of Neurophysiology*, 74(6), 2685–2706.

Wang, X., Qi, Q., Huang, C., Chomiak, T., & Luo, F. (2016). Duration sensitivity of neurons in the primary auditory cortex of albino mouse. *Hearing Research*, 332, 160–169.

West, C. D. (1985). The relationship of the spiral turns of the cochlea and the length of the basilar membrane to the range of audible frequencies in ground dwelling mammals. *The Journal of the Acoustical Society of America*, 77(3), 1091–1101.

Wigderson, E., Nelken, I., & Yarom, Y. (2016). Early multisensory integration of self and source motion in the auditory system. *Proceedings of the National Academy of Sciences of the Unites States of America*, 113(29), 8308–8313.

Wilczynski, W., & Ryan, M. J. (2010). The behavioral neuroscience of anuran social signal processing. *Current Opinion in Neurobiology*, 20(6), 754–763.

Wild, J. M., Li, D., & Eagleton, C. (1997). Projections of the dorsomedial nucleus of the intercollicular complex (DM) in relation to respiratory-vocal nuclei in the brainstem of pigeon (*Columba livia*) and zebra finch (*Taeniopygia guttata*). *The Journal of Comparative Neurology*, 377, 392–413.

Willard, F. H., & Ryugo, D. K. (1983). Anatomy of the central auditory system. In J. F. Willott (Ed.), *The auditory psychobiology of the mouse* (pp. 201–304). Springfield, MA: Charles C. Thomas.

Winer, J. A. (1992). The functional architecture of the medial geniculate body and the primary auditory cortex. In D. B. Webster, A. N. Popper, & R. R. Fay (Eds.), *The mammalian auditory pathway: Neuroanatomy* (pp. 222–409). New York: Springer-Verlag.

Winer, J. A. (2006). Decoding the auditory corticofugal systems. *Hearing Research*, 212(1-2), 1–8.

Winer, J. A., & Schreiner, C. E. (2005). The central auditory system: A functional analysis. In J. A. Winer & C. E. Schreiner (Eds.), *The inferior colliculus* (pp. 1–68). New York: Springer Science+Media Publishing.

Wöhr, M., & Schwarting, R. K. W. (2010). Activation of limbic system structures by replay of ultrasonic vocalization in rats. In S. M. Brudzynski (Ed.), *Handbook of behavioral neuroscience* (pp. 113–124). New York: Elsevier.

Wöhr, M., Oddi, D., & D'Amato, F. R. (2010). Effect of altricial pup ultrasonic vocalization on maternal behavior. In S. M. Brudzynski (Ed.), *Handbook of behavioral neuroscience* (pp. 159–166). New York: Elsevier.

Woolley, S. M. N. (2012). Early experience shapes vocal neural coding and perception in songbirds. *Developmental Psychobiology*, 54, 612–631.

Woolley, S. M. N., & Portfors, C. V. (2013). Conserved mechanisms of vocalization coding in mammalian and songbird auditory midbrain. *Hearing Research*, 305, 45–56.

Xia, Y. F., Qi, Z. H., & Shen, J. X. (2000). Neural representation of sound duration in the inferior colliculus of the mouse. *Acta Oto-Laryngologica*, 120(5), 638–643.

Zhang, S. P., Davis, P. J., Bandler, R., & Carrive, P. (1994). Brain stem integration of vocalization: Role of the midbrain periaqueductal gray. *Journal of Neurophysiology*, 72(3), 1337–1356.

Chapter 7
Lessons from Rodent Models for Genetic and Age-Related Hearing Loss

Kevin K. Ohlemiller

Abstract This chapter compares the six most heavily studied rodent models with regard to hearing-in-aging and the availability of mutant lines that recapitulate human genetic hearing loss. Four of the six models are available only as outbreds, and much of that work has been based on genetically nonstandard animals of unclear origin. Some of these (guinea pigs and chinchillas) may no longer resemble their wild counterparts. Some results from outbred models may not be reproducible, since it may be impossible for experimenters to know if they are testing the same genetic models. Likewise, engineered or induced mutations onto outbred lines may not be productive because characterization can be confounded by variable and unknown modifier genes. Naturally arising coat color-related mutations may influence hearing through an absence of melanin or melanocytes. These lines may not be commercially available, however. Hamsters are not well described with respect to detailed hearing or aging studies. Gerbils, guinea pigs, and chinchillas are well explored both as general hearing models and as aging models. Inbred mice and rats have become the primary models for most research over the last 20 years. Inbred models offer a high degree of genetic standardization and reproducibility of results. Their short lifespans and the availability of lines with progressive hearing loss have made mice and rats popular for aging research. They also foster transgenic methods and gene discovery, but mice and rats may not be optimal for studies that require low-frequency hearing or readily accessible inner ear fluid spaces.

Keywords Auditory neuron · Chinchilla · Cochlea · Endocochlear potential · Gerbil · Guinea pig · Hair cells · Hamster · Inbred stock · Mouse · Mutation · Outbred stock · Presbycusis · Rat · Stria vascularis

---

K. K. Ohlemiller (✉)
Fay and Carl Simons Center for Biology of Hearing and Deafness, Central Institute for the Deaf at Washington University, Saint Louis, MO, USA

Department of Otolaryngology, Washington University, School of Medicine, Saint Louis, MO, USA
e-mail: kohlemiller@wustl.edu

M. L. Dent et al. (eds.), *Rodent Bioacoustics*, Springer Handbook of Auditory Research 67, https://doi.org/10.1007/978-3-319-92495-3_7

7.1 Introduction

In this chapter, rodents are considered in the context of genetic and age-related pathology. In doing so, the focus has been narrowed to the similarities of rodents to humans and what rodents may reveal about human hearing loss. Relatively few rodent species have been studied in this regard and coverage has been very uneven. This chapter covers the same six species highlighted in Dent, Screven, and Kobrina (Chap. 4): hamsters (*Mesocricetus auratus*), Mongolian gerbils (*Meriones unguiculatus*), guinea pigs (*Cavia porcellus*), chinchillas (*Chinchilla sp.*), mice (*Mus musculus*), and rats (*Rattus norvegicus*). Among these, only mice and rats are available commercially in many genetically standardized varieties, a vital feature for identifying hearing loss genes. In nature, it is an open question as to how often rodents survive to old age and whether there is a re-deployment of resources from healthy aging to basic survival. Whether or not this is the case, rodents show similar age-related pathology of hearing as humans and they appear to carry many of the same deafness genes.

7.2 Philosophy of a Comparative Approach

Why adopt a comparative approach when considering hearing or any other function common to most organisms? First, a comparative approach is inherently valuable for the perspectives gained, regardless of any potential clinical benefit. Over the course of evolution, the neural mechanisms involved in capturing and interpreting acoustic energy co-opted structures and mechanisms that evolved to support mechanoreception in organisms as diverse as insects and fish (e.g., French 1992). Some of the same genes and developmental patterns were put to work. Even if the interest were narrowly clinical, ignoring genes that were first discovered in drosophila (*Drosophila melanogaster*) and zebrafish (*Danio rerio*) that have also been identified as human deafness genes (Whitfield 2002; Senthilan et al. 2012) would be counterproductive. Some aspects of age-associated pathology may begin during development, which involves some evolutionarily very old and highly conserved genes. Nevertheless, there are about as many differences in cochlear anatomy and function between humans and rodents as between humans and macaques. These differences take the form of variations in which cells express which genes and isoforms (splice variants) rather than in what genes are expressed (Ohlemiller et al. 2016). This means that some cell types may perform slightly different functions in different species, even while the overall functions of their respective epithelia are the same. Yet there exists tremendous overlap in the known genes that cause deafness in rodents and humans and in the basic anatomic aspects of human cochlear pathology related to genetics, noise, and aging. This includes pathology of cochlear hair cells, neurons, and stria vascularis, the three major players in cochlear function. For the rodent species considered in this chapter, unknown comparative details derive primarily from the fact that so much more is known about how these species lose their hearing than can be gleaned from relatively few, poorly preserved, human temporal bones.

7.3 Inbred Versus Outbred Characters of Different Models

Other than inbred mice and rats, the research models one may purchase commercially are outbred (http://isogenic.info/html/outbred_stocks.html). Outbred stocks utilize closed rotational breeding schemes intended to keep the number of homozygous loci to less than 1% per generation. The breeding is *closed* only with respect to a particular supplier. Across breeders of outbred lines, there is little standardization, little genetic monitoring, and few claims are made about the genetic similarity within a named outbred line.

Outbreeding means that animals are assumed to be heterozygous at many unknown loci, and the number of alleles in play is not constrained. One implication is that one should not be too rigid in stating *the* phenotype of a particular outbred line, since the same line purchased from a different vendor may have a different phenotype. Another implication is that, even within a vendor, the variance of any data obtained from the animals will be relatively large compared to inbred mice or rats. Outbred lines are argued to yield a closer animal equivalent for the human population that the animals are intended to model (Festing 2010). After all, humans are not typically highly inbred and most genes have many variants. The posited advantages of outbred models have been challenged as a false economy, however (Festing 2010). Outbred animals are not particularly similar to randomly bred humans, and introducing unknown genetic variance into any study simply adds data variance than can obscure the answer to the experimental question. Whether this poses a problem depends on the question.

The experimenter should always beware of making sweeping statements about all members of a species based on a single strain or population. Certainly, there are some features that genetic background seems to affect very little, like the very different roles of outer and inner hair cells. Other features, like noise vulnerability and aging characteristics, are classic complex features that will be highly subject to genetic and environmental modulation. In between lie a host of features for which the sources of variance are simply not known. If the goal is to detect genetic background effects and to identify the underlying genes, this task will be approachable (though often still difficult) only in inbred models. The potential problem of overgeneralizing findings can be minimized by testing 4–5 genetically divergent inbred strains (see Petkov et al. 2004).

7.4 Key Metrics for Comparison

How does one decide how well a species models human hearing and hearing loss? It may first be useful to lay to rest the potential issue of frequency range. Should it be disqualifying that many rodents possess ultrasonic hearing (>20 kHz) and often have poor low-frequency hearing (see Dent, Screven, and Kobrina Chap. 4)? Does this perhaps make their cochlea too different from the human cochlea for modeling disease? Other than some echolocating bats and aquatic mammals (Ketten 1992; Pollak 1992), there are no striking qualitative operational differences between the cochleas of humans and other mammals. That said, some rodents show high-frequency sensitivity peaks in their behavioral audiograms or physiological responses that suggest specializations for detecting ultrasonic communication sounds (see Kubke and Wild, Chap. 6; Hurley and Kalcounis-Rueppel, Chap. 8). Notably, laboratory mice hear poorly in the frequency range where neural phase-locking can be observed in cats, chinchillas, guinea pigs, and gerbils. Nevertheless, mouse cochlear neurons phase lock to intense tones in the low-frequency tails of their frequency tuning curves (Taberner and Liberman 2005). The responses to such stimuli, which may never occur in nature, suggest that key timing mechanisms are conserved.

Ultrasonic hearing is believed to have arisen early in mammalian evolution when there were no large mammals (Masterton et al. 1969). Small mammals of the day were able to expand their hearing range upward due to the innovation of three middle ear bones balanced around a single axis of rotation (Rosowski 1992). The dominant driver in this advance was likely the need to localize sounds, and interaural intensity would have been the major cue available to these animals (Masterton et al. 1969). As mammals grew in size, interaural timing cues became available, relaxing the selective pressure to maintain high-frequency hearing (Heffner and Masterton 1980). Many small mammals have expanded their lower frequency range by expanding the air space surrounding the cochlea to form a Helmholtz resonator (the bulla), although one may wonder why this feature is more exaggerated in some species than in others (Ravicz and Rosowski 1997). The innovations that facilitate low- and high-frequency hearing are somewhat independent of each other (but see Heffner and Masterton 1980), which is why some small mammals (e.g., Mongolian gerbils and chinchillas) can reasonably match the hearing range of humans.

A second point of emphasis is that no species can model humans in any universal sense. Each model may usefully mimic particular disease features of humans, but even that will likely only apply to particular humans, namely the carriers of particular modifier genes. The real question will always be: Which members of the species may best model which humans? Since humans are genetically heterogeneous, it is best if the experimenter has some choice of genetic backgrounds for any species studied. In any case, human versus animal comparisons require numerical descriptors of normal and pathologic cochlear anatomy, function, and behavior.

7.4.1 Requisites for Normal Hearing

The cochlea contains a variety of cell types, many whose primary function is not clear (see Kubke and Wild, Chap. 6). Certainly, hearing requires functioning inner hair cells (IHCs) for sound coding and outer hair cells (OHCs) for amplification. For any cochlear frequency place, some minimal number of hair cells are needed to support normal detection thresholds, frequency discrimination, and intensity

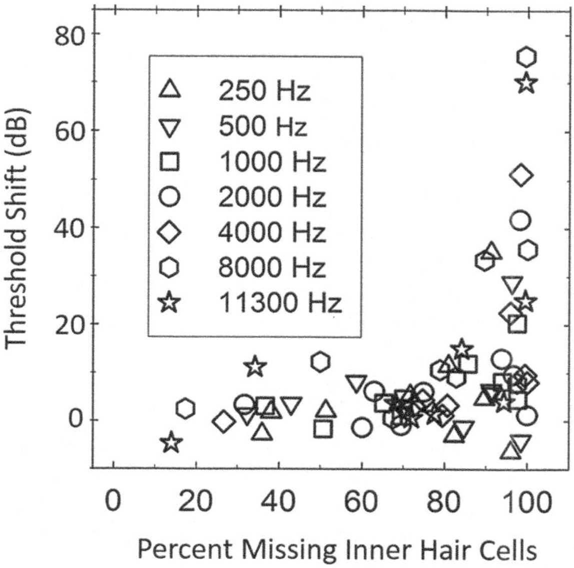

Fig. 7.1 Behavioral threshold shifts in chinchillas for frequencies from 250 to 11,300 Hz as a function of inner hair cell loss following application of the ototoxin carboplatin. Threshold shifts ranged from 10 dB to 80 dB only after IHC loss exceeded 80%, suggesting large IHC losses are needed to produce threshold shifts greater than 10 dB. (Modified from Lobarinas et al. (2013a) with permission)

discrimination. However, the steepness of the dependence of thresholds on OHC function is greater in the high-frequency base of the cochlea than in the low-frequency apex (Prosen et al. 1990). Also, detection thresholds are surprisingly resistant to moderate IHC loss (Lobarinas et al. 2013a) (Fig. 7.1). Nevertheless, it is still common and nearly universal to count hair cells as an essential metric of cochlear and hearing health. Isolated IHC loss is less common, and selective inner hair cell pathology is more difficult to detect. Each IHC may provide the sole driving force to ~20 radial afferent neurons, so that every millimeter of mammalian cochlear length typically includes about 100 IHCs and up to 2000 afferents. This number of afferents is not needed to detect sounds in quiet; rather, they appear necessary to separate and encode sound sources by spectral shape in noisy backgrounds.

Cochlear hair cells rely on transducer currents whose electromotive force is partly set by the stria vascularis, which generates the endocochlear potential (EP). Response thresholds of both single neurons and cochlear potentials, such as the compound action potential and Wave I of the auditory brainstem response (ABR), depend on the EP with a slope of 1–2 mV/dB (Sewell 1984; Ohlemiller 2009). The EP, in turn, requires a normally functioning stria and spiral ligament, although the dependence of the EP on percent strial function is surprisingly forgiving: a normal EP (80–110 mV, depending on species) may be maintained with only half the stria functioning (Schulte and Schmiedt 1992). While the hallmarks of an abnormal stria are not difficult to spot by light microscopy (thinning, cyst formation, loss of blood vessels, fewer and larger marginal cells), one cannot judge the EP based solely on the appearance of the stria. The EP depends, in part, on an ion barrier that normally lines the cochlear scala media, which is the large fluid space where the EP can be recorded. This barrier is formed by the membranes of the cells that line the compartment and by tight junctions between these cells. Notably, what does *not* seem to be a feature of aging or most genetic hearing loss (as judged by mouse models) is a degradation of this barrier sufficient to "short out" the EP (Ohlemiller 2009). Age-associated loss of hair cells, strial marginal cells, or other cells of the lining does not seem to leave holes that allow dissipation of key ion gradients (mostly K^+).

A host of cochlear cell types beyond hair cells, neurons, and strial cells must be doing something critical, but their unique roles are not yet known, nor is it possible to relate their health or numbers to hearing (Slepecky 1996). Why does the organ of Corti require Boettcher cells or Hensen's cells? What do Claudius cells do that these do not? What are the unique functions of the multiple types of fibrocytes in the spiral ligament? The organ of Corti is highly susceptible to fixation artifacts that distort its appearance, so it is not typical to emphasize subtle differences in the appearance of supporting cells or to count them.

Often the first guess about the critical role of a cell type comes from the discovery of natural mutations in inbred models that impact only that cell. Most identified deafness genes are primarily expressed in hair cells, stria, or a combination of cells (e.g., connexins), such that one cannot assign unique roles to specific supporting cells or cells of the spiral limbus or ligament. A recent exception is "root cell" pathology that has been reported in some aging mouse strains (Ohlemiller et al. 2010). Follow-up studies have revealed channels and pumps unique to these cells and the assignment of a unique role (Jagger et al. 2010), although their functions may be divided among neighboring cells in other species (Ohlemiller et al. 2016).

In summary, scientists lack proper metrics for addressing the health of many cell types necessary for hearing. Counting hair cells and neurons or recording distance metrics (e.g., height of the organ of Corti, thickness of the stria and ligament) are useful, but those metrics will not always indicate where a problem originates or the proximate cause of hearing loss. Classic light microscope histopathology of well-fixed tissue has become a lost art, but it is as vital as ever. The absence of a favorite antigen in immunocytochemical analyses is no substitute for microscopic histopathology, since the expression of many proteins will be altered in sick cells.

7.4.2 Diagnosing Tinnitus in Rodents

Rodent models are frequently considered for the study of tinnitus, so that particular consideration is merited here. Tinnitus, the perception of sound that is not physically present, frequently accompanies hearing loss (Eggermont and Roberts 2015). Treatments for tinnitus have proven elusive, and good animal models for studying the condition are not abundant since it is hard to get an animal to tell the experimenter when it is experiencing a phantom sound. This has been unhelpfully equated with the inability of an animal with tinnitus to recognize silence. This is an important distinction, because the latter is taken to imply that the animal cannot tell its own tinnitus from an external sound presented by the experimenter. Thus, if played a tone or noise with an embedded silent period (gap), the animal ostensibly cannot detect the gap. Moreover, it has been argued that the frequency of the tone or noise band in which the gap is embedded can provide information about the frequency band of the animal's tinnitus (e.g., Turner et al. 2012). This is a staggering leap that is only addressable in experiments where the same assumptions are tested in humans. Such experiments do not clearly support these assertions (Fournier and Hébert 2013; Galazyuk and Hébert 2015).

Essentially, all animal-based experiments studying and treating tinnitus rely on the animal's confusion of internal and external sounds. In the same way that a tone presented just before a loud, startle-evoking stimulus can attenuate the startle response, a silent gap embedded in an ongoing tone or noise is also inhibitory. Particularly in rats and mice, so called "gap detection" has been used to diagnose and characterize tinnitus (e.g., Longenecker and Galazyuk 2011; Turner et al. 2012). Another problem is that noise-induced hearing loss that is often inflicted to trigger tinnitus will interfere with the task unless it is unilateral. Salicylate or other ototoxic protocols that cause bilateral hearing loss are inescapably confounded. Yet it is not clear that unilateral hearing loss caused by unilateral noise exposure is any less confounding (Lobarinas et al. 2013b). This typically requires that the animals be anesthetized during exposure, an unnatural condition that likely alters the nature of any central auditory effects of the noise, including the nature of any tinnitus. Tinnitus has many manifestations (e.g., tonal, pulsatile, clicking, buzzing), and there is no guarantee that identically treated animals will acquire tinnitus that is similarly conducive to diagnosis by gap detection. Finally, comparison of data across animals requires normalization to a control startle condition. Interpretation may be complicated by any change to the startle response. Ratios are highly sensitive to changes in the denominator. In summary, applying rodent models to the study of tinnitus has not proven straightforward, and it is not clear how to get around the current limitations.

7.5 The Nature of Cochlear Pathology Due to Aging and Genetics

The discussion of how rodents can model human hearing loss can be streamlined by summarizing how major forms of hearing loss affect the inner ear of humans. Since all aspects of the study of rodent models can be optimized and human temporal bones are often poorly fixed with unknown histories, much more is known about cochlear pathology in some rodents than is known for humans. Thus, much of what has been reported in rodent studies, including highly detailed characterizations of some cellular changes, presently has no counterpart in human studies. For this reason, it is unclear whether some reported aspects of rodent pathology are broadly applicable.

7.5.1 Aging and Presbycusis

The generally accepted framework for characterizing presbycusis is one championed by H. Schuknecht in a series of human temporal bone studies (Schuknecht 1993; Schuknecht and Gacek 1993). In his attempts to identify the primary cause of hearing loss using often-compromised specimens, Schuknecht proposed that the initial effects of aging can appear in hair cells, afferent neurons, or stria vascularis, so that pathology of one of these can drive the life-long pattern of hearing loss and determine its essential features. As a corollary, he proposed that the pathology of these cells/structures can arise independently. That is, organ of Corti pathology need not cause strial pathology, and although the stria is needed for normal hearing, strial pathology need not cause hair cell loss, per se. In humans and animals, unfortunately, multiple pathologies typically occur at once. In animals, at least, it is possible to dissociate these and isolate their causes by appropriate model choices.

Neither human nor animal observations have proven which type of presbycusis is most prevalent. Audiogram shape is not definitive in human diagnoses since a given pattern can be produced by different combinations of underlying factors (Allen and Eddins 2010). Schuknecht termed age-related primary hair cell pathology as *sensory presbycusis*. The audiogram is very sensitive to OHC loss. The steepness of the relation depends on stimulus frequency, with high frequency hearing more readily affected. It is usually not possible in humans to separate hair cell loss caused by aging from that caused by noise. Yet a popular view of aging is that of accumulated injury (Ohlemiller and Frisina 2008; Ohlemiller 2015), and some alleles that promote presbycusis and other age-associated pathology appear to influence cellular responses to stress (Fransen et al. 2003, 2015). For this reason, there may be no clear delineation between pure aging effects and injury.

Schuknecht distinguished primary neural pathology/loss (*neural presbycusis*) from that resulting secondarily from hair cell loss. This was taken to affect not thresholds but supra-threshold sound identification, particularly speech. Indeed, loss of speech-in-noise decoding ability, the so-called *cocktail party effect*, is the most common hearing complaint in aging. Studies in mice and guinea pigs have given impetus to this area of inquiry (Kujawa and Liberman 2009; Furman et al. 2013). This work suggests that primary pathology of afferent synapses on IHCs (with eventual loss of neural cell bodies) can result from even mild noise exposure, so that apparent neural presbycusis also may often be a consequence of noise injury. Finally, Schuknecht proposed that the stria vascularis is also a frequent target of aging, giving rise to *strial* or *metabolic presbycusis*. Oddly, human *strial presbycusis* remains largely theoretical, since recording of the EP is highly invasive and has only been performed anecdotally in humans (Tran Ba Huy et al. 1989; Kobayashi et al. 1996). Nevertheless, analysis of human audiograms and chance features of a particular animal model, the Mongolian gerbil (Sect. 7.6.2), has led to the suggestion that strial presbycusis represents the major form of presbycusis. Since so many unknowns surround the features and risk factors for human strial presbycusis, rodent work has been particularly important for its study. At present, it does not appear that noise exposure typically promotes isolated strial injury or EP decline in animals, although this conclusion is based on a single intense exposure (Hirose and Liberman 2003; Ohlemiller et al. 2011). A chronic exposure paradigm might indicate otherwise.

While the general form of presbycusis appears similar in humans and rodents, it may be more problematic to study the influence of co-morbidities, such as cardiovascular disease, diabetes, hypertension, or any of a number of conditions that can magnify hearing loss. There is no single normalization for all aspects of aging for any two species (Ohlemiller et al. 2016). That is, different tissues may age at different rates across species, so that the same factors may not interact at a given age. Thus, to study the effects of type 2 diabetes on presbycusis in a rodent model, for example, diabetes may have to be triggered by chemicals or diet.

A potential related issue is whether species with a life span of 1–3 years can capture the same aspects of aging as would be found after 70 years in a human. For example, if all hair cells are the same and have the same intrinsically set life span, then an old mouse or Mongolian gerbil should have a young cochlea in human terms. This is not the case. Old mouse and gerbil cochleas can look very much like old human cochleas with respect to degenerative changes (e.g., Tarnowski et al. 1991; Ohlemiller 2006). What may not be possible to normalize is the Mongolian gerbil life span to that of a typical human with the expectation that progressive cochlear changes will fall on the same curve. The rate of degeneration will depend on higher order cell–cell and system–system interactions that may not translate across species.

7.5.2 Genetically Related Pathology

As of this writing, well over 100 human deafness genes have been identified and another 100+ quantitative trait loci (QTLs) have been identified that contain one or more unknown deafness genes (Steel 2014). Steel (2014) has estimated that more than 1000 genes may be associated with hearing loss, either in an isolated manner or as part of a syndrome. Genetically related cochlear pathology encompasses genes related to both early Mendelian hearing loss and to hearing loss acquired over time. *Mendelian hearing loss* is hearing loss that is deterministic and influenced little by environment: the genetic contribution is nearly 100%. *Acquired hearing loss* is heavily influenced by environment, and the genetic contribution may be less than 5% (Parker and Palmer 2011). That 5%, moreover, may represent the aggregate effects of multiple small-effect alleles.

Mendelian hearing loss can be studied within pedigrees in linkage studies, while acquired hearing loss requires large scale genetic association studies. Mendelian hearing loss can be caused by alleles that act either dominantly or recessively. Dominant acting alleles (only one copy required to cause a hearing loss phenotype) also tend to be associated with progressive hearing loss (Dawes and Payton 2016), so that the tendency toward presbycusis may reflect the action of dominant Mendelian-acting alleles, multiple small-effect alleles, or both.

The earliest genes found to be associated with early recessive Mendelian hearing loss often encoded critical proteins expressed by hair cells, such as stereocilia proteins, while those associated with acquired hearing loss tended to encode protective or homeostatic factors. As more genes have been identified, however, such broad statements are harder to justify. Many newly discovered genes are not particularly intuitive or obvious candidates for causing hearing loss, and different alleles of the same gene may cause different forms of hearing loss.

Since it is easiest to detect and isolate new hearing loss mutations in inbred models like rats and mice, most animal work in the area of genetic hearing loss is performed in these species. This work has been fostered by the rapid growth of DNA sequencing methods and the parallel application of these to humans and other species. Syntenic regions of human, mouse, and rat genomes can be compared, and candidate deafness genes can be knocked out and/or subjected to expression profiling.

Parallel study of humans, rats, and mice has greatly expedited the search for human deafness genes and potential therapies. How well complex hearing traits can be modeled in rodents remains to be seen, as these will depend on subtle genetic and environmental interactions that may be fundamentally different from humans. Some higher order gene–gene interactions may be species dependent and may require genetic variants that not all species possess. Even if the right variants do exist within a species naturally, they may be missing in commercially bred models.

7.6 The Models

The rodent models that have been explored in any detail regarding hearing have been studied very unevenly. Fewer specifics can be stated about hamsters, guinea pigs, or chinchillas in the realm of aging, or certainly genetics, than for mice and rats. Mongolian gerbils and chinchillas have been popular subjects due to their human-like hearing range and large auditory bullae, while guinea pigs offer useful varieties and easy surgical access to four cochlear turns. Prior to the molecular revolution of the 1990s, the vast majority of hearing research was directed at basic aspects of cochlear function and employed guinea pigs, chinchillas, and Mongolian gerbils (in addition to cats, primates, and bats). Now all research is molecular, in that the basic molecules of hearing—and thus the bases for all physical processes in hearing—can best be identified using molecular methods. For this reason, mouse and rat work has supplanted much work in the other models, even though they are not ideally suited in terms of frequency range of hearing, cochlear size, or cochlear access.

The high degree of genetic standardization in rats and mice is essential for gene identification and manipulation that provide the power of molecular methods. Many of the varieties of rodent models arose from the interest of hobbyists in varied coat colors, which in turn led to more rigorous control of breeding. Indeed, many of the classic methods of genetic analysis arose from studies of the inheritance of coat color in mice (Silver 1995). Since the primary natural hair pigment, melanin, is involved in cochlear function, some hearing defects are linked to coat or eye color, so that the earliest models with hearing or balance defects also happened to be those with particular coloration. These animals were typically first identified by their circling behavior, which only accompanies some hearing loss.

7.6.1 Hamsters

The most popular hamster breed used for research in the United States is the Syrian or golden hamster (Pritt et al. 2012). Laboratory golden hamsters are thought to have originated from a handful of animals captured in Syria in the 1930s. Thus, they may carry very little of the extant genetic variation that exists in nature, and they have been domesticated for a relatively short time in comparison with mice, rats, and guinea pigs. While potentially useful coat color variants have been described in hamsters (Alizadeh et al. 2009), presently two color variants are known to affect hearing. The black tremor (*bt*) mutation, reportedly arose in a research colony in Japan (Naito et al. 1999). This mutation, which has not been identified, causes demyelination of cochlear nerve fibers and prolonged response latencies. Another mutation, anophthalmic white (*Wh*), causes dominantly inherited white coat color with dark eyes along with hearing loss and poorly characterized cochlear degeneration (Amedofu et al. 1999). Sánchez-Benito and colleagues (2017) suggested that the *Wh* mutation is homologous to one causing Waardenburg Syndrome in humans.

7.6.2 Mongolian Gerbils

The only gerbilline species used extensively in laboratory research is the Mongolian gerbil (Pritt et al. 2012). The entire population of gerbils used for research in the United States is genetically constrained since they all arose from 20 breeding pairs brought to the U. S. in 1954, so the animals available may lack useful and informative natural variation. All commercially available research lines of gerbils appear to be outbred. Some gerbil strains carry pigmentation mutations as reflected by coat and eye color, although no hearing phenotype has been reported (e.g., Ukaji et al. 2016). There are also seizure-sensitive and seizure-resistant lines of gerbils (Lee et al. 1987). None of these models can be directly compared since they are not congenic.

Gerbils possess a broad range of hearing (0.1–60 kHz) (Ryan 1976) and a large auditory bulla that provides ease of access for cochlear recording. Their short lifespan (about 3 years) also makes them excellent for aging research. While aging gerbils show some loss of cochlear hair cells and neurons with aging, the driving influence on age-related threshold shifts in these animals appears to be degeneration of stria vascularis and associated EP reduction (Tarnowski et al. 1991; Spicer and Schulte 2005). Gerbils, therefore, appear to model human strial presbycusis in an isolated form, which has enhanced their value given how difficult this condition is to study. Although no genetic analysis has been attempted (and would be difficult to perform in this genetically heterogeneous model) Mongolian gerbils may concentrate multiple alleles that promote age-associated strial degeneration. Among key attributes of strial presbycusis that Mongolian gerbils have helped to establish are a weak relation between the EP and endolymphatic K^+ and the potential importance of strial capillary loss (Gratton and Schulte 1995; Schmiedt 2010).

Gerbils also are prone to brain cysts that resemble encephalopathies like those associated with prion diseases (Faddis and McGinn 1997), although in the case of Mongolian gerbils, the causes are not well understood. They are also susceptible to cholesteatomas that can be induced by surgical ligation of the external ear canal (Kim and Chole 1998). Gerbils are commonly used for study of otitis media.

Mongolian gerbils have served as a model for a broad array of studies examining basic cochlear function and noise injury. Among their advantages are ease of access to all three cochlear turns (though not all scala) (Schmiedt and Zwislocki 1977; Ohlemiller and Siegel 1992) and a unique approach to the auditory nerve (Schmiedt et al. 1980; Ohlemiller and Siegel 1994). Mongolian gerbils show pronounced cochlear place-dependence of spontaneous firing rates (SRs) that is not as clear in other published species (Ohlemiller and Siegel 1994; Bourien et al. 2014), whereby average SRs are lower in neurons tuned to high characteristic frequencies (CFs) (Fig. 7.2). Such variation is predicted to magnify neural firing rate responses at frequencies where phase locking does not occur in response to CF tones (Patuzzi 2011).

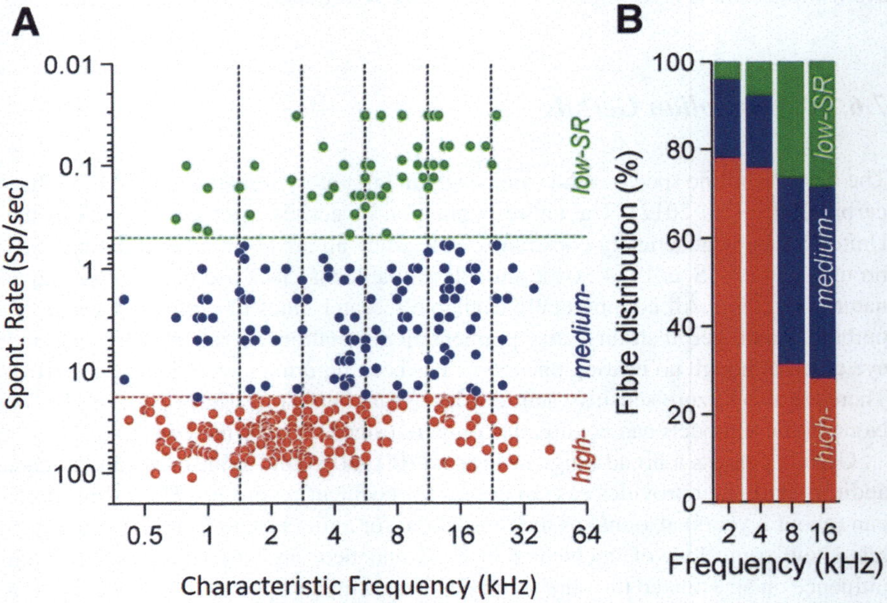

Fig. 7.2 (**a**) Spontaneous firing rates (SR, in spikes/sec) of single cochlear neurons in gerbils as a function of characteristic frequency (CF). (**b**) Proportion of neurons with *low* (green), *medium* (blue), and *high* (red) SRs in different CF bins. Average SR decreases at high CFs. (Modified from Bourien et al. (2014) with permission)

7.6.3 Guinea Pigs

Guinea pigs, notably unrelated to either pigs or the Guinea region of Africa, were one of the earliest popular models for hearing and other types of research. Laboratory guinea pigs largely derive from an outbred albino strain begun by Dunkin and Hartley in 1926 (Pritt et al. 2012). Most studies characterize their animals simply as pigmented or albino without giving the source or lineage. Domestic guinea pigs are so different from their nearest wild ancestor that they do not exist in the wild. Similar to other laboratory rodent models, hobbyist interest in novel coat colors facilitated the development of many breeds. Few breeds, however, are inbred, so that comparison of hearing characteristics in different breeds may be fraught with interpretation problems as they may reflect many unknown genes. The large auditory bulla of guinea pigs facilitated all types of cochlear recordings. Most of what we know about hair cell responses comes from work in guinea pigs from two laboratories working at opposite ends of the cochlea (e.g., Russell and Sellick 1978; Dallos and Cheatham 1992). Guinea pigs have also been used extensively for auditory nerve and basilar membrane recordings (e.g., Winter et al. 1990; Cooper and Yates 1994) and continue to be used heavily for studies of noise injury, therapeutic testing, cochlear fluid dynamics, and cochlear biochemistry.

The life expectancy of guinea pigs is 4–7 years (Pritt et al. 2012). Hearing-in-aging research in guinea pigs has utilized pigmented and albino guinea pigs of unclear origin. Most quantitative studies were restricted to cursory physiology and hair cell counts (e.g., Covell and Rogers 1957; Coleman 1976). The primary changes noted are typical: hair cell loss, some neuronal loss, and modest strial degeneration. What may differentiate guinea pigs—or at least the types studied—from other published models is a reported predominant loss of apical hair cells. However, the pathologies described at present poorly account for the published changes in hearing with age in guinea pigs. Comparison of pigmented and albino strains has led to inferences about the role of cochlear melanin for maintaining hearing sensitivity with age (Conlee et al. 1986, 1988), although the strains studied were not congenic.

There has been little work aimed at manipulating specific hearing genes in guinea pigs, but useful spontaneous mutations have appeared. Waltzing guinea pigs have a dominant-acting mutation (*W*) causing early atrophy of the organ of Corti that results in deafness and circling behaviors (Ernstson 1971; Festing 1976). Homozygosity for the mutation leads to perinatal mortality. The mutation in German waltzing guinea pigs (*gw*) appeared spontaneously in a breeding facility in Germany in 1996 (Jin et al. 2006, 2007). Affected animals exhibit deafness, balance disorders, and circling from birth. Unlike the *W* mutation, the *gw* mutation acts in a recessive manner to cause a collapse of the Reissner's membrane and an absence of the scala media. The mutation is a pigmentation type considered in other models, although coat color is unaffected. Heterozygotes reportedly have normal hearing with an increased resistance to noise-induced trauma and an exacerbated response to the ototoxic drug cisplatin (Skjönsberg et al. 2014; Skjönsberg and Mannström 2015).

German waltzing guinea pigs serve as an animal model of human vestibular atelectasis, since the morphology caused by the mutation and the human disease appear identical (Kawaguchi et al. 2010). As of this writing, the causative genetic mutation is not known for either mutant model.

7.6.4 Chinchillas

Like guinea pigs, long-tailed chinchillas originated in the Andes of South America (Pritt et al. 2012). Those used for research in this country remain the most poorly genetically defined rodent species used for hearing research and may represent a hybrid (mixing *Chinchilla laniger* and *C. chinchilla*) that does not exist in the wild (Pritt et al. 2012). Chinchillas grew in popularity for hearing research due to the superior match of their behavioral audiogram with humans versus other models plus other features that include economy, a large middle ear, resistance to middle ear infections, relative ease of training, and tolerance for anesthesia (Miller 1970). The suitability of chinchillas for behavioral paradigms facilitated the early assemblage of a large body of comparative psychoacoustic data (e.g., Clark et al. 1974, 1987). Most notably, compared to humans, they have relatively poor frequency resolution both behaviorally and at the level of cochlear neurons (Long and Clark 1984).

Chinchillas are also popular models for auditory nerve recording, basilar membrane studies, and noise injury studies (e.g., Ruggero and Rich 1983; Hamernik et al. 1989). According to one investigator (Bohne 1972; Ahmad et al. 2003), they are prone to a mode of rapid OHC loss following moderate noise exposure that leaves holes in the reticular lamina. These holes ostensibly permit mixing of endolymph and perilymph and may exacerbate OHC loss. Whether such holes are common to other models, and simply have been missed, is a matter of controversy. Chinchillas are moderately long lived, with some estimates ranging up to 20 years (Bohne et al. 1990), and they have been the subject of several aging studies (e.g., Bohne et al. 1990; McFadden et al. 1997). While characteristics of all of Schuknecht's presbycusis types have been noted, the major driver of age-related threshold shifts appears to be hair cell loss.

At present there are no commercial chinchilla genetic variants that might facilitate the study of mutations affecting hearing. A recent effort to sequence the chinchilla genome has led to an online resource for comparison with human DNA sequences (Shimoyama et al. 2016). However, without multiple, genetically well-characterized chinchilla lines, it may be difficult to capitalize on this information.

7.6.5 Mice

The majority of mechanistic questions in physiology can now best be addressed using molecular methods, which is why inbred laboratory mice and rats have become the dominant model for much of biomedical research (Ohlemiller et al.

2016). Not all commercial mice are inbred; outbred mice, the genetic equivalent of outbred guinea pigs or gerbils, are popular for some kinds of work and can be a powerful tool for gene mapping (Yalcin and Flint 2012). It may not be profitable, however, to devote a study to nuances of hearing or cochlear phenotype in outbred CD-1 mice, for example, since the precise phenotype can be expected to vary from mouse to mouse, time to time, and breeder to breeder.

Most experimental requirements will favor inbred mice, yet there is always the danger that results will depend on the inbred strain picked. One solution is simply to avoid sweeping language about what inbred mice do, instead including a caveat that the results might have been different in a different inbred strain. A more useful approach would be to include three or four distantly related inbred strains in experiments (Ohlemiller et al. 2016). Even then, results may depend on mouse age, sex, or the particular paradigm (e.g., noise exposure type and level). One strength of using inbred mice to study hearing is the wealth of information written on this topic, including books edited by J. F. Willott that cover a wide range of topics (Willott 1983, 2001).

The first inbred mice were developed by the eventual founders of the Jackson Laboratory (JAX) around 1910 (Silver 1995). Most mouse-based research focuses on relatively few of the now more than 400 commercial inbred strains. Each strain is demarcated by its precise name, which includes a major strain designation and potentially multiple substrain names reflecting each successive breeder. Thus, each segment of what may be a long strain name indicates part of a strain's history that might have facilitated additional genetic drift. Only when the strain names match *exactly* are inbred mice truly genetically similar. Some strains are so popular throughout the world that commercial breeders have set up a system for breeding exact matches to each other's mice. For example, C57BL/6 J mice can be obtained from several commercial breeders, even though they were first bred at JAX. Notably, inbreeding does *not* guarantee identical mice. Recently, geneticists have come to appreciate how omnipresent and unrelenting are the processes that alter DNA. Even nominally identical inbred mice will differ with respect to copy numbers of genes, somatic mutations, and epigenetic changes caused by the uterine and postnatal environment (Provenzano and Domann 2007).

For much research, it is not important how far removed the mice are from their evolutionary antecedents, nor is it important to attempt to place any results into an evolutionary context. For research into normative hearing processes, it is simply necessary to have mice that hear normally. Popular strains that fit this requirement include CBA/CaJ, CBA/J, and FVB/J. Several strains are popular despite—or sometimes because of—progressive hearing loss that may model aspects of aging (e.g., DBA/2J, C57BL/6J, BALB/cJ) (Ohlemiller et al. 2016). The major commercial strains overlap genetically a great deal, having been derived from a limited number of progenitors. For some kinds of research the experimenter may wish to recoup less common alleles. For this purpose, wild-derived inbred strains (e.g., MOLF/EiJ, CAST/EiJ) can be purchased. These were developed deliberately to capture more of the genetic variety that exists across *Mus musculus* in nature. Note that wild-derived mice are still inbred and are not the equivalent of wild-caught mice. Nevertheless, the wild-derived inbred strains, having been subjected to selective pressures more recently than many of the more frequently used strains and exhibit good hearing and noise resistance, while many popular strains show progressive hearing loss and relative noise vulnerability.

7.6.5.1 Mouse Aging Models

The approximate 2 year lifespan of most commercial inbred mouse strains makes them relatively economical for aging research. Still, only a few strains are well characterized with respect to aging. As canonical good hearing, healthy aging strains, CBA/CaJ and CBA/J mice are popular (Ohlemiller et al. 2010). These strains differ usefully in that only CBA/CaJs feature strial degeneration after about 1 year, while CBA/J mice show mostly OHC loss. The rates of hearing loss in DBA/2J, C57BL/6J (B6J), and BALB/cJ differ owing to different age-related hearing loss alleles plus some of the same mutations. All three strains carry *Cdh23*^{753A}, a mutation that destabilizes hair cell stereocilia, thereby promoting OHC death, progressive hearing loss, and noise-induced hearing loss (Johnson et al. 1997, 2000). The *Cdh23*^{753A} allele promotes sensory presbycusis-like pathology by exacerbating injury, which fits with a notion of aging as cumulative injury. BALB/cJ mice also add a component of strial degeneration after about 1 year (Ohlemiller et al. 2006) (Fig. 7.3). Surprisingly, B6J mice show very little EP reduction over a typical life span, while albino congenics to B6J tend toward EP decline (Ohlemiller et al. 2009) (Fig. 7.3). The albino B6J findings support the idea that melanin in the stria vascularis is important for preserving strial function with age.

Mouse strial presbycusis models have added key details to the conception of strial presbycusis. First, EP decline is highly variable: it does not appear in all mice, suggesting environmental or stochastic influences. As a corollary, EP decline is not omnipresent or inevitable, as also indicated by inbred strains showing no EP decline with age (CBA/J, B6J) (Fig. 7.3). Absence of EP decline in B6J mice further argues that moderate hair cell loss need not be associated with strial dysfunction. Second, EP decline occurs more readily in female mice, matching some suggestions for humans. Third, EP decline in females accelerates after menopause (or estropause in animals), again matching reports for humans. Fourth, the primary pathology impacts strial marginal cells, matching reports in gerbils and Schuknecht's characterization of human temporal bones. Finally, EP decline in mice corresponds with a shift to fewer, larger, strial capillaries. This may ultimately provide clues about underlying vascular processes.

B6J mice, often used as rapid aging models, may be problematic due to a null mutation of the *nnt* gene that may impart atypical stress responses (Ohlemiller et al. 2016). One alternative is to use C57BL/6N mice instead, which carry a functional *nnt* allele. These mice carry the same *Cdh23*^{753A} allele as B6J and show similar aging characteristics, although their vulnerability to noise is not yet well characterized.

Most mouse hearing-in-aging models resemble sensory presbycusis, wherein hair cell loss seems to drive hearing loss. The "cleanest" mouse strial presbycusis

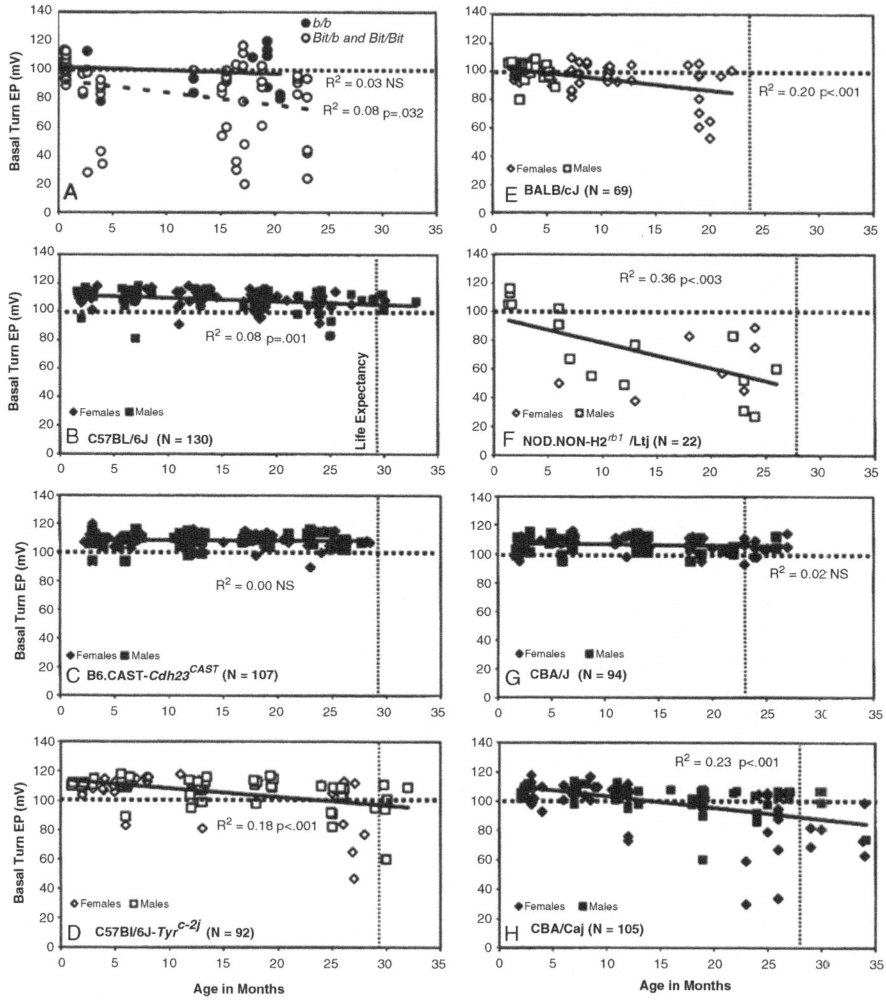

Fig. 7.3 (**a–h**) Scatter plots of age versus endocochlear potential (EP) in eight different mouse models (strain identified on plot). *Horizontal dotted line* in each graph indicates 100 mV for reference. *Vertical dotted lines* indicate life expectancy averaged for males and females of each strain. Genders (where known) are also indicated (females, *diamonds*; males, *squares*). All strains compared are inbreds or their congenics except for panel **a**. Carriers of the *Tyrp1*$^{B\text{-}lt}$ mutation in panel **a** (replotted from Cable et al. 1993) were generated on a mixed background and are compared with *Tyrp1*^{b}/*Tyrp1*^{b} littermates. R^2, correlation coefficient; NS, not significant. (Figure taken from Ohlemiller (2009) with permission)

model (defined by an initially normal EP that declines after 1 year, but only mild hair cell or neuronal loss) is presently CBA/CaJ. Human neural presbycusis, defined by primary neuronal loss with minimal hair cell loss, is poorly understood in terms of genetic versus environmental origins. There are, however, both mouse genetic models and noise-exposure models that mimic the key elements of this condition. Inactivating mutations of *NFκB*, *Bsn*, and *VgluT3* cause primary neural loss or dysfunction in mice (Khimich et al. 2005; Lang et al. 2006; Moser and Starr 2016). As referenced in Sect. 7.5.1, a recent finding that has generated much discussion is that of primary neural pathology after a single modest noise exposure, in fact, an exposure that causes no permanent threshold shift (Kujawa and Liberman 2006, 2009). The initial change after noise, first reported in CBA/CaJ mice, involves only loss of afferent synapses with eventual loss of neural cell bodies. If this finding translates to humans, it suggests that cochlear neurons may be at risk from noise exposures once thought harmless. Some amount of neural loss reliably accompanies human aging (about 10% per decade) (Makary et al. 2011), a loss assumed to be unavoidable. But if that loss is caused in part by continual modest noise exposure, it might be appropriate to rethink what exposures are advisable in entertainment and occupational settings. There are anatomical differences between human and essentially all animal cochlear neurons examined, and not all inbred mouse strains or species show the dramatic effect found in CBA/CaJ mice (Ohlemiller et al. 2016). To date, human studies do not show as clear an effect (e.g., Stamper and Johnson 2015), although these are limited to proxy physiological measures such as ABR Wave I amplitude. It remains to be seen how clinically relevant this finding in mice will be.

As mentioned in Sect.7.5.1, there is a convergence of concepts surrounding aging and injury. Accordingly, there are many genes that encode protective and repair factors that, when mutated, accelerate aging-like pathology (Bowl and Dawson 2015). These encode antioxidants, DNA repair enzymes, heat shock proteins, and a host of homeostatic factors. Many of these, when rendered inactive by genetic engineering, also accelerate age-associated hearing loss in mice. The cochlea appears to be particularly sensitive to perturbations of homeostatic machinery, perhaps because the organ of Corti directly receives mechanical input that may be inherently traumatic and also because of the metabolic demands on sensory cells (Schuknecht 1964; Schuknecht and Gacek 1993). Some mutations (e.g., those affecting mitochondria) drive a broader, accelerated aging phenotype so that many tissues are affected at the same time (Kujoth et al. 2005). This latter class of genes and alleles are of broad interest for aging research; however, they may be of less interest in understanding why some people lose more hearing with age than others with a similar health profile.

7.6.5.2 Mouse Genetic Deafness Models

Over 400 deafness genes have been identified in mice (Ohlemiller et al. 2016). Progress in mice has been more rapid than in other models due to the advantages offered by inbred strains. It is likely that most human deafness genes yet to be discovered will also show extensive overlap with known mouse genes (Steel 2014). Where there is no overlap, this may reflect the expression of different protein isoforms in humans and mice or expression of the same genes but in different cells. The cochlea contains unique cell types with unusual structures, such as stereocilia.

As pointed out in Sect. 7.2, mechanoreception constitutes an innovation with evolutionarily old roots. Key proteins evolved and were conserved, apparently with few compensatory genes that might alleviate the effects of mutations. This may be why so many deafness genes encode stereociliary proteins.

Steel and colleagues (Steel and Brown 1994; Steel 1995) assigned the effects of deafness genes in mice and humans to three general categories. *Neuroepithelial defects* are associated primarily with hair cell loss. The vast majority of deafness genes identified in mice and humans cause pathology of this type. *Cochleosaccular defects* primarily affect cochlear and saccular function, as the name suggests, through impaired strial function. The ~100 mV EP generated by the mouse cochlear stria vascularis also provides much of the electrochemical drive for currents through saccular hair cells, as the saccule possesses no epithelium for producing an endolymphatic potential on its own. In contrast, the semicircular canals do not require a large endolymphatic potential, and function normally with the typically <5 mV potential generated by dark cells adjacent to the cristae. The particular type of strial defect in cochleosaccular mutants is a lack of intermediate cells, which are a type of melanocyte that is critical for EP generation. These cells migrate into the cochlea during development. This migration can be disrupted by mutations, leaving some regions and tissues without melanocytes. For this reason, cochleosaccular defects may involve pigmentation anomalies of skin, hair, and eyes and are the type of defect underlying pigmentation-related hearing loss in many models considered in this chapter.

Mouse genes known to be associated with cochleosaccular hearing loss include *c-kit*, *Trp1*, and *Pax3*, which is also the gene affected in human Waardenburg syndrome type 1. For reasons that are unclear, hair cells also die in these models even though the mutations primarily impact the stria. Apparently there are limits to the independence of the fates of the stria and hair cells. Albinism, in which otherwise normal melanocytes produce no melanin, appears to be associated with only a modest cochlear phenotype, although this may depend on genetic background. This is not true for the visual system, where melanin plays a more important role during development.

Morphogenetic defects, Steel's third major type of genetic hearing loss, alter the normal developmental layout of the inner ear and can be associated with significant alterations of the sensory epithelia or fluid spaces. In extreme cases, the normal boney spaces may be absent along with the absence of entire sensory epithelia. Such cases in humans are incompatible with cochlear implantation. Since a thorough treatment of mutant mouse hearing models is beyond the scope of this chapter, the reader is directed to comprehensive tables by Ohlemiller and colleagues (2016).

7.6.6 Rats

Much of what was stated in Sect. 7.5 regarding mice in hearing research also applies to rats. Only rats approach laboratory mice in the number of inbred and outbred varieties. While there are many fewer knockout and transgenic rat models than exist for mice, inbred rats have been more heavily used for studies of auditory central nervous system function and for many studies of cochlear biochemistry, noise injury, and pharmacotherapies. Genetic engineering has been revolutionized by efficient and relatively cheap methods such as CRISPR/Cas9 (Zou et al. 2015) that will likely lead to a wave of engineered mutations in many species. This line of experimentation will still favor inbred models because only in inbreds will subsequent breeding yield genetically uniform animals. In some cases, however, the larger size of rat cochleas, middle ears, and brains may make them the preferred model. One weakness of mouse models for cochlear injury and remediation studies is that noise-induced hearing loss in mice is not closely linked to hair cell loss. Instead, mice are more likely to show spatial alterations within the organ of Corti (Wang et al. 2002). Rats, by contrast, appear more like chinchillas, guinea pigs, and potentially humans in the amount of hearing loss that can be explained by hair cell loss (Ohlemiller 2012).

Rats and mice have similar life spans (2–3 years). Multiple inbred rat strains are used as models of aging, but in no case are the causative genes currently known. These models remain useful for their general similarity to human presbycusis and for mechanistic and therapeutic studies. The most-studied rat presbycusis models are Fischer 344, Wistar, and Sprague Dawley. All are albino, which could influence how age-associated cochlear pathologies are manifested. Descriptions of these models in aging emphasize hair cell loss and are consistent with sensory presbycusis, although moderate pathology of neurons, stria vascularis, and spiral ligament have been noted (Chen et al. 2009). No rat genetic model of primary neural presbycusis has been reported. The phenotype of Fischer 344 rats may depend on the substrain. Fischer 344/NHsd rats show mostly OHC pathology and loss (Bielefeld et al. 2008), while Fischer 344/DuCrl rats may include components of middle ear changes and strial presbycusis, although no EP data have been reported (Buckiova et al. 2007). The latter may be related to collagen deposition in the spiral ligament. Wistar rat studies have emphasized central auditory changes with age. The most typically used rat model with good hearing is the outbred Long-Evans hooded (pigmented) rat.

7.7 Summary

This chapter assembles information about genetic aspects of hearing and gene/environment interactions for the six rodent species most extensively covered in the literature. Each model species is inherently valuable for what can be learned about how its inner ear is adapted to survival needs. One caveat is that some models (guinea pigs and chinchillas) may be too genetically different from wild ancestors to be useful for neuroethological research. Inbred rats and mice likewise may represent unnatural populations for study in some neuroethological contexts. Nevertheless, all six species reasonably model human inner ear function and dysfunction, although detailed information is very unevenly distributed. Except for inbred rats and mice, the models are outbred, and most inbred lines are not commercially available. Most of the species include models that vary in coat or eye color, including albinism and pigmentation defects. Melanin and melanocytes are widely present in the cochlea, especially in the lateral wall, so that these models are relevant to the genetics of hearing. Pigmentation mutations have appeared spontaneously on outbred lines that are available only from pet stores or academic sources. These lines could be used to study albinism or cochleosaccular mutations, although only in congenic lines can such mutations be studied without the potentially confounding effects of other alleles. Interesting hearing phenotypes appearing in outbred lines of any species could reflect the effects of multiple genes and could prove difficult to study.

Rat and mouse models are available with known deafness mutations that appeared spontaneously or were created using mutagens or genetic engineering. In many cases, these correspond to known human deafness genes, and mutations of these produces human-like cochlear phenotypes. Although *circling* phenotypes have been identified in mutant lines of guinea pigs, the causative genes remain unknown. With time, the genomes of most mammalian species will be sequenced. However, without multiple genetically standardized subpopulations, it will be difficult to use these models to identify deafness genes.

Compliance with Ethics Requirements Kevin Ohlemiller declares that he has no conflict of interest.

References

Ahmad, M., Bohne, B. A., & Harding, G. W. (2003). An in vivo tracer study of noise-induced damage to the reticular lamina. *Hearing Research*, 175, 82–100.

Alizadeh, A., Hong, L. Z., Kaelin, C. B., Raudsepp, T., et al. (2009). Genetics of sex-linked yellow in the Syrian hamster. *Genetics*, 181, 1427–1436.

Allen, P. D., & Eddins, D. A. (2010). Presbycusis phenotypes form a heterogeneous continuum when ordered by degree and configuration of hearing loss. *Hearing Research*, 264, 10–20.

Amedofu, G. K., Gopal, K. V., Asher, Jr., J. H., Ahmadizadeh, M., & Moore, E. J. (1999). Auditory brainstem responses in golden Syrian hamsters (*Mesocricetus auratus*) affected with the Wh gene. *Comparative Medicine*, 49, 173–178.

Bielefeld, E., Coling, D. E., Chen, G.-D., Li, M., et al. (2008). Age-related hearing loss in the Fischer 344/NHsd rat substrain. *Hearing Research*, 241, 26–33.

Bohne, B. A. (1972). Location of small cochlear lesions by phase contrast microscopy prior to thin sectioning. *The Laryngoscope*, 82, 1–16.

Bohne, B. A., Gruner, M. M., & Harding, G. W. (1990). Morphological correlates of aging in the chinchilla cochlea. *Hearing Research*, 48, 79–91.

Bourien, J., Tang, Y., Batrel, C., Huet, A., et al. (2014). Contribution of auditory nerve fibers to compound action potential of the auditory nerve. *Journal of Neurophysiology*, 112, 1025–1039.

Bowl, M. R., & Dawson, S. J. (2015). The mouse as a model for age-related hearing loss—a mini-review. *Gerontology*, 61, 149–157.

Buckiova, D., Popelar, J., & Syka, J. (2007). Aging cochleas in the F344 rat: Morphological and functional changes. *Experimental Gerontology*, 42, 629–638.

Cable, J., Jackson, I. J., & Steel, K. P. (1993). Light (B^{lt}), a mutation that causes melanocyte death, affects stria vascularis function in the mouse inner ear. *Pigment Cell Research*, 6, 215–225.

Chen, G. D., Li, M., Tanaka, C., Bielefeld, E. C., et al. (2009). Aging outer hair cells (OHCs) in the Fischer 344 rat cochlea: Function and morphology. *Hearing Research*, 248, 39–47.
Clark, W. W., Clark, C. S., Moody, D. B., & Stebbins, W. C. (1974). Noise-induced hearing loss in the chinchilla as determined by a positive-reinforcement technique. *The Journal of the Acoustical Society of America*, 56, 1202–1209.
Clark, W. W., Bohne, B. A., & Boettcher, F. A. (1987). Effect of periodic rest on hearing loss and cochlear damage following exposure to noise. *The Journal of the Acoustical Society of America*, 82, 1253–1263.
Coleman, J. W. (1976). Hair cell loss as a function of age in the normal cochlea of the guinea pig. *Acta Otolaryngologica*, 82, 33–40.
Conlee, J. W., Abdul-Baqi, K. J., McCandless, G. A., & Creel, D. J. (1986). Differential susceptibility to noise-induced permanent threshold shift between albino and pigmented guinea pigs. *Hearing Research*, 23, 81–91.
Conlee, J. W., Abdul-Baqi, K. J., McCandless, G. A., & Creel, D. J. (1988). Effects of aging on normal hearing loss and noise-induced threshold shift in albino and pigmented guinea pigs. *Acta Otolaryngologica*, 106, 64–70.
Cooper, N. P., & Yates, G. K. (1994). Nonlinear input-output functions derived from the responses of guinea-pig cochlear nerve fibres: Variations with characteristic frequency. *Hearing Research*, 78, 221–234.
Covell, W. P., & Rogers, J. B. (1957). Pathologic changes in the inner ear of senile guinea pigs. *Laryngoscope*, 67, 118–129.
Dallos, P., & Cheatham, M. A. (1992). Cochlear hair cell function reflected in intracellular recordings in vivo. *Society of General Physiologists,* Series 47, 371–393.
Dawes, P., & Payton, A. (2016). Genetics of age-related hearing loss. In B. Vona & T. Haaf (Eds.), *Genetics of deafness* (pp. 84–96). New York: Karger.
Eggermont, J. J., & Roberts, L. E. (2015). Tinnitus: Animal models and findings in humans. *Cell and Tissue Research*, 36, 311–336.
Ernstson, S. (1971). Cochlear morphology in a strain of the waltzing guinea pig. *Acta Otolaryngologica*, 71, 469–482.
Faddis, B. T., & McGinn, M. D. (1997). Spongiform degeneration of the gerbil cochlear nucleus: An ultrastructural and immunohistochemical evaluation. *Journal of Neurocytology*, 26, 625–635.
Festing, M. F. (1976). The guinea pig. In *Universities Federation for Animal Welfare Handbook on the Care and Management of Laboratory Animals* (pp. 229–247). Edinburgh: Churchill Livingstone.
Festing, M. F. (2010). Inbred strains should replace outbred stocks in toxicology, safety testing, and drug development. *Toxicological Pathology*, 38, 681–690.
Fournier, P., & Hébert, S. (2013). Gap detection deficits in humans with tinnitus as assessed with the acoustic startle paradigm: Does tinnitus fill in the gap? *Hearing Research*, 295, 16–23.
Fransen, E., Lemkens, N., Van Laer, L., & Van Camp, G. (2003). Age-related hearing impairment (ARHI): Environmental risk factors and genetic prospects. *Experimental Gerontology*, 38, 353–359.
Fransen, E., Bonneux, S., Corneveaux, J. J., Schrauwen, I., et al. (2015). Genome-wide association analysis demonstrates the highly polygenic character of age-related hearing impairment. *European Journal of Human Genetics*, 23, 110–115.
French, A. S. (1992). Mechanotransduction. *Annual Review of Physiology*, 54, 135–152.
Furman, A. C., Kujawa, S. G., & Liberman, M. C. (2013). Noise-induced cochlear neuropathy is selective for fibers with low spontaneous rates. *Journal of Neurophysiology*, 110, 577–586.
Galazyuk, A., & Hébert, S. (2015). Gap-prepulse inhibition of the acoustic startle reflex (GPIAS) for tinnitus assessment: Current status and future directions. *Frontiers in Neurology*, *6*. https://doi.org/10.3389/fneur.2015.00088
Gratton, M. A., & Schulte, B. A. (1995). Alterations in microvasculature are associated with atrophy of the stria vascularis in quiet-aged gerbils. *Hearing Research*, 82, 44–52.

Hamernik, R. P., Patterson, J. H., Turrentine, G. A., & Ahroon, W. A. (1989). The quantitative relation between sensory cell loss and hearing thresholds. *Hearing Research*, 38, 199–212.
Heffner, H., & Masterton, B. (1980). Hearing in glires: Domestic rabbit, cotton rat, feral house mouse, and kangaroo rat. *The Journal of the Acoustical Society of America*, 68, 1584–1599.
Hirose, K., & Liberman, M. C. (2003). Lateral wall histopathology and endocochlear potential in the noise-damaged mouse cochlea. *Journal of the Association for Research in Otolaryngology*, 4, 339–352.
Jagger, D. J., Nevill, G., & Forge, A. (2010). The membrane properties of cochlear root cells are consistent with roles in potassium recirculation and spatial buffering. *Journal of the Association for Research in Otolaryngology*, 11, 435–448.
Jin, Z., Mannström, P., Skjönsberg, Å., Järlebark, L., & Ulfendahl, M. (2006). Auditory function and cochlear morphology in the German waltzing guinea pig. *Hearing Research*, 219, 74–84.
Jin, Z., Mannström, P., Järlebark, L., & Ulfendahl, M. (2007). Malformation of stria vascularis in the developing inner ear of the German waltzing guinea pig. *Cell and Tissue Research*, 328, 257–270.
Johnson, K. R., Erway, L. C., Cook, S. A., Willott, J. F., & Zheng, Q. Y. (1997). A major gene affecting age-related hearing loss in C57BL/6J mice. *Hearing Research*, 114, 83–92.
Johnson, K. R., Zheng, Q. Y., & Erway, L. C. (2000). A major gene affecting age-related hearing loss is common to at least 10 inbred strains of mice. *Genomics*, 70, 171–180.
Kawaguchi, S., Hultcrantz, M., Jin, Z., Ulfendahl, M., & Suzuki, M. (2010). Vestibular morphology in the German waltzing guinea pig. *Journal of Otolaryngology--Head & Neck Surgery*, 39(2), 115–121.
Ketten, D. R. (1992). The marine mammal ear: Specializations for aquatic audition and echolocation. In D. B. Webster, R. R. Fay, & A. N. Popper (Eds.), *The evolutionary biology of hearing* (pp. 717–750). New York: Springer-Verlag.
Khimich, D., Nouvian, R., Pujol, R., tom Dieck, S., et al. (2005). Hair cell synaptic ribbons are essential for synchronous auditory signalling. *Nature*, 434, 889–894.
Kim, H. J., & Chole, R. A. (1998). Experimental models of aural cholesteatomas in Mongolian gerbils. *Annals of Otology, Rhinology & Laryngology*, 107, 129–134.
Kobayashi, T., Aslan, A., Chiba, T., Takasaka, T., & Sanna, M. (1996). Measurement of endocochlear DC potentials in ears with acoustic neuromas: A preliminary report. *Acta Oto-Laryngologica*, 116, 791–795.
Kujawa, S. G., & Liberman, M. C. (2006). Acceleration of age-related hearing loss by early noise: Evidence of a misspent youth. *The Journal of Neuroscience*, 26, 2115–2123.
Kujawa, S. G., & Liberman, M. C. (2009). Adding insult to injury: Cochlear nerve degeneration after 'temporary' noise-induced hearing loss. *The Journal of Neuroscience*, 29, 14077–14085.
Kujoth, G. C., Hiona, A., Pugh, T. D., Someya, S., et al. (2005). Mitochondrial DNA mutations, oxidative stress, and apoptosis in mammalian aging. *Science*, 309, 481–484.
Lang, H., Schulte, B. A., Zhou, D., Smythe, N., et al. (2006). Nuclear factor κB deficiency is associated with auditory nerve degeneration and increased noise-induced hearing loss. *The Journal of Neuroscience*, 26, 3541–3550.
Lee, R. J., Hong, J. S., McGinty, J. F., & Lomax, P. (1987). Increased enkephalin and dynorphin immunoreactivity in the hippocampus of seizure sensitive Mongolian gerbils. *Brain Research*, 401, 353–358.
Lobarinas, E., Salvi, R., & Ding, D. (2013a). Insensitivity of the audiogram to carboplatin induced inner hair cell loss in chinchillas. *Hearing Research*, 302, 113–120.
Lobarinas, E., Hayes, S. H., & Allman, B. L. (2013b). The gap-startle paradigm for tinnitus screening in animal models: Limitations and optimization. *Hearing Research*, 295, 150–160.
Long, G. R., & Clark, W. W. (1984). Detection of frequency and rate modulations by the chinchilla. *The Journal of the Acoustical Society of America*, 75, 1184–1190.
Longenecker, R. J., & Galazyuk, A. V. (2011). Development of tinnitus in CBA/CaJ mice following sound exposure. *Journal of the Association for Research in Otolaryngology*, 12, 647–658.

Makary, C. A., Shin, S., Kujawa, S. G., Liberman, M. C., & Merchant, S. N. (2011). Age-related primary cochear neuronal degeneration in human temporal bones. *Journal of the Association for Research in Otolaryngology*, 12, 711–717.
Masterton, B., Heffner, H., & Ravizza, R. (1969). The evolution of human hearing. *The Journal of the Acoustical Society of America*, 45, 966–985.
McFadden, S. L., Quaranta, N., & Henderson, D. (1997). Suprathreshold measures of auditory function in the aging chinchilla. *Hearing Research*, 111, 127–135.
Miller, J. D. (1970). Audibility curve of the chinchilla. *The Journal of the Acoustical Society of America*, *48*, 513–523.
Moser, T., & Starr, A. (2016). Auditory neuropathy—neural and synaptic mechanisms. *Nature Reviews Neurology*, 12, 135–149.
Naito, R., Murofushi, T., Mizutani, M., & Kaga, K. (1999). Auditory brainstem responses, electrocochleograms, and cochlear microphonics in the myelin deficient mutant hamster 'bt'. *Hearing Research*, 136, 44–48.
Ohlemiller, K. K. (2006). Contributions of mouse models to understanding of age- and noise-related hearing loss. *Brain Research*, 1091, 89–102.
Ohlemiller, K. K. (2009). Mechanisms and genes in human strial presbycusis from animal models. *Brain Research*, 1277, 70–83.
Ohlemiller, K. K. (2012). Current issues in noise exposure. In K. L. Tremblay & R. F. Burkard (Eds.), *Translational perspectives in auditory neuroscience*. Vol. 3 (pp. 1–34). San Diego: Plural Publishing.
Ohlemiller, K. K. (2015). A question of balance: Free radicals in inner ear homeostasis. In J. Miller & C. G. Le Prell (Eds.), *Free radicals in ENT medicine* (pp. 21–55). New York: Springer Science+Business Media.
Ohlemiller, K. K., & Frisina, R. D. (2008). Age-related hearing loss and its cellular and molecular bases. In J. Schacht, A. N. Popper, & R. R. Fay (Eds.), *Auditory trauma, protection, and repair* (pp. 145–194). New York: Springer Science+Business Media.
Ohlemiller, K. K., & Siegel, J. H. (1992). Effects of cooling on gross cochlear potentials in the gerbil: Basal and apical differences. *Hearing Research*, 63, 79–89.
Ohlemiller, K. K., & Siegel, J. H. (1994). Cochlear basal and apical differences reflected in the effects of cooling on responses of single auditory nerve fibers. *Hearing Research*, 80, 174–190.
Ohlemiller, K. K., Lett, J. M., & Gagnon, P. M. (2006). Cellular correlates of age-related endocochlear potential reduction in a mouse model. *Hearing Research*, 220, 10–26.
Ohlemiller, K. K., Rice, M. R., Lett, J. M., & Gagnon, P. M. (2009). Absence of strial melanin coincides with age-associated marginal cell loss and endocochlear potential decline. *Hearing Research*, 249, 1–14.
Ohlemiller, K. K., Dahl, A. R., & Gagnon, P. M. (2010). Divergent aging characteristics in CBA/J and CBA/CaJ mouse cochleae. *Journal of the Association for Research in Otolaryngology*, 11, 605–623.
Ohlemiller, K. K., Rosen, A. R., Rellinger, E. A., Montgomery, S.C, & Gagnon, P. M. (2011). Different cellular and genetic basis of noise-related endocochlear potential reduction in CBA/J and BALB/cJ mice. *Journal of the Association for Research in Otolaryngology*, 12, 45–58.
Ohlemiller, K. K., Jones, S. M., & Johnson, K. R. (2016). Application of mouse models to research in hearing and balance. *Journal of the Association for Research in Otolaryngology*, 17, 1–31.
Parker, C. C., & Palmer, A. A. (2011). Dark matter: Are mice the solution to missing heritability? *Froniters in Genetics*, 2, 32. https://doi.org/10.3389/fgene.2011.00032
Patuzzi, R. (2011). Ion flow in cochlear hair cells and the regulation of hearing sensitivity. *Hearing Research*, 280, 3–20.
Petkov, P. M., Ding, Y., Cassell, M. A., Zhang, W., et al. (2004). An efficient SNP system for mouse genome scanning and elucidating strain relationships. *Genome Research*, 14, 1806–1811.
Pollak, G. D. (1992). Adaptations of basic structures and mechanisms in the cochlea and central auditory pathway of the mustache bat. In D. B. Webster, R. R. Fay, & A. N. Popper (Eds.), *The evolutionary biology of hearing* (pp. 751–778). New York: Springer-Verlag.

Pritt, S., Hargaden, M., Singer, L., Smith, G. D., et al. (2012). *The laboratory rabbit, guinea pig, hamster, and other rodents*. San Diego: Academic Press.
Prosen, C. A., Moody, D. B., Stebbins, W. C., Smith, D. W., et al. (1990). Apical hair cells and hearing. *Hearing Research*, 44, 179–193.
Provenzano, M. J., & Domann, F. E. (2007). A role for epigenetics in hearing: Establishment and maintenance of auditory specific gene expression patterns. *Hearing Research*, 233, 1–13.
Ravicz, M. E., & Rosowski, J. J. (1997). Sound-power collection by the auditory periphery of the Mongolian gerbil *Meriones unguiculatus*: III. Effect of variations in middle-ear volume. *The Journal of the Acoustical Society of America*, 101, 2135–2147.
Rosowski, J. J. (1992). Hearing in transitional mammals: Predictions from the middle-ear anatomy and hearing capabilities of extant mammals. In D. B. Webster, R. R. Fay, & A. N. Popper (Eds.), *The evolutionary biology of hearing* (pp. 615–631). New York: Springer-Verlag
Ruggero, M. A., & Rich, N. C. (1983). Chinchilla auditory-nerve responses to low-frequency tones. *The Journal of the Acoustical Society of America*, 73, 2096–2108.
Russell, I. J., & Sellick, P. M. (1978). Intracellular studies of hair cells in the mammalian cochlea. *Journal of Physiology*, 284, 261–289.
Ryan, A. (1976). Hearing sensitivity of the mongolian gerbil, *Meriones unguiculatis*. *The Journal of the Acoustical Society of America*, 59, 1222–1226.
Sánchez-Benito, D., Gómez-Nieto, R., Hernández-Noriega, S., Murashima, A. A. B., et al. (2017). Morphofunctional alterations in the olivocochlear efferent system of the genetic audiogenic seizure-prone hamster GASH: Sal. *Epilepsy & Behavior*, 71, 193–206.
Schmiedt, R. A. (2010). The physiology of cochlear presbycusis. In S. Gordon-Salant, R. D. Frisina, A. N. Popper, & R. R. Fay (Eds.), *The aging auditory system* (pp. 9–38). New York: Springer Science+Business Media.
Schmiedt, R. A., & Zwislocki, J. J. (1977). Comparison of sound-transmission and cochlear-microphonic characteristics in Mongolian gerbil and guinea pig. *The Journal of the Acoustical Society of America*, 61, 133–149.
Schmiedt, R. A., Zwislocki, J. J., & Hamernik, R. P. (1980). Effects of hair cell lesions on responses of cochlear nerve fibers. I. Lesions, tuning curves, two-tone inhibition, and responses to trapezoidal wave patterns. *Journal of Neurophysiology*, 43, 1367–1389.
Schuknecht, H. F. (1964). Further observations on the pathology of presbycusis. *Archives of Otolaryngology*, 80, 369–382.
Schuknecht, H. F. (1993). *Pathology of the ear*. 2nd ed. Philadelphia: Lea and Febiger.
Schuknecht, H. F., & Gacek, M. R. (1993). Cochlear pathology in presbycusis. *Annals of Otology, Rhinology and Laryngology*, 102, 1–16.
Schulte, B. A., & Schmiedt, R. A. (1992). Lateral wall Na, K-ATPase and endodochlear potentials decline with age in quiet-reared gerbils. *Hearing Research*, 61, 35–46.
Senthilan, P. R., Piepenbrock, D., Ovezmyradov, G., Nadrowski, B., et al. (2012). Drosophila auditory organ genes and genetic hearing defects. *Cell*, 150, 1042–1054.
Sewell, W. (1984). The effects of furosemide on the endocochlear potential and auditory nerve fiber tuning curves in cats. *Hearing Research*, 14, 305–314.
Shimoyama, M., Smith, J. R., De Pons, J., Tutaj, M., et al. (2016). The chinchilla research resource database: Resource for an otolaryngology disease model. *Database: The Journal of Biological Databases and Curation*, 2016. https://academic.oup.com/database/article/doi/10.1093/database/baw073/2630430
Silver, L. M. (1995). *Mouse genetics*. Oxford, UK: Oxford Press.
Skjönsberg, Å., & Mannström, P. (2015). Reduced noise susceptibility in littermate offspring from heterozygous animals of the German waltzing guinea pig. *Neuroreport*, 26, 593–597.
Skjönsberg, Å., Duan, M., Johnson, A. C., & Ulfendahl, M. (2014). Effect of auditory stress agents on heterozygous German waltzing guinea pigs. *Journal of Otology*, 9, 179–190.
Slepecky, N. B. (1996). Structure of the mammalian cochlea. In P. Dallos, A. N. Popper, & R. D. Fay (Eds.), *The cochlea* (pp. 44–129). New York: Springer-Verlag.

Spicer, S. S., & Schulte, B. A. (2005). Pathologic changes of presbycusis begin in secondary processes and spread to primary processes of strial marginal cells. *Hearing Research*, 205, 225–240.

Stamper, G. C., & Johnson, T. A. (2015). Auditory function in normal-hearing, noise-exposed human ears. *Ear and Hearing*, 36, 172.

Steel, K. P. (1995). Inherited hearing defects in mice. *Annual Review of Genetics*, 29, 675–701.

Steel, K. P. (2014). What's the use of genetics? *Perspectives on Auditory Research* (pp. 569–584). New York: Springer Science+Business Media.

Steel, K. P., & Brown, S. D. M. (1994). Genes and deafness. *Trends in Genetics*, 10, 428–435.

Taberner, A. M., & Liberman, M. C. (2005). Response properties of single auditory nerve fibers in the mouse. *Journal of Neurophysiology*, 93, 557–569.

Tarnowski, B. I., Schmiedt, R. A., Hellstrom, L. I., Lee, F. S., & Adams, J. C. (1991). Age-related changes in cochleas of mongolian gerbils. *Hearing Research*, 54, 123–134.

Tran Ba Huy, P., Ferrary, E., & Roinel, N. (1989). Electrochemical and clinical observations in 11 cases of Meniere's disease. In J. B. Nadol (Ed.), *Meniere's disease* (pp. 241–246). Amsterdam: Kugler and Ghedini.

Turner, J., Larsen, D., Hughes, L., Moechars, D., & Shore, S. (2012). Time course of tinnitus development following noise exposure in mice. *Journal of Neuroscience Research*, 90, 1480–1488.

Ukaji, T., Iwasa, M. A., & Kai, O. (2016). Tyrosinase (Tyr) gene mutation in albino Mongolian gerbil (*Meriones unguiculatus*). *Open Journal of Animal Sciences*, 6, 259. https://doi.org/10.4236/ojas.2016.64031

Wang, Y., Hirose, K., & Liberman, M. C. (2002). Dynamics of noise-induced cellular injury and repair in the mouse cochlea. *Journal of the Association for Research in Otolaryngology*, 3, 248–268.

Whitfield, T. T. (2002). Zebrafish as a model for hearing and deafness. *Journal of Neurobiology*, 53, 157–171.

Willott, J. F. (1983). Introduction: *Mus musculus*. In J. F. Williot (Ed.), *Auditory psychobiology of the mouse* (pp. 3–12). Springfield, IL: Charles C. Thomas.

Willott, J. F. (Ed.). (2001). *Handbook of mouse auditory research: From behavior to molecular biology*. New York, NY: CRC Press.

Winter, I. M., Robertson, D., & Yates, G. K. (1990). Diversity of characteristic frequency rate intensity functions in guinea pig auditory nerve fibers. *Hearing Research*, 45, 191–202.

Yalcin, B., & Flint, J. (2012). Association studies in outbred mice in a new era of full-genome sequencing. *Mammalian Genome*, 23, 719–726.

Zou, B., Mittal, R., Grati, M. H., Lu, Z., et al. (2015). The application of genome editing in studying hearing loss. *Hearing Research*, 327, 102–108.

Chapter 8
State and Context in Vocal Communication of Rodents

Laura M. Hurley and Matina C. Kalcounis-Rueppell

Abstract Context adds meaning to vocal communication. For an individual that is encountering conspecifics, sensitivity to factors such as the identity of a social partner and presence of eavesdroppers (the external scene) or reproductive state and experience (the internal scene) may be essential for effective communication. Although the external and internal scenes are often categorized as separate, they are functionally entangled because they often interact to influence communication and they converge at the physiological level. The external and internal scenes have well-documented effects on both vocal production by signalers and the responses to vocal signals by receivers. This commonality supports a view of context as an emergent phenomenon with individuals acting as both senders and receivers during a communication event and contributing to the tone of the interaction via feedback to social partners. Amid this complexity, an operational view defines context as the set of factors that influences communication within a given interaction. This definition can be used as a simplifying tool for exploring both the functions and mechanisms of context sensitivity. Since both the external and internal scenes affect physiological systems involved in the internal representations of qualities such as stress, reward, and positive or negative valence, an integrated concept of context also unites ecological and biomedical perspectives.

Keywords Abiotic · Behavioral context · Biotic · Environment · Individual experience · Internal state · Signal receiver · Signal sender · Social physiology · Ultrasonic · Vocalization

---

L. M. Hurley (✉)
Department of Biology, Indiana University, Bloomington, IN, USA
e-mail: lhurley@indiana.edu

M. C. Kalcounis-Rueppell
Biology Department, University of North Carolina at Greensboro, Greensboro, NC, USA
e-mail: mckalcou@uncg.edu

M. L. Dent et al. (eds.), *Rodent Bioacoustics*, Springer Handbook of Auditory Research 67, https://doi.org/10.1007/978-3-319-92495-3_8

8.1 A Broad Definition of Context

To paraphrase evolutionary biologist Theodosius Dobzhansky (1973), nothing in communication makes sense except in the light of context. Understanding the function of a vocal signal requires knowing the characteristics of the signaler and the receiver and the circumstances surrounding them. For an individual participating in a communication event, for example, a nocturnal "mouse" (a descriptive term applied to a wide range of rodent species), a wealth of biotic and abiotic factors could be relevant. The mouse may exist in a complex social environment, possessing a home range that overlaps with home ranges of female or male conspecifics or heterospecifics. The individual could be in a reproductive phase that may include caring for juveniles in a nest or interacting with sub-adults within its home range. The mouse may be in marginal habitat with very little cover from predators or harsh weather, or the mouse may be in high-quality habitat protected by a dense understory of vegetation, or in a solid nest that would be difficult for a predator to access. The developmental history or adult experience of the mouse, including whether it has completed a recent period of plentiful or scarce resources, may influence current interactions. Finally, the mouse may interact with other individuals possessing their own distinctive combinations of these same factors, which provide social feedback that shapes the course and outcome of a communicative event. Thus, when considering how vocal signals (as behavioral traits) have evolved and are maintained and mediated, it is important to consider context from the perspectives of both signalers and receivers.

For the purposes of this chapter, *context* will be defined very broadly as a series of factors that describe the communicative environment, reflecting previous ethological approaches (e.g., Smith 1965). These factors can be separated into the *external scene* and the *internal scene*. The external scene reflects factors originating outside of an individual (e.g., abiotic conditions or the identity of a partner). It incorporates the concept of an "auditory scene" (Bregman 1990), but also extends to cues from other sensory modalities, including chemical or tactile cues. The internal scene includes factors such as self-identity and internal state. The external and internal scenes can be subdivided further into: (1) the nature of external events, such as the presence of conspecific individuals or predators; (2) the identities of social partners participating in a communicative exchange, including their sex, age, or genetic makeup; (3) the current state of the individuals, including reproductive, nutritional, or affective state; (4) the social history of the individuals, including juvenile and adult experiences, and familiarity with social partners or past roles in social hierarchies (Fig. 8.1).

Many of these factors are interrelated; for example, history with an event often influences acute responses during a similar event (Maggio et al. 1983). Diverse contextual factors may also converge on the same physiological pathways that represent the salience of events, including neurochemical signals (Hall et al. 2011; Hanson and Hurley 2014) and neural spike trains (Lin et al. 2013). It is the interdependence

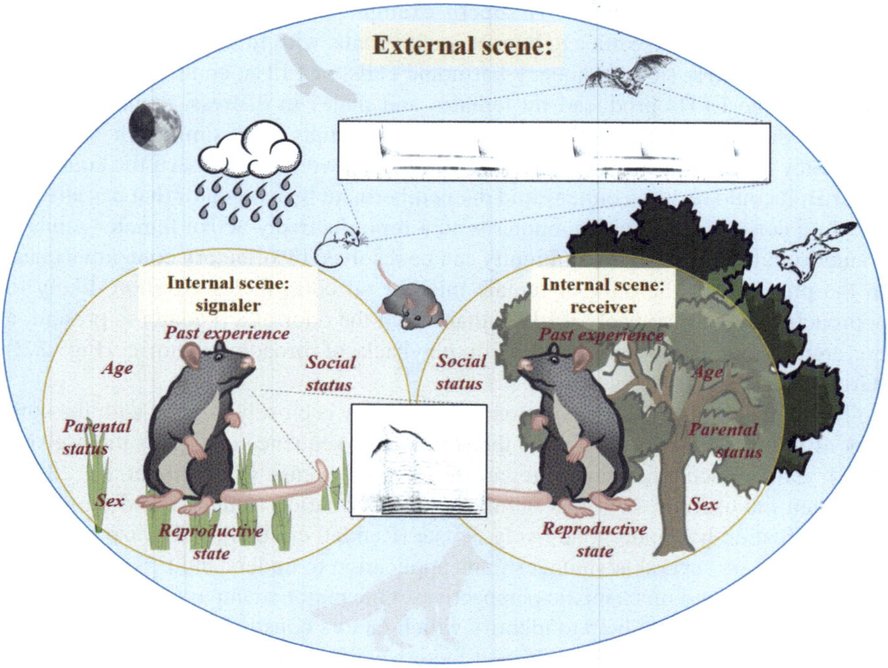

Fig. 8.1 Considerations of context in the study of rodent vocalizations. The internal scene can have effects on both the production and perception of a vocal signal. The internal scene can be mediated by age, sex, developmental history, experience, reproductive state, social status and parental status, as well as other physiological properties of the individual. The external scene is relevant to both the signaler and the receiver and includes both abiotic and biotic factors. Particular habitat types and weather conditions/moon phase not only influence how the signal is propagated in the environment but also how risk-averse the senders or receivers are (e.g., perceived risk). Social properties, like the presence of pups, other family members (e.g., juveniles that have not dispersed), and conspecific audiences, also influence signal production and reception. The presence of heterospecifics, including predators, may shape signals and responses. Importantly, conspecifics and heterospecifics also communicate, causing the soundscape to reflect ultrasonic vocalizations (*USVs*) from other rodents, including mice, rats, and squirrels, and echolocation signals and social calls from flying or roosting bats

of these different factors that makes it conceptually coherent to include them all within a broad framework of context.

8.1.1 Context as Information

Sensitivity to contextual factors can inform decisions about communication. Given a range of potential signals that can be produced at a given time, or a range of potential responses to a given signal, context can reduce the ambiguity about the most appropriate behavioral choice. A superb example of this phenomenon is in the responses of male house mice to human-audible calls with fundamental frequencies well below 20 kHz (low-frequency harmonic calls, or LFHs, commonly known as squeaks). The LFHs produced by females and males in distress, or produced by females during rejection of male courtship attempts, are similar in structure (Grimsley et al. 2013). Therefore, if a male mouse were to approach the source of an LFH, it could result in either rapid dismemberment by a predator that has already caught a conspecific or an encounter with a reproductively active female—quite a conundrum for a male! This ambiguity can be resolved by olfactory contextual cues. In the presence of the odor of female mice or no odor, males are more likely to approach during a playback of LFHs than when the odor of a predator is present, a preference that is not observed with playback of broadband noise (Fig. 8.2) (Grimsley et al. 2013).

The idea of context as a resolution for ambiguity is closely aligned with the concept of information represented by the structure or sequence of signals themselves, which can reduce uncertainty regarding related events (Seyfarth et al. 2010). Although the usefulness of this definition of information to understanding animal communication has undergone lively debate (Rendall et al. 2009; Carazo and Font 2010), there are several advantages in its application to understanding context from both functional and mechanistic perspectives. One major advantage is that this concept of information can help to identify which factors constitute context for a given situation. Thus, contextually important factors can be defined as any set of circumstances that reduce ambiguity about the appropriate signal or response in each social interaction. Like the definition of "functional information" in signal structures described by Carazo and Font (2010), this view incorporates the advantages that sensitivity to context confers on the signaler or receiver.

In contrast to the definition of Carazo and Font, the idea of context broadens the scope of relevant information to factors outside of specific communication signals in time, space, or modality. For the previous example of male mice listening to LFHs, it would be advantageous to reduce the chance of encountering a predator of known location as signaled by the distress of a conspecific or, alternatively, to increase the chances of encountering a reproductively available female as signaled by olfactory cues of a female's identity and estrous state (Hayashi and Kimura 1974). In this vein, hypotheses on the adaptiveness of context sensitivity would predict that individuals are most likely to be influenced by the specific features of the internal and external scenes that favor survival or reproduction but be unresponsive to features that do not.

Finally, the concept of context as information can extend to mechanisms of context-dependent responses, as for the information content of neural spike trains. For the previous example of a male mouse listening to LFHs, the firing patterns of neurons in the basolateral amygdala in response to LFHs are different when no odor is present versus when predator odor is present, but the same distinction does not occur for broadband noise (Fig 8.2b) (Grimsley et al. 2013). In this situation, neural firing patterns are influenced by behaviorally relevant combinations of contextual

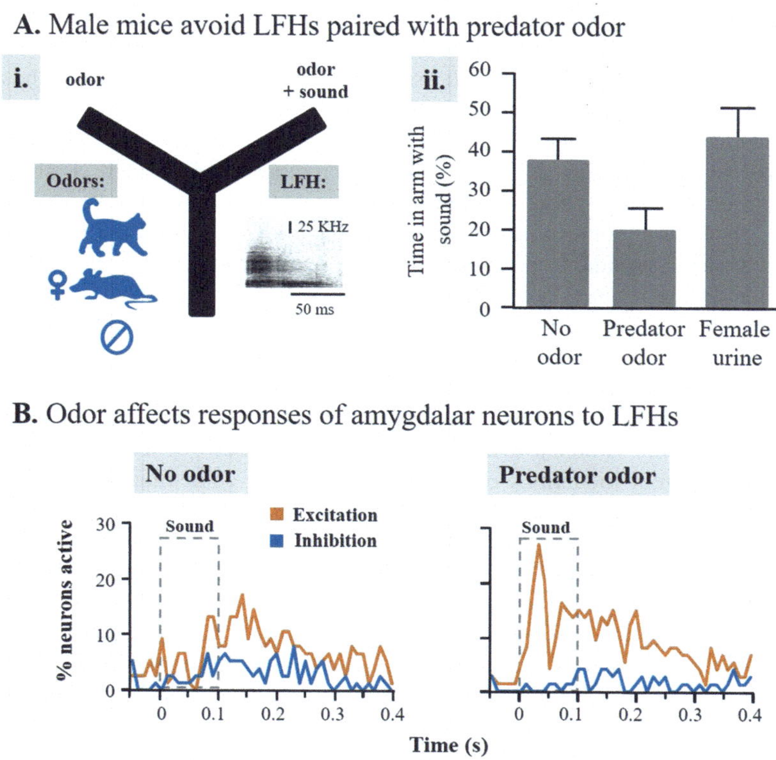

Fig. 8.2 Olfactory context decreases ambiguity about the most appropriate response to a vocal signal. (**a**) Behavioral responses to a call are influenced by odor. (**i**) Y-maze used to assess approach or avoidance of low-frequency harmonic calls (*LFH*s). Recorded LFHs were played through one speaker while no odor or odor sources consisting of cat fur or mouse urine were placed at the end of each arm of the maze. (**ii**) The presence of predator odor (cat fur) caused male mice to avoid LFH playback more than when LFHs were paired with no odor or female urine. (**b**) Responses of neurons in the basolateral amygdala to LFHs also were influenced by the presence of predator odor. In the presence of predator odor, more neurons showed excitatory responses to LFHs and fewer showed inhibitory responses than when LFHs were not paired with predator odor. (Adapted from Grimsley et al. (2013))

features. The idea of context as information can thus link functional and mechanistic levels of analysis.

This chapter, like other chapters in this volume (Okanoya and Screven, Chap. 2), raises issues that are common to acoustic communication across vertebrate groups. Just as for rodents, acoustic communication in fish, amphibians, and birds is shaped by such salient factors as the presence and identity of conspecifics (Woolley and Doupe 2008; Halfwerk et al. 2014). Internal factors such as motivation or hormonal state also shape communication behavior (Sisneros et al. 2004; Maney 2013). Finally, experience may influence how these external and internal factors come to bear on communication events (Poremba et al. 2013). Rodents as a group are a rich proving ground for general hypotheses related to context-dependent communication because of the diversity of natural histories and acoustic ecologies. Some rodent species produce and receive acoustic signals in ways that overlap with those of humans (e.g., chinchilla, *Chinchilla lanigera*) (Shofner 2000; Moreno-Gómez et al. 2015). Other rodent species access an exceptionally wide range of frequencies, similarly to echolocating bats (suborder Microchiroptera) (Bauer et al. 2002; Gadziola et al. 2016). These species include one of the most commonly studied rodents, the house mouse (*Mus musculus*), which produces and receives a variety of distinct acoustic signals, from human-audible vocalizations (Finton et al. 2017) to ultrasonic vocalizations (USVs) (Nyby et al. 1977).

An example of the usefulness of rodents in testing broad theories on the importance of context in communication is a consideration of how the complexity of social interactions drives the evolution of social signals (Freeberg et al. 2012). Among a wide range of animals, the number of categories of communication signals, the variability of signal structure and signal sequences, or the combination of signals in different modalities can be associated with contextual features of the social environment. These features include the size and relatedness of social groups, the diversity of individual social roles, the degree of hierarchy, and the sizes and proximity of territories. In a rodent-specific application of this hypothesis, Pollard and Blumstein (2012) have made the additional important point that particular features of social context may be associated with specific types of acoustic variation. For example, large group size corresponds to high individual acoustic variability for the alarm calls of sciurid rodents, while the diversity of social roles corresponds to alarm call repertoire size. Although the main emphasis in this chapter will be plasticity in acoustic signaling rather than signal evolution, this example illustrates how rodents as a whole provide the scope for testing a wide range of concepts.

8.1.2 Signalers Versus Receivers?

This chapter first considers contextual influences on senders and receivers separately (Sects. 8.2 and 8.3) in line with studies that isolate the influence of specific contextual factors on the production of, or responses to, vocalizations. These studies portray rodent vocalizations as versatile signals, capable of conveying nuanced information that represents multiple time scales. The structure or use of vocalizations may correspond to highly stable characteristics like the sex of a signaler (Keesom et al. 2015) or to characteristics that change relatively slowly or predictably, like reproductive phase for seasonal breeders (Rendon et al. 2015). Rapidly changing factors, such as the behavior of a social partner, can also influence the production of vocalizations (Lupanova and Egorova 2015). When combined with additional signaling modalities that are prevalent in rodents, such as olfaction, vocalizations are potentially rich in information (Asaba et al. 2014b; Okanoya and Screven, Chap. 2; Schleich and Francescoli, Chap. 3).

As some authors have noted, however, extrapolating from variation in signal production to variation in signal perception is not necessarily straightforward, particularly since behavioral or neural responses to vocal signals may be more cryptic than the signals themselves (Freeberg et al. 2012). Section 8.3, therefore, focuses on several excellent examples of the context dependence of signal perception in rodents and expands on physiological and neural mechanisms that may underlie this dependence. Section 8.4 acknowledges the artificiality of separating senders and receivers by exploring studies that take emergent approaches to sender–receiver interactions, and it provides a broad view of context as a joint construct that can link multiple actively contributing social partners.

8.2 The Influence of Context on Signalers

The context experienced by signalers shapes the course and outcome of a communication event. Hypothetically, there are many factors that could be involved in defining external and internal facets of context (Fig. 8.1). Factors might be biotic and/or abiotic, external and/or internal, and all could be potentially important and competing stimuli for the signaler (see Sect. 8.1.1). In many cases, cues indicating the presence of conspecifics and predators would demand different and opposite responses from a signaler (and a receiver). For example, in response to either concomitant acoustic or olfactory signals of a potential mate and a predator, a signaler may simultaneously respond acoustically to attract a mate or not respond at all to avoid being heard by the predator. There may also be influences of abiotic factors. Consider, for example, a mouse with only a limited number of biotic stimuli in a situation with heavy rain or wind. In heavy wind, the effective distance of an acoustic signal, especially in the ultrasonic frequency range, may be too low to justify vocalizing, given the costs of signal production in other animals such as birds (Oberweger and Goller 2001; Ward et al. 2003).

The majority of what is known about mice and rats with regard to how signaler context shapes a communication event comes from experiments with a single mouse presented with a stimulus. In these experiments the signaler, usually a male, is stimulated by the presence of a female. The female may be anesthetized or awake or represented by an olfactory cue in the form of either urine or urine-soaked bedding (e.g., Chabout et al. 2015). There are studies, however, that bring a more emergent and realistic social and physical environment to bear. These studies either use groups of mice to better simulate a natural social setting (multiple individuals living in groups or family groups) within a lab environment (e.g., Hoier et al. 2016; Neunuebel et al. 2015; Pultorak et al. 2015) or examine signalers in completely natural contexts by eavesdropping on them in the wild (e.g, Briggs and Kalcounis-Rueppell 2011; Petric and Kalcounis-Rueppell 2013).

Four model systems are highlighted to examine how the social and genetic context of conspecifics primarily influences the signals that senders produce. Section 8.2.1 describes a study that used different stimuli presented to male house mice

(B6D2F1/J) to simulate different contexts associated with mating (Chabout et al. 2015). Section 8.2.2 outlines a series of papers with pairs of mice that used lab-bred progeny of wild-caught house mice from Germany and France, which allowed for manipulations that varied context with respect to familiarity, gender, and population origin (von Merten et al. 2014, 2015; Hoier et al. 2016). Section 8.2.3 describes a study in which house mice (C57Bl/6J) were tested in a large group enclosure to allow competing external biotic stimuli of both sexes (groups of males and females) to determine whether the context of group size and membership influenced the sender (Neunuebel et al. 2015). Section 8.2.4 focuses on two species of mice in the genus *Peromyscus*, the brush mouse (*P. boylii*) and the California mouse (*P. maniculatus*), each with a different mating system, that were studied in the lab and the wild. *Peromyscus* are New World relatives of house mice, the species from which lab mice were derived (for relationships between *Peromyscus* and *Mus* see Bedford and Hoekstra 2015). For the monogamous California mouse, an important aspect of the internal state of a male sender is whether he is mated or unmated (Pultorak et al. 2015), and the external context may or may not include a female mate or a neighbor (Briggs and Kalcounis-Rueppell 2011). For the polygynous brush mouse, the external context may include a same-sex or opposite-sex conspecific or no conspecific at all (Petric and Kalcounis-Rueppell 2013).

8.2.1 Stimuli

In a study to determine if signals were context-specific in an *individual versus stimulus paradigm*, Chabout et al. (2015) presented male lab mice (B6D2F1/J) with a series of different stimuli that included either an olfactory signal from urine or an actual mouse (anaesthetized female, active female, or an anaesthetized male). Male mice vocalized at a higher rate in response to fresh female urine as well as to an active female or an anaesthetized female when compared to fresh male urine. This result suggested to the authors that the female need not reciprocate in social interactions for the communication event to occur (this is different from the outcome from Neunuebel et al. 2015; see Sect. 8.2.3).

In contrast, the spectral characteristics and the syntax of the sounds associated with USVs varied with the different stimuli (Neunuebel et al. 2015). For example, syllable duration, amplitude, and frequency were higher for awake than anaesthetized females, and these characteristics also differed between urine and mouse-present stimuli. In addition, the syntax of the sequences differed among stimuli with simpler sequences produced in response to the presence of females as opposed to complex sequences that were produced in response to urine. This is curious because in this study, females showed a preference for complex songs based on playbacks, and the authors argue that the complex songs in response to urine serve to lure females in, whereas the simple songs in response to females serve to facilitate copulation. Here the female is only considered as the receiver, but it would be very interesting for this experiment to be repeated from the perspective of the female as the sender/signaler, as some studies have demonstrated that females are active senders in dyadic interactions (see Sects. 8.2.2, 8.2.3, and 8.2.4). Chabout and colleagues (2015) showed that the nature of the stimulus influences the spectral characteristics, syntax, and complexity of the signals produced.

8.2.2 *Gender, Familiarity, and Population of Origin*

In a series of studies to assess the effect of divergence of two populations of wild-caught house mice on USVs, von Merten and colleagues (2014) placed lab-bred offspring of wild-caught house mice into two different pair contexts: both members of the pair were either from the same country of origin (Germany or France) or not. Pairs were either male-female from the same population or male-female from different populations, or were female-female or male-male from the same population. Thus, there were two contextual differences: the origin and sex of social partners. These factors allowed the authors to assess mating signal relevance in different populations and to assess whether USVs were used in nonmating contexts (e.g., during same sex interactions). Not surprisingly, German and French mice differed in terms of the type and syntax of the USVs, which can be explained by the genetic divergence between these populations.

Importantly, and more germane to the discussion of context, von Merten et al. (2014) set up a recording chamber (Fig. 8.3) where both individuals in the pair were awake and were recorded separately. Additionally, familiarity could be investigated because the recordings were done over two nights, so the mice were familiar on night two but not on night one. Signals were emitted by both sexes in all contexts, demonstrating that USVs are not just emitted by males to facilitate mating with females. Context mattered in terms of the number of calls and the spectral features of calls, but the main influence was from sex rather than familiarity (but see Hammerschmidt et al. 2012). There were higher rates of calling between female-male pairs, and males had high syllable rates with USVs at higher frequencies when they were interacting with females from the same populations (i.e., with putative mates).

Even though rates of calling between female–male pairs were higher than same sex interactions, there was calling during same sex interactions. Female–female interactions yielded higher rates of calling, with long and structured USVs, when compared to male–male interactions (von Merten et al. 2014). Thus, female–female interactions are important. These female–female interactions were further examined by stimulating females with unfamiliar females and monitoring, over time, both direct (at a contact window) and indirect (not at the contact window) interactions with and without the presence of male bedding (i.e., chemical cue of potential mate) (Hoier et al. 2016). Female production of USVs was highest when the signaler was first presented with a new female but production decreased over subsequent nights of interaction. Additionally, the spectral characteristics of USVs differed depending on whether the interaction between the stimulus and the signaler was direct or

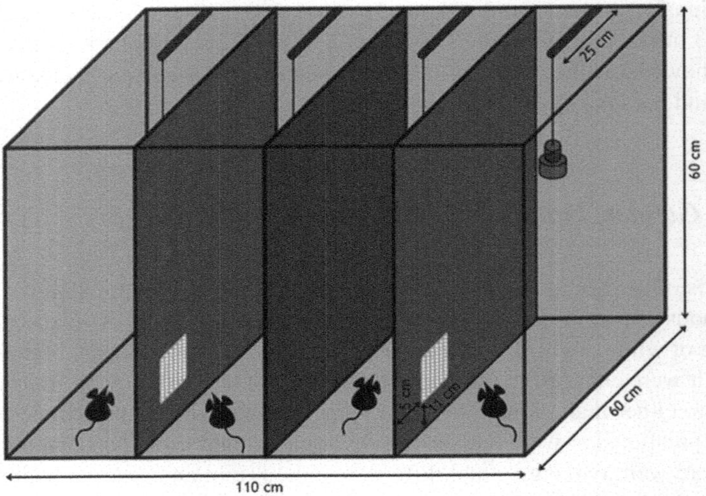

Fig. 8.3 Experimental setup for recording all dyad types as in von Merten et al. (2014). The box allows for recording two dyads (one dyad in the left two compartments and the other in the right two compartments). Although the box is not drawn to scale, there are several features of the design that are important. *First*, each compartment has its own microphone so that both the signaler and the receiver (or both members of the dyad) can be independently recorded. *Second*, there is a mesh wire contact window that, when open, allows for direct contact between interacting mice. *Third*, the setup is large enough that the mice can remain in the setup for up to four days (with food, water, and bedding) so that individuals can become familiar with the environment and the other member of the dyad for two nights before the window is opened, and recording begins on the 3rd night. (Work originally published in von Merten et al. (2014); used with permission of S. von Merten)

indirect. These two studies showed that: (1) context matters in the sense that mice only communicate with relevant conspecifics with whom they are likely to mate, in the case of male–female interactions; (2) female signalers respond to both indirect and direct stimuli in the form of other females, and familiarity modulates the communication event (von Merten et al. 2014, 2015; Hoier et al. 2016).

8.2.3 Signaling in Complex Social Environments

In a study designed to better approximate a social environment that a mouse might encounter in nature, Neunuebel and colleagues (2015) followed lab mice (C57Bl/6J) that were in small groups to determine when USVs were produced and how interactions with social partners corresponded to production of USVs. The authors showed several important influences of context. First, most USVs were produced when males and females were together, as opposed to when two males were together or two females were together. Second, USV occurrence was more likely when pairs of animals were closer together than further apart in the group. Third, the likelihood of a USV was highest when the pair of mice was moving relatively quickly or when there was a chase occurring. Lastly, when the female was involved in the USV interaction, her movement speed slowed (even during a chase), suggesting a correlation between movements and vocalizations, which may be related to the male assessing her receptivity. Thus, proximity is important in determining the rate of vocalization production by senders. Whether the syntax or the spectral characteristics of individual vocalizations also vary with proximity has yet to be determined within this system. However, importantly, this study highlighted the importance of spatial scale, movement dynamics associated with signals, and social complexity.

8.2.4 Using *Peromyscus* to Study Signalers in the Field and the Laboratory

The study of context with the most emergent approach was performed with wild mice in the genus *Peromyscus*. The following studies are the closest approximation to understanding context on the signaler as depicted in Fig. 8.1. Kalcounis-Rueppell and colleagues examined how the identity of conspecifics influenced both rate and spectral structure of USVs (Briggs and Kalcounis-Rueppell 2011; Petric and Kalcounis-Rueppell 2013). In the monogamous California mouse, males and females have a lasting pair bond and production of USVs depends on presence and proximity of the mate. Signals were produced when the mate was more than roughly 1 m away, both in the presence and absence of a nonmate (Briggs and Kalcounis-Rueppell 2011). In the brush mouse, which has a mating system that is common among rodents, females are generally solitary and share smaller territories within a male's larger territory. In the brush mouse, signals were produced when females were alone, when females and males were in the presence of the opposite sex, and when females were near other neighboring females at territory boundaries (Petric and Kalcounis-Rueppell 2013).

In both studies, spectral characteristics and rate of calling did not differ markedly between the identity of a partner or sex, suggesting that USVs are general signals used in multiple contexts. However, caution is warranted in comparing the results of these emergent field studies with all the preceding lab studies because the types of USVs recorded on the scale of meters in the field (more likely to record loud, long, and directional USVs) can be different from those recorded on the scale of centimeters in the laboratory (more likely to record quiet, very short, highly frequency-modulated USVs). Regardless, results from emergent field studies support the influence of context on USVs produced by rodents in the lab. Another important outcome of these field studies is the discovery that the signals themselves (e.g., loud, long, and directional USVs) show general similarities in structure whether they are recorded in the lab or in the field (Kalcounis-Rueppell et al. 2010).

A comprehensive study of the internal scene and its effects on USV production in mice also was done using the monogamous California mouse (Pultorak et al.

2015). In this laboratory study, both the reproductive history (sexually naive or mated) and the hormonal status (pulse of testosterone or pulse of vehicle) of signaling males in the presence of an unknown female were manipulated. A testosterone pulse decreased the total number of USVs produced by mated, but not sexually naive, males in the presence of an unknown female. These results suggest that steroid hormones may be responsible for rapidly mediating and suppressing USVs in mated males, providing a mechanism for mate fidelity in this species of mouse.

The approaches exemplified in these five studies are all important because they provide insight about both the external and internal context and how those contexts relate to USV production by signalers. Signalers are responding with USVs only to certain stimuli (Chabout et al. 2015), and the USVs differ depending on stimulus. Sex, identity, and familiarity of the intended receivers are all important parts of the external scene that mediate both USV production and structure (von Merten et al. 2014, 2015; Hoier et al. 2016). Moreover, even within complex spatial and social conditions in the external scene, there are specific patterns associated with USV production and structure (Briggs and Kalcounis-Rueppell 2011; Petric and Kalcounis-Rueppell 2013; Neunuebel et al. 2015). Lastly, the internal context for the signaler is important, and both the reproductive experience and hormonal state influence the production of USVs (Pultorak et al. 2015).

An interesting note in several of these studies (e.g. Briggs and Kalcounis-Rueppell 2011; Petric and Kalcounis-Rueppell 2013) is that not all mice vocalize to the same extent, and in some cases, there is more variation among individuals than between contexts (e.g., von Merten et al. 2014), a result that had also been highlighted by Hoffmann et al. (2012). Individual differences were especially pronounced in female house mice, both in terms of the amount of calling and the spectral structure of the calls (e.g., slope, number of turns, and number of jumps) (von Merten et al. 2014). This suggests that independent of internal and external scene, there may be intrinsic individual differences in vocal signals that set up an important component of the external scene: individual identity of conspecifics.

8.3 The Influence of Context on Receivers

8.3.1 The External Scene

Having a range of potential responses to vocal signals that depend on the external scene can be a great advantage to receivers. In some cases, relevant cues of context are encoded by salient differences in the structure of vocalizations (Matrosova et al. 2011; Blumstein et al. 2013). As one example, sciurid rodents (in the ground squirrel family) produce relatively well-studied alarm calls. In addition to alerting conspecifics of danger, these calls may be influenced by individual identity or kinship (Matrosova et al. 2011; Blumstein et al. 2013), stress levels (Blumstein and Chi 2012), proximity of a predator (Furrer and Manser 2009), age (Schneiderová et al.

2015), or the type of predatory threat (Slobodchikoff and Placer 2006). Call structure may also be used by Richardson's ground squirrels (*Urocitellus richardsonii*) to match the perceptual abilities of conspecific listeners while excluding eavesdroppers. In this species, audible vocal signals are used as general alarms, but when conspecifics are nearby, signalers send highly directional ultrasonic alarm calls that cannot be heard by predators (Wilson and Hare 2004, 2006).

Listeners respond in appropriate ways to this varied content. For example, Gunnison's prairie dogs (*Cynomys gunnisoni*) respond to predator-specific alarm calls with predator-specific defensive behavior (Kiriazis and Slobodchikoff 2006), while Cape ground squirrels (*Xerus inauris*) respond more actively to calls of high urgency than to calls of lower urgency (Furrer and Manser 2009). Acoustic structure also distinguishes the alarm calls of young yellow-bellied marmot pups (*Marmota flaviventris*), which produce alarm calls that are longer than those of adults and contain more segments with noise-like structure; these calls are termed "screams" due to their rough acoustic quality to human listeners (Blumstein et al. 2008). Adult marmots are more responsive to the playback of screams than to playback of adult calls, but adults are not as responsive to artificially shortened calls. The addition of synthetic noise to calls during playbacks also enhances adult responses (Fig. 8.4) (Blumstein and Récapet 2009), suggesting that both duration and spectral structure may contribute to the specific behavioral response to screams. Being able to distinguish and differentially respond to the alarm calls of juveniles in contrast to those of adults could contribute to the survival of young offspring or other kin (Blumstein et al. 2004).

Another example of call structure containing contextual cues is found in the males of several species of cricetid rodents in the genus *Scotinomys* (singing mice), which perform songs that consist of a series of repeated downward frequency sweeps (Campbell et al. 2010). These songs are thought to function in both attraction of females and repulsion of males. Male *Scotinomys* listening to the songs of other males may perform their own song in reply (Pasch et al. 2011). For *Scotinomys teguina* (Alston's singing mice), these replies depend on whether the perceived song is produced by a conspecific or by a congener, *Scotinomys xerampelinus* (Chiriquí singing mice) (Pasch et al. 2013). The congener is larger and can exclude the smaller Alston's singing mouse males from sites that they would normally inhabit. Correspondingly, the response of male Alston's singing mice to playbacks of the songs of male Chiriquí singing mice is the suppression of their own calls and greater avoidance of a playback speaker relative to control. In contrast, songs of their own species elicit an overall increase in response songs and greater proximity to a playback speaker relative to control. Because replies are elicited solely by playback of the songs of conspecific or congeneric individuals, the cues that lead to these differential responses must be in the acoustic structures of the stimuli.

A wide range of hormonal and neurochemical systems are responsive to changes in the external scene and could potentially mediate context-dependent behavioral responses of listeners. Two examples of such physiological systems are signaling molecules involved in responses to stressors (e.g., glucocorticoids) and socially activated neuromodulators such as dopamine and serotonin. A general rule for these

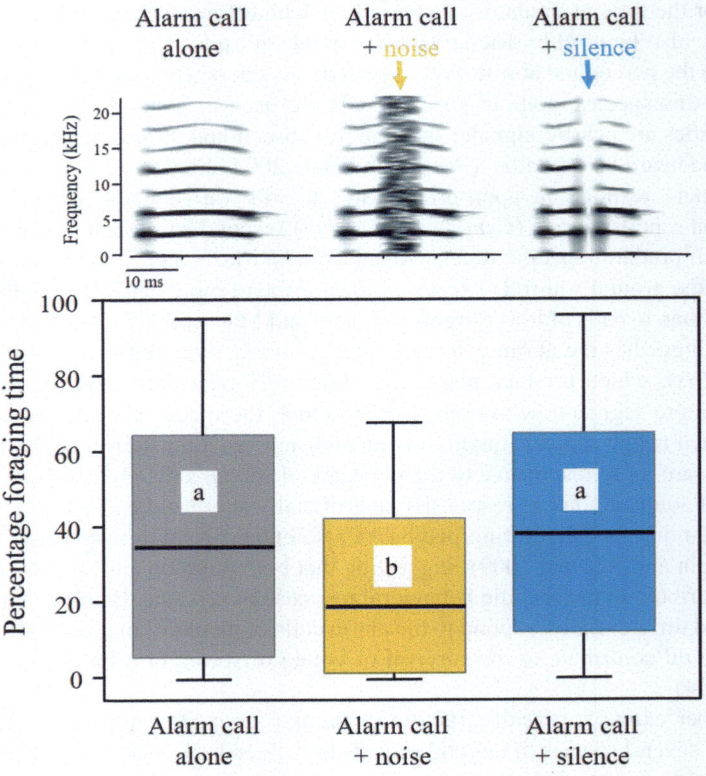

Fig. 8.4 Call structure influences responses of receivers. Playbacks of alarm calls with inserted noise (*middle*) causes receivers to forage less than unmanipulated alarm calls (*left*) and alarm calls with inserted silence (*right*). Different *letters* indicate statistically distinct groups. (Adapted from Blumstein and Récapet (2009))

agents is that they not only affect the processing of acoustic signals both peripherally and centrally (Bartolomé and Gil-Loyzaga 2005; Meltser and Canlon 2011), but they also act broadly through neural and even somatic systems to influence behavior (de Kloet et al. 2006).

For example, predatory threats may engage stress systems that release glucocorticoid hormones. In some rodents, hormone levels are highly responsive to cues of predators (Mateo 2010; Harris and Saltzman 2013). Cortisol levels in juvenile Belding's ground squirrels even respond differently to alarm calls that indicate different types of predatory threats (Mateo 2010). In turn, glucocorticoids are capable of altering auditory processing via receptors located in the auditory periphery and the central auditory system (Mazurek et al. 2010; Meltser and Canlon 2011). Related to social contexts, the release of neuromodulators, such as dopamine and serotonin, may be sensitive to specific aspects of social encounters, such as positive or negative social valence (Gittelman et al. 2013; Keesom and Hurley 2016), and can influence the responses of auditory neurons to neutral sounds and vocalizations alike (Hurley and Pollak 2005; Gittelman et al. 2013).

Important to a broad model of context, some physiological signals can integrate multiple features of the external and internal scenes. For example, serotonin levels in the auditory midbrains of female mice increase in the presence of male partners, and experience with a previous stressor elevates the serotonergic increases in response to an acute stressor (Hanson and Hurley 2014). Further, the immediate early gene responses of midbrain auditory neurons to serotonin depend on both the presence of a social partner and estrous state (Hanson and Hurley 2016). In addition to integrating disparate features of context, modulatory signaling systems target diverse brain regions (Lee et al. 2008; Vasudeva et al. 2011). By coordinating responses in sensory, affective, and motor pathways, these systems can facilitate context-dependent behavioral responses to stimuli.

8.3.2 The Internal Scene

Both internal state and previous experience have strong effects on the acute responses of rodents to vocal signals. These two factors also interact with each other, as well as with external cues, to produce appropriate behavior. Two such examples occur in the responses of females to male courtship vocalizations and the responses of parents to the calls of pups.

Males of many rodent species produce vocalizations during their courtship of females (Holy and Guo 2005; Pultorak et al. 2015; Okanoya and Screven, Chap. 2). Just as for songbirds, a functional prediction is that the responses of females to these calls depend in part on whether females are in reproductive condition (Maney 2013; Yoder et al. 2015). In two well-studied rodent model systems, rats (*Rattus norvegicus*) and house mice, female reproductive state does influence responses to male courtship calls but in some rather nuanced ways. Female house mice maintain more proximity to speakers playing recorded ultrasonic vocalizations of the kind that male mice produce during courtship than to silent speakers (Hammerschmidt et al. 2009; Shepard and Liu 2011). Females also prefer intact males to males that have been surgically devocalized (Pomerantz et al. 1983a). In none of these preference paradigms do females in estrus versus other reproductive phases show a difference in their preference for vocalizations versus silence. On the other hand, ovariectomy decreases the preference for intact males over devocalized males, and the deficit is rescued by treatment with estradiol and progesterone (Pomerantz et al. 1983a). Thus, a milieu of ovarian hormones seems to be necessary for females to express a preference for male calls over an alternative.

Experience plays a role in some aspects of these preferences. Interaction with a male restores female preference for male vocalizations over silence after females have habituated to the vocalizations, potentially by imparting salience to the vocalizations (Shepard and Liu 2011). Vocalization structure may also complement olfactory cues in allowing female mice to assess the relatedness of potential mates

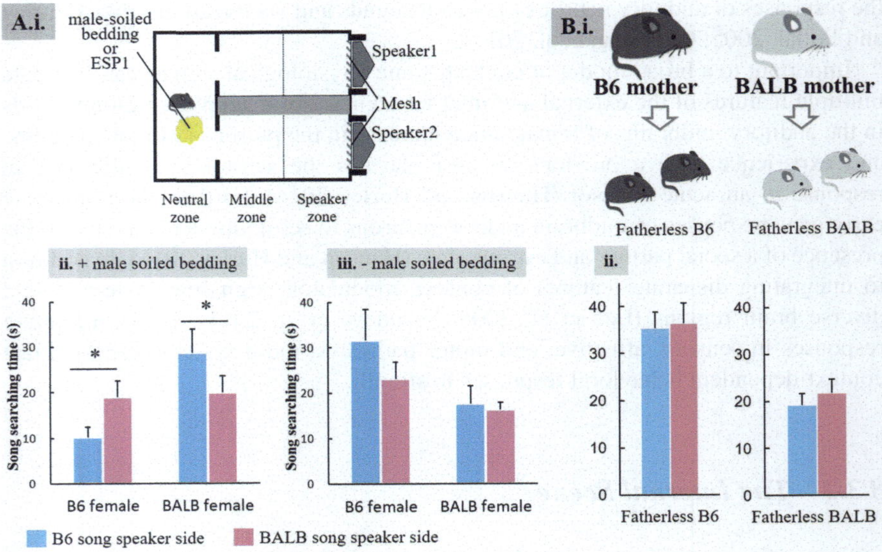

Fig. 8.5 Responses of female mice to playback of male USVs depend on multiple features of context. (**a**) Olfactory cues gate the expression of female auditory preference. (**i**) Experimental setup used to test the preferences of female B6 and BALB females for recorded male USVs. (**ii**) Females prefer song recorded from males of different strains; B6 females prefer USVs of BALB males and BALB females prefer USVs of B6 males. (**iii**) The expression of female preferences is gated by olfactory cues; in the absence of bedding soiled by males, females do not express preferences for song from a different strain. (**b**) Female preference for male song depends on juvenile experience. (**i**) Experimental design used to prevent experience of juvenile females with adult males. (**ii**) Females raised in the absence of males express no preference for USVs from a particular strain. (Modified from Asaba et al. (2014a); original figures provided by A. Asaba)

(Penn and Potts 1998). Vocalizations of male house mice show consistent differences in structure at the level of species (Musolf et al. 2015). Strains (Asaba et al. 2014a), kin groups, and individual differences (Hoffmann et al. 2012) also play a role. Females may use this information to choose a mate that is not too distantly related (different species) (Musolf et al. 2015) but also not too closely related (siblings) (Musolf et al. 2010).

Juvenile experience is important for avoiding close relatives; females prefer calls produced by males from lab strains that are different from their own with significantly different vocal characteristics (Asaba et al. 2014a). Cross-fostering mice from different strains reverses female preferences, however, so that adult females now prefer vocalizations from males of their own strain, and preferences are removed by raising females without fathers (Fig. 8.5). This experience-based preference is only observed in diestrous females that have been primed by exposure to bedding soiled by males (not other females) or to a male pheromone. The conclusion of these studies is that female mice imprint on the calls of their fathers as juveniles and avoid males with similar calls as adults, but estrous state and the presence of appropriate multimodal stimuli are necessary to express this experience. These studies are an excellent demonstration that juvenile experience interacts with both the internal scene (estrous phase) and the external scene (olfactory cues of males) to shape female preference for male vocalizations.

A second example of the interaction of experience and state on signal perception relates to maternal responses to pup calls. The pups of many different rodent species produce calls under different circumstances that promote caring behaviors from their mothers (Hahn and Lavooy 2005; Shair 2014). In some cases it also promotes caring behaviors from their fathers (Wright and Brown 2004; Robison et al. 2016). These parental behaviors are likely to be highly adaptive in natural settings, resulting in increased survival and optimal development of pups (Curley and Champagne 2016).

An assortment of the behavioral, hormonal, and neural mechanisms contributing to maternal responsiveness to pup calls has been intensively studied in female house mice. Ultrasonic vocalizations are produced by pups when they are isolated from their litter, which may prompt the mother to search, resulting in retrieval of pups to the nest (Ehret 2005; Hahn and Lavooy 2005). When in the nest, pups of some species may also produce harmonic wriggling calls (with most energy below 10 kHz) that prompt mothers to show caring behaviors, such as licking, changing position, or nest building (Ehret and Bernecker 1986). Mothers who have undergone gestation and birth perform these behaviors without prior experience, while pup-naive females show few of these behaviors (Ehret 2005; Ehret and Schmid 2009). When naive females are given experience with pups by cohabiting with females who have litters, however, they develop enhanced rates of maternal behavior that increase over time (Ehret et al. 1987; Geissler et al. 2013). Such experience-dependent caring for pups does not fully mimic maternal behavior, indicating that experience with pups and maternal state do not influence the responses to calls identically. Maternal state and experience also interact with features of the external scene, notably the characteristics of pup calls. Construction of synthesized models of pup calls shows that the frequency structure of pup calls (Ehret and Haack 1981; Ehret and Koch 1989) must fall within the natural range to evoke the maximum responses of females. Repetition rate is also vital for female response (Gaub and Ehret 2005).

The changes in behavior of both mothers and cohabiting females who care for pups (co-carers) relative to pup-naive females are paralleled by changes in neural processing within the auditory systems of female house mice from the levels of the auditory brainstem through the cortex (Miranda and Liu 2009; Miranda et al. 2014). Overall, maternal state and experience with pups both enhance the representation of pup isolation calls in the auditory system. Some changes in response to playbacks of pup calls are observed across populations of neurons. For example, neurons in relatively low-frequency auditory cortical fields show greater increases in the inhibitory responses to pup calls than do neurons in auditory cortical fields that are more sensitive to the ultrasonic frequencies contained in pup calls. This selective pattern may emphasize the neural representation of pup calls in both mothers and co-carers, and mirrors more pronounced phonotactic behavior of mothers and co-carers to playback of pup calls (Lin et al. 2013). Interestingly, these changes deteriorate in co-carers but not in mothers soon after weaning, suggesting important differences in the influence of maternal state versus experience on neural response properties relevant to pup retrieval.

High rates of pup care in mothers are also associated with shifts in the timing of auditory cortical neuron responses to increase the neural information available for both the detection and discrimination of pup isolation calls (Liu and Schreiner 2007). Further, maternal state increases the ability of auditory cortical neurons to respond to calls at repetition rates that better encompass the natural range of pup calls (Liu et al. 2006). Not just primary auditory areas, but also limbic areas, are likely to be involved in neural responses of mothers to pup-related stimuli. In regions including the amygdala, bed nucleus of the stria terminalis, and medial preoptic area, immediate early gene activity increases more in mother mice exposed to both auditory and olfactory cues of pups relative to unimodal stimuli, which is in parallel with the behavioral attraction to these stimuli (Okabe et al. 2013). Therefore, maternal responses to pups involve coordination among sensory regions and regions of the brain involved in affect.

Hormonal mechanisms underlie at least some of the behavioral and neural changes to call responses that occur when female mice become mothers (Miranda and Liu 2009). The estrous phases of virgin females influence responses to artificial wriggling calls. Females in diestrus show higher rates of pup care to artificial calls that are better approximations of pup calls than to poorer call models. Females in estrus show relatively high rates of response to all call models, and females in metestrus show relatively low rates of response to all call models (Ehret and Schmid 2009). Pup-naive ovariectomized females with implants of estradiol show preferential responses to more accurate ultrasonic pup call models after several days of experience with pups. Ovariectomized females without estradiol implants must have extensive pup experience before accurate call recognition, and experience also increases the maximum rate of call recognition (Ehret and Koch 1989; Koch and Ehret 1989).

Oxytocin, a hormone released during birth in mammals, is also central to orchestrating parental responses to infant-related stimuli (Rilling and Young 2014). Injecting virgin females with oxytocin or optogenetically activating a population of oxytocinegic neurons accelerates the development of retrieval responses relative to controls even when drug administration is limited to a single hemisphere of the auditory cortex (Marlin et al. 2015). Pairing oxytocin with exposure to pup calls in virgins also replicates many of the characteristics of the responses of auditory cortical neurons of mothers to pup calls.

The examples in the previous paragraphs highlight two important points. First, behaviors with high fitness consequences, such as choosing a mate or providing parental care, are likely to be influenced by multiple contextual factors. Such context-sensitivity may robustly increase the chances of showing a behavior when it is advantageous. From the perspective of a female mouse, there may be several possible behavioral options when hearing a pup call (Liu and Schreiner 2007). A female may perform pup care and/or retrieval, or not perform these activities; for some rodents, eating pups is an additional possibility (Day et al. 2002). Having experience with pups, becoming a mother, and hearing the correct acoustic structure of pup calls increase the probability that females will appropriately care for pups. A second important point is that what appear to be clearly separate classes of contextual features (e.g., experience, external events, and internal state) may be entangled with each other through their conditional and interdependent effects on behavior and through their convergence on the same hormonal and neural mechanisms.

8.4 Context Links Signalers and Receivers

Up to now, this chapter has presented a broad view of context that is inclusive of multiple internal states and external events. This section extends these ideas to a model of context as a joint dynamic construct linking the behaviors of both senders and receivers. Two major points in service of this model are that: (1) both senders and receivers respond to some of the same contextual factors, thus coordinating their behaviors; and (2) mutual interaction between social participants contributes to creating a specific context.

The first point, that senders and receivers are coordinated through context, is easily illustrated through examples already described in Sects. 8.2 and 8.3. For example, the presence of predators may equally influence the production of alarm calls (Slobodchikoff and Placer 2006) and responses to distress calls (Grimsley et al. 2013). In sexual interactions, a factor that influences the behavior of both males and females is the estrous state of the female partner. In house mice, males may produce vocalizations expected to require greater effort when female partners are in receptive reproductive stages (Hanson and Hurley 2012). From a female perspective, estrous state influences preferences for males of differing relatedness (Asaba et al. 2014a) and the immediate early gene activation of auditory neurons (Hanson and Hurley 2016). Thus, common contextual factors may be relevant to both senders and receivers of vocal signals, who adjust their behavior accordingly.

Second, the outcomes of social interactions are often highly dependent on exchanges of signals and cues between participants. Thus, individuals are both senders and receivers and do not simply respond to context but also play a role in establishing it. A prime example of this is in the isolation calls of the pups of many rodent species (Motomura et al. 2002), although pups of some rodent species produce isolation calls at greatly reduced rates relative to others (Shapiro and Insel 1990). Pup calls are sensitive to a range of contextual factors, such as pup age (Szentgyörgyi et al. 2008; Grimsley et al. 2011) and environmental temperature (Blumberg et al. 1992), they but are also influenced by interactions with nearby adults.

These effects are species-specific in ways that may reflect aspects of natural history (e.g., communal nesting) or whether pups are relatively altricial or precocial (Shair 2014). For example, if pups are briefly reunited with their mothers after a period of isolation, then re-isolated, the maternal contact potentiates calling or prevents a decline in calling in mice, rats, guinea pigs, and multiple species of voles

(Hennessy et al. 2006; Shair 2014). Pups may even change the sound frequencies of their calls when exposed to the odor of the home nest relative to odors of unfamiliar adults (Kapusta et al. 1995). In rats, the potentiation of calls is enhanced when pups interact with active, as opposed to anesthetized, mothers (Hofer et al. 1996). There is also a role for paternal contact in some species. In rats, the presence of males, even fathers, suppresses pup calls, which is a sensible response to potential cannibalism, but rearing experience with a father changes suppression to potentiation (Brunelli et al. 1998; Wiedenmayer et al. 2003). On the other hand, in prairie voles (*Microtus ochrogaster*), a species in which paternal care of offspring is common, calls of pups are potentiated only by contact with mothers and not by contact with fathers (Robison et al. 2016). Overall, interactions with parents fine-tune pup calls, which in turn likely influences parental responses to these signals, but there is a great deal of variation in the relevant details of context among species (Brunelli et al. 1994; Yu et al. 2011).

The exchange of multimodal signals between males and females during sexual interaction is also mutual. Courtship vocalizations produced by males have been studied in a range of species, are initiated in response to cues indicating the presence of females, and may be greater in number if males have had prior experience with female interactions (Lepri et al. 1988; Roullet et al. 2011). The production of vocalizations by males is highly dependent on gonadal hormones in a wide range of species, including singing mice (Pasch et al. 2011), deer mice (*Peromyscus maniculatis bairdi*) (Pomerantz et al. 1983b), house mice (Nunez and Tan 1984), and golden hamsters (*Mesocricetus auratus*) (Floody et al. 1979). Females not only act as receivers in using calls as a signal for approach or as the basis for potential mate preference (rat: Willadsen et al. 2014; mouse: Pomerantz et al. 1983a), or in facilitating lordosis (Floody and Bauer 1987), but females also produce a variety of vocal signals during interactions with males.

Some of these signals relate to rejection. In response to initial investigation by males, female house mice kick at or dart away from males, or they lunge at males with open mouths, often in conjunction with the production of very loud low-frequency multiharmonic calls or squeaks (Sugimoto et al. 2011; Lupanova and Egorova 2015). These vocalizations may function to pace sexual interactions (Johansen et al. 2008) and correspond to the reduction of courtship effort by males (Finton et al. 2017). Squeaks also correlate with neurochemical changes in the male auditory system (Keesom and Hurley 2016). In males that are interacting with females, levels of the neuromodulator serotonin increase on average in the auditory midbrain, but the change from baseline is inversely correlated with female rejection behavior, including the number of squeaks. Males with squeaky female partners show much more modest increases in serotonin than males with less squeaky partners.

The females of multiple species of rodents also produce USVs during sexual interactions, and these correspond positively to sexual behaviors or preference for males in house mice (Neunuebel et al. 2015), rats (Börner et al. 2016), and golden hamsters (Fernández-Vargas and Johnston 2015). Thus, both male and female rodents produce and respond to multimodal signals that incorporate vocalizations from their partners during an opposite-sex encounter and thereby determine the outcome of a specific interaction from a range of possibilities.

8.5 Context Sensitivity Is Widespread

As demonstrated in Sects. 8.2, 8.3, and 8.4, sensitivity of signalers and receivers to context is a robust phenomenon in rodents. Specific types of context sensitivity, such as the modulation of isolation calls by pups in response to conspecifics, occur across a range of taxonomic families of rodents (Hennessy et al. 2006; Shair 2014). More broadly, sensitivity to context has been documented in studies taking psychological or ecological approaches in both field and lab studies. The ubiquity of context sensitivity in vocal communication is not surprising when viewed through the ethological lens of functionality (Tinbergen 1963; Liu et al. 2003). With this view, a fundamental question is whether the individuals that modify their acoustic signal production or modify their behavior following signal reception in response to context are likelier to survive or reproduce than individuals with inflexible signaling. Context sensitivity is therefore often framed as a type of communicative plasticity that provides an advantage to individuals.

This basic principle of context sensitivity occurs in additional vertebrate groups, including a classic group of model systems for acoustic signaling: oscine songbirds. Male song is sensitive to factors such as the presence of female conspecifics (Sakata and Vehrencamp 2012) and juvenile acoustic and nonacoustic social experience (Bottjer and Arnold 1997; Cousillas et al. 2006). Female receivers are also sensitive to context, including the characteristics of male song (Woolley and Doupe 2008). Like female house mice, female zebra finches (*Taeniopygia guttata*) show responses to male song that are influenced by interacting internal and external features of context that include juvenile experience, song structure, and hormonal state (Vyas et al. 2008). How neurochemical systems modulate signal production and reception according to behavioral salience has been relatively well-explored in songbirds (Riters 2012; Maney 2013). How context influences neural activity and behavior in songbirds is also known (Castelino and Schmidt 2010; Chen et al. 2017). Studies in songbirds, therefore, can provide some guiding principles for exploring context sensitivity in rodents. Even in nonvertebrate species, however, acoustic communication is influenced by contextual information such as nonacoustic cues that indicate the potential presence of a predator (Jacob and Hedwig 2015). Therefore, context sensitivity in acoustic communication is a fundamental phenomenon that is extremely widespread across animal taxa.

8.6 Chapter Summary

This chapter has described some of the ways in which vocal communication in rodents is sensitive to context. These examples serve to highlight omissions in the current understanding of vocal production relating to a practical consideration of experimental work and to the creation of additional conceptual frameworks. Of immediate practicality, some of the contextual factors that have a demonstrated effect on vocal communication are often not considered in experimental work. This is particularly true of the social history of individuals and is especially relevant given the emphasis on social housing in emerging standards of lab animal care. Assessing the influence of all of the historical social variables important to vocal communication would be an undue burden for much experimental work. Developing standards for reporting on variables such as the nature of social housing and experience, as well as relevant nonsocial conditions (e.g., exposure to stressors) would be an excellent start in documenting the contextual factors that are associated with specific experimental results.

There is also a need for more coherent theories of context as guides for experimentation. For example, an operational definition of context would be of use in creating logical predictions regarding which components of the internal and external scenes should influence vocal signaling or behavioral responses to signals in different situations. Such a definition of context could guide experiments in areas like studies of multimodal integration or neural information content. Novel experimental approaches can likewise expand the range of phenomena under study. For example, the development of technology allows more emergent approaches in field studies (Briggs and Kalcounis-Rueppell 2011). The continued development of ethological and psychophysical assays allows for measurement of more aspects of vocal signal perception in the lab (Pomerantz et al. 1983a; Neilans et al. 2014).

Concepts of context sensitivity can also play a role in studies of signal evolution. For example, a case could be made for context sensitivity in receivers that would reduce the need for variation in signals. In the earlier example of male house mice responding to audible LFHs or squeaks of females, the degree of approach to the same signal depends on whether contextual odor cues arise from potential mates or potential predators (Grimsley et al. 2013). This finding suggests that context can allow the same acoustic signal to serve multiple behavioral functions. On the other hand, a more complex signal repertoire could reflect the nuances of different contexts. For male house mice, sensitivity to the estrous state of their female partners is associated with greater structural variation in vocalizations, and statistically distinct call structures are associated with different estrous states (Hanson and Hurley 2012).

From a mechanistic perspective, much remains to be understood regarding physiological representations of context, both in terms of how the internal and external scenes influence physiological systems and how these systems in turn coordinate signal production with signal perception and behavioral responses. Broadly projecting neurochemical systems could play an important role in the coordination of signaling and receiving, since they have access to sensory, motor, and integrative brain regions. This is encapsulated in the example of single serotonergic neurons sending collaterals to whisker-related sensory and motor regions in rats (Lee et al. 2008). Although such neurochemical systems are usually studied in isolation, contexts related to vocal communication typically engage multiple interconnected physiological systems, including broadly projecting neuromodulatory systems (Nevue et al. 2016). Smaller populations of neurons producing peptides (e.g., oxytocin or kisspeptin) also have a widespread effect on activity at neural sites relevant to communication (Marlin and Froemke 2016; Parhar et al. 2016). Excitatory–inhibitory macrocircuits and microcircuits may be modulated by these and by intrinsic circuit mechanisms (Newman 1999; Voytenko and Galazyuk 2011). Therefore, one pressing question is: How do these different physiological mechanisms coordinate to produce appropriate internal representations of context such as reward or valence (Nieh et al. 2013; Petersen and Hurley 2017)?

In conclusion, the idea of context is "organism centered", drawing from classic ethological and semiotic work on the perceptual worlds of animals (von Uexküll 1934; von Uexküll et al. 2010; Lettvin et al. 1968). At the same time, contextual factors in the external and internal scene act through systems related to mental health that underlie reward, stress, and social motivation. The idea of context therefore can be an informative lens for viewing both function and dysfunction in communication behavior.

Acknowledgments The work described here that was completed in the authors' laboratories was supported by the National Institute on Deafness and Other Communication Disorders award R01DC008963 and National Science Foundation awards 1460949 and 1456298 (Hurley lab) and National Science Foundation grants IOS-1132419, IOS-1355163, and IOS-0641530 (Kalcounis-Rueppell Lab). M. C. Kalcounis-Rueppell acknowledges her collaborator and co-PI on IOS-1355163 and IOS-1132419, Cathy Marler, for important discussions in developing her thoughts and ideas on contexts and USVs. The content is solely the responsibility of the authors and does not necessarily represent the official views of the National Institutes of Health. The authors wish to thank Drs. Jasmine Grimsley, Daniel Blumstein, Sophie von Merten, and Akari Asaba for generously sharing data for particular illustrations.

Compliance with Ethics Requirements Laura Hurley declares that she has no conflicts of interest.

Matina Kalcounis-Rueppell declares that she has no conflicts of interest.

References

Asaba, A., Okabe, S., Nagasawa, M., Kato, M., et al. (2014a). Developmental social environment imprints female preference for male song in mice. *PLoS One*, 9, e87186. https://doi.org/10.1371/journal.pone.0087186

Asaba, A., Hattori, T., Mogi, K., & Kikusui, T. (2014b). Sexual attractiveness of male chemicals and vocalizations in mice. *Frontiers in Neuroscience*, 8, 231. https://doi.org/10.3389/fnins.2014.00231

Bartolomé, M. V., & Gil-Loyzaga, P. (2005). Serotonergic innervation of the inner ear: Is it involved in the general physiological control of the auditory receptor? *International Tinnitus Journal*, 11, 119–125.

Bauer, E. E., Klug, A., & Pollak, G. D. (2002). Spectral determination of responses to species-specific calls in the dorsal nucleus of the lateral lemniscus. *Journal of Neurophysiology*, 88, 1955–1967.

Bedford, N. L., & Hoekstra, H. E. (2015). Peromyscus mice as a model for studying natural variation. *Elife* 4, e06813. https://doi.org/10.7554/eLife.06813

Blumberg, M. S., Efimova, I. V., & Alberts, J. R. (1992). Ultrasonic vocalizations by rat pups: The primary importance of ambient temperature and the thermal significance of contact comfort. *Developmental Psychobiology*, 25, 229–250. https://doi.org/10.1002/dev.420250402

Blumstein, D. T., & Chi, Y. Y. (2012). Scared and less noisy: Glucocorticoids are associated with alarm call entropy. *Biological Letters*, 8, 189–192. https://doi.org/10.1098/rsbl.2011.0832

Blumstein, D. T., & Récapet, C. (2009). The sound of arousal: The addition of novel non-linearities increases responsiveness in marmot alarm calls. *Ethology*, 115, 1074–1081. https://doi.org/10.1111/j.1439-0310.2009.01691.x

Blumstein, D. T., Verneyre, L., & Daniel, J. C. (2004). Reliability and the adaptive utility of discrimination among alarm callers. *Proceedings of the Royal Society B: Biological Sciences*, 271, 1851–1857. https://doi.org/10.1098/rspb.2004.2808

Blumstein, D. T., Richardson, D. T., Cooley, L., Winternitz, J., & Daniel, J. C. (2008). The structure, meaning and function of yellow-bellied marmot pup screams. *Animal Behaviour*, 76, 1055–1064. https://doi.org/10.1016/j.anbehav.2008.06.002

Blumstein, D. T., Nguyen, K. T., & Martin, J. G. (2013). Ontogenetic variation of heritability and maternal effects in yellow-bellied marmot alarm calls. *Proceedings of the Royal Society B: Biological Sciences*, 280, 20130176. https://doi.org/10.1098/rspb.2013.0176

Börner, A., Hjemdahl, R., Götz, T., & Brown, G. R. (2016). Ultrasonic vocalizations of female Norway rats (*Rattus norvegicus*) in response to social partners. *Journal of Comparative Psychology*, 130, 76–80. https://doi.org/10.1037/com0000017

Bottjer, S. W., & Arnold, A. P. (1997). Developmental plasticity in neural circuits for a learned behavior. *Annual Review of Neuroscience*, 20, 459–481. https://doi.org/10.1146/annurev.neuro.20.1.459

Bregman, A. S. (1990). *Auditory scene analysis: The perceptual organization of sound.* Cambridge, MA: The MIT Press.

Briggs, J. R., & Kalcounis-Rueppell, M. C. (2011). Similar acoustic structure and behavioural context of vocalizations produced by male and female California mice in the wild. *Animal Behaviour*, 82, 1263–1273. https://doi.org/10.1016/j.anbehav.2011.09.003

Brunelli, S. A., Shair, H. N., & Hofer, M. A. (1994). Hypothermic vocalizations of rat pups (*Rattus norvegicus*) elicit and direct maternal search behavior. *Journal of Comparative Psychology*, 108, 298–303.

Brunelli, S. A., Masmela, J. R., Shair, H. N., & Hofer, M. A. (1998). Effects of biparental rearing on ultrasonic vocalization (USV) responses of rat pups (*Rattus norvegicus*). *Journal of Comparative Psychology*, 112, 331–343.

Campbell, P., Pasch, B., Pino, J. L., Crino, O. L., et al. (2010). Geographic variation in the songs of neotropical singing mice: Testing the relative importance of drift and local adaptation. *Evolution*, 64, 1955–1972. https://doi.org/10.1111/j.1558-5646.2010.00962.x

Carazo, P., & Font, E. (2010). Putting information back into biological communication. *Journal of Evolutionary Biology*, 23, 661–669. https://doi.org/10.1111/j.1420-9101.2010.01944.x

Castelino, C. B., & Schmidt, M. F. (2010). What birdsong can teach us about the central noradrenergic system. *Journal of Chemical Neuroanatomy*, 39, 96–111. https://doi.org/10.1016/j.jchemneu.2009.08.003

Chabout, J., Sarkar, A., Dunson, D. B., & Jarvis, E. D. (2015). Male mice song syntax depends on social contexts and influences female preferences. *Frontiers in Behavioral Neuroscience*, 9. Artn 76 https://doi.org/10.3389/Fnbh.2015.00076

Chen, Y., Clark, O., & Woolley, S. C. (2017). Courtship song preferences in female zebra finches are shaped by developmental auditory experience. *Proceedings of the Royal Society B: Biological Sciences*, 284(1855), 20170054. https://doi.org/10.1098/rspb.2017.0054

Cousillas, H., George, I., Mathelier, M., Richard, J.-P., et al. (2006). Social experience influences the development of a central auditory area. *Naturwissenschaften*, 93, 588–596. https://doi.org/10.1007/s00114-006-0148-4

Curley, J. P., & Champagne, F. A. (2016). Influence of maternal care on the developing brain: Mechanisms, temporal dynamics and sensitive periods. *Frontiers in Neuroendocrinology*, 40, 52–66. https://doi.org/10.1016/j.yfrne.2015.11.001

Day, D. E., Mintz, E. M., & Bartness, T. J. (2002). Diet choice exaggerates food hoarding, intake and pup survival across reproduction. *Physiology and Behavior*, 75, 143–157.

de Kloet, C. S., Vermetten, E., Geuze, E., Kavelaars, A., et al. (2006). Assessment of HPA-axis function in posttraumatic stress disorder: Pharmacological and non-pharmacological challenge tests, a review. *Journal of Psychiatric Research*, 40, 550–567. https://doi.org/10.1016/j.jpsychires.2005.08.002

Dobzhansky, T. (1973). Nothing in biology makes sense except in the light of evolution. *The American Biology Teacher*, 35, 125–129.

Ehret, G. (2005). Infant rodent ultrasounds—a gate to the understanding of sound communication. *Behavioral Genetics*, 35, 19–29. https://doi.org/10.1007/s10519-004-0853-8

Ehret, G., & Bernecker, C. (1986). Low-frequency sound communication by mouse pups (*Mus musculus*): Wriggling calls release maternal behavior. *Animal Behaviour*, 34, 821–830. https://doi.org/10.1016/S0003-3472(86)80067-7

Ehret, G., & Haack, B. (1981). Categorical perception of mouse pup ultrasound by lactating females. *Naturwissenschaften*, 68, 208–209.

Ehret, G., & Koch, M. (1989). Ultrasound-induced parental behavior in house mice is controlled by female sex hormones and parental experience. *Ethology*, 80, 81–93.

Ehret, G., & Schmid, C. (2009). Reproductive cycle-dependent plasticity of perception of acoustic meaning in mice. *Physiology and Behavior*, 96, 428–433. https://doi.org/10.1016/j.physbeh.2008.11.005

Ehret, G., Koch, M., Haack, B., & Markl, H. (1987). Sex and parental experience determine the onset of an instinctive behavior in mice. *Naturwissenschaften*, 74, 47.

Fernández-Vargas, M., & Johnston, R. E. (2015). Ultrasonic vocalizations in golden hamsters (*Mesocricetus auratus*) reveal modest sex differences and nonlinear signals of sexual motivation. *PLoS One*, 10, e0116789. https://doi.org/10.1371/journal.pone.0116789

Finton, C. J., Keesom, S. M., Hood, K. E., & Hurley, L. M. (2017). What's in a squeak? Female vocal signals predict the sexual behaviour of male house mice during courtship. *Animal Behaviour*, 126, 163–175.

Floody, O. R., & Bauer, G. B. (1987). Selectivity in the responses of hamsters to conspecific vocalizations. *Hormones and Behavior*, 21, 522–527.

Floody, O. R., Walsh, C., & Flanagan, M. T. (1979). Testosterone stimulates ultrasound production by male hamsters. *Hormones and Behavior*, 12, 164–171.

Freeberg, T. M., Dunbar, R. I. M., & Ord, T. J. (2012). Social complexity as a proximate and ultimate factor in communicative complexity. *Philosophical Transactions of the Royal Society B: Biological Sciences*, 367, 1785–1801. https://doi.org/10.1098/rstb.2011.0213

Furrer, R. D., & Manser, M. B. (2009). The evolution of urgency-based and functionally referential alarm calls in ground-dwelling species. *American Naturalist*, 173, 400–410. https://doi.org/10.1086/596541

Gadziola, M. A., Shanbhag, S. J., & Wenstrup, J. J. (2016). Two distinct representations of social vocalizations in the basolateral amygdala. *Journal of Neurophysiology*, 115, 868–886. https://doi.org/10.1152/jn.00953.2015

Gaub, S., & Ehret, G. (2005). Grouping in auditory temporal perception and vocal production is mutually adapted: The case of wriggling calls of mice. *Journal of Comparative Physiology A: Neuroethology, Sensory, Neural, and Behavioral Physiology*, 191, 1131–1135. https://doi.org/10.1007/s00359-005-0036-y

Geissler, D. B., Schmidt, H. S., & Ehret, G. (2013). Limbic brain activation for maternal acoustic perception and responding is different in mothers and virgin female mice. *Journal of Physiology-Paris*, 107, 62–71. https://doi.org/10.1016/j.jphysparis.2012.05.006

Gittelman, J. X., Perkel, D. J., & Portfors, C. V. (2013). Dopamine modulates auditory responses in the inferior colliculus in a heterogeneous manner. *Journal of the Association for Research in Otolaryngology*, 14, 719–729. https://doi.org/10.1007/s10162-013-0405-0

Grimsley, J. M., Monaghan, J. J., & Wenstrup, J. J. (2011). Development of social vocalizations in mice. *PLoS One*, 6, e17460. https://doi.org/10.1371/journal.pone.0017460

Grimsley, J. M. S., Hazlett, E. G., & Wenstrup, J. J. (2013). Coding the meaning of sounds: Contextual modulation of auditory responses in the basolateral amygdala. *Journal of Neuroscience*, 33, 17538–17548. https://doi.org/10.1523/Jneurosci.2205-13.2013

Hahn, M. E., & Lavooy, M. J. (2005). A review of the methods of studies on infant ultrasound production and maternal retrieval in small rodents. *Behavioral Genetics*, 35, 31–52. https://doi.org/10.1007/s10519-004-0854-7

Halfwerk, W., Jones, P. L., Taylor, R. C., Ryan, M. J., & Page, R. A. (2014). Risky ripples allow bats and frogs to eavesdrop on a multisensory sexual display. *Science*, 343, 413–416. https://doi.org/10.1126/science.1244812

Hall, I. C., Sell, G. L., & Hurley, L. M. (2011). Social regulation of serotonin in the auditory midbrain. *Behavioral Neuroscience*, 125, 501–511. https://doi.org/10.1037/a0024426

Hammerschmidt, K., Radyushkin, K., Ehrenreich, H., & Fischer, J. (2009). Female mice respond to male ultrasonic 'songs' with approach behaviour. *Biological Letters*, 5, 589–592. https://doi.org/10.1098/rsbl.2009.0317

Hammerschmidt, K., Radyushkin, K., Ehrenreich, H., & Fischer, J. (2012). The structure and usage of female and male mouse ultrasonic vocalizations reveal only minor differences. *PLoS One*, 7, e41133. https://doi.org/10.1371/journal.pone.0041133

Hanson, J. L., & Hurley, L. M. (2012). Female presence and estrous state influence mouse ultrasonic courtship vocalizations. *PLoS One*, 7(7), e40782.

Hanson, J. L., & Hurley, L. M. (2014). Context-dependent fluctuation of serotonin in the auditory midbrain: The influence of sex, reproductive state and experience. *Journal of Experimental Biology*, 217, 526–535. https://doi.org/10.1242/jeb.087627

Hanson, J. L., & Hurley, L. M. (2016). Serotonin, estrus, and social context influence c-Fos immunoreactivity in the inferior colliculus. *Behavioral Neuroscience*, 130, 600–613. https://doi.org/10.1037/bne0000165

Harris, B. N., & Saltzman, W. (2013). Effect of reproductive status on hypothalamic-pituitary-adrenal (HPA) activity and reactivity in male California mice (*Peromyscus californicus*). *Physiology and Behavior*, 112, 70–76. https://doi.org/10.1016/j.physbeh.2013.02.016

Hayashi, S., & Kimura, T. (1974). Sex-attractant emitted by female mice. *Physiology and Behavior*, 13, 563–567.

Hennessy, M. B., Miller, E. E., & Shair, H. N. (2006). Brief exposure to the biological mother "potentiates" the isolation behavior of precocial Guinea pig pups. *Developmental Psychobiology*, 48, 653–659. https://doi.org/10.1002/dev.20180

Hofer, M. A., Brunelli, S. A., Masmela, J., & Shair, H. N. (1996). Maternal interactions prior to separation potentiate isolation-induced calling in rat pups. *Behavioral Neuroscience*, 110, 1158–1167.

Hoffmann, F., Musolf, K., & Penn, D. J. (2012). Spectrographic analyses reveal signals of individuality and kinship in the ultrasonic courtship vocalizations of wild house mice. *Physiology and Behavior*, 105, 766–771. https://doi.org/10.1016/j.physbeh.2011.10.011

Hoier, S., Pfeifle, C., von Merten, S., & Linnenbrink, M. (2016). Communication at the garden fence: Context dependent vocalization in female house mice. *PLoS One*, 11(3), e0152255. https://doi.org/10.1371/journal.pone.0152255

Holy, T. E., & Guo, Z. S. (2005). Ultrasonic songs of male mice. *Plos Biology*, 3, 2177–2186. https://doi.org/10.1371/journal.pbio.0030386

Hurley, L. M., & Pollak, G. D. (2005). Serotonin modulates responses to species-specific vocalizations in the inferior colliculus. *Journal of Comparative Physiology A: Neuroethology, Sensory, Neural, and Behavioral Physiology*, 191, 535–546. https://doi.org/10.1007/s00359-005-0623-y

Jacob, P. F., & Hedwig, B. (2015). Impact of cercal air currents on singing motor pattern generation in the cricket (*Gryllus bimaculatus* DeGeer). *Journal of Neurophysiology*, 114, 2649–2660. https://doi.org/10.1152/jn.00669

Johansen, J. A., Clemens, L. G., & Nunez, A. A. (2008). Characterization of copulatory behavior in female mice: Evidence for paced mating. *Physiology and Behavior*, 95, 425–429. https://doi.org/10.1016/j.physbeh.2008.07.004

Kalcounis-Rueppell, M. C., Petric, R., Briggs, J. R., Carney, C., et al. (2010). Differences in ultrasonic vocalizations between wild and laboratory California mice (*Peromyscus californicus*). *PLoS One*, 5, e970510. https://doi.org/10.1371/journal.pone.0009705

Kapusta, J., Marchlewskakoj, A., & Sales, G. D. (1995). Home bedding modifies ultrasonic vocalization of infant bank voles. *Journal of Chemical Ecology*, 21, 577–582. https://doi.org/10.1007/Bf02033702

Keesom, S. M., & Hurley, L. M. (2016). Socially induced serotonergic fluctuations in the male auditory midbrain correlate with female behavior during courtship. *Journal of Neurophysiology*, 115, 1786–1796. https://doi.org/10.1152/jn.00742.2015

Keesom, S. M., Rendon, N. M., Demas, G. E., & Hurley, L. M. (2015). Vocal behaviour during aggressive encounters between Siberian hamsters, *Phodopus sungorus*. *Animal Behaviour*, 102, 85–93. https://doi.org/10.1016/j.anbehav.2015.01.014

Kiriazis, J., & Slobodchikoff, C. N. (2006). Perceptual specificity in the alarm calls of Gunnison's prairie dogs. *Behavioral Processes*, 73, 29–35. https://doi.org/10.1016/j.beproc.2006.01.015

Koch, M., & Ehret, G. (1989). Estradiol and parental experience, but not prolactin are necessary for ultrasound recognition and pup-retrieving in the mouse. *Physiology and Behavior*, 45, 771–776.

Lee, S. B., Lee, H. S., & Waterhouse, B. D. (2008). The collateral projection from the dorsal raphe nucleus to whisker-related, trigeminal sensory and facial motor systems in the rat. *Brain Research*, 1214, 11–22. https://doi.org/10.1016/j.brainres.2008.04.003

Lepri, J. J., Theodorides, M., & Wysocki, C. J. (1988). Ultrasonic vocalizations by adult prairie voles, *Microtus ochrogaster*. *Experientia*, 44, 271–273.

Lettvin, J. Y., Maturana, H. R., McCulloch, W. S., & Pitts, W. H. (1968). What the frog's eye tells the frog's brain. In W. C. Corning & M. Balaban (Eds.), *The mind: Biological approaches to its functions* (pp. 233–258). New York: Interscience Publishers.

Lin, F. G., Galindo-Leon, E. E., Ivanova, T. N., Mappus, R. C., & Liu, R. C. (2013). A role for maternal physiological state in preserving auditory cortical plasticity for salient infant calls. *Neuroscience*, 247, 102–116. https://doi.org/10.1016/j.neuroscience.2013.05.020

Liu, R. C., & Schreiner, C. E. (2007). Auditory cortical detection and discrimination correlates with communicative significance. *PloS Biology*, 5, e173. https://doi.org/10.1371/journal.pbio.0050173

Liu, R. C., Miller, K. D., Merzenich, M. M., & Schreiner, C. E. (2003). Acoustic variability and distinguishability among mouse ultrasound vocalizations. *The Journal of the Acoustical Society of America*, 114, 3412–3422. https://doi.org/10.1121/1.1623787

Liu, R. C., Linden, J. F., & Schreiner, C. E. (2006). Improved cortical entrainment to infant communication calls in mothers compared with virgin mice. *European Journal of Neuroscience*, 23, 3087–3097. https://doi.org/10.1111/j.1460-9568.2006.04840.x

Lupanova, A. S., & Egorova, M. A. (2015). [Translation: Vocalizations of sex partners in the house mouse (*Mus musculus*)]. *Zhurnal Evoliutsionnoi Biokhimii I Fiziologii*, 51, 283–289.

Maggio, J. C., Maggio, J. H., & Whitney, G. (1983). Experience-based vocalization of male mice to female chemosignals. *Physiology and Behavior*, 31, 269–272. https://doi.org/10.1016/0031-9384(83)90186-5

Maney, D. L. (2013). The incentive salience of courtship vocalizations: Hormone-mediated 'wanting' in the auditory system. *Hearing Research*, 305, 19–30. https://doi.org/10.1016/j.heares.2013.04.011

Marlin, B. J., & Froemke, R. C. (2016). Oxytocin modulation of neural circuits for social behavior. *Developmental Neurobiology*, 77, 169–189. https://doi.org/10.1002/dneu.22452

Marlin, B. J., Mitre, M., D'amour, J. A., Chao, M. V., & Froemke, R. C. (2015). Oxytocin enables maternal behaviour by balancing cortical inhibition. *Nature*, 520, 499–504. https://doi.org/10.1038/nature14402

Mateo, J. M. (2010). Alarm calls elicit predator-specific physiological responses. *Biological Letters*, 6, 623–625. https://doi.org/10.1098/rsbl.2010.0118

Matrosova, V. A., Blumstein, D. T., Volodin, I. A., & Volodina, E. V. (2011). The potential to encode sex, age, and individual identity in the alarm calls of three species of Marmotinae. *Naturwissenschaften*, 98, 181–192. https://doi.org/10.1007/s00114-010-0757-9

Mazurek, B., Haupt, H., Joachim, R., Klapp, B. F., et al. (2010). Stress induces transient auditory hypersensitivity in rats. *Hearing Research*, 259, 55–63. https://doi.org/10.1016/j.heares.2009.10.006

Meltser, I., & Canlon, B. (2011). Protecting the auditory system with glucocorticoids. *Hearing Research*, 281, 47–55. https://doi.org/10.1016/j.heares.2011.06.003

Miranda, J. A., & Liu, R. C. (2009). Dissecting natural sensory plasticity: Hormones and experience in a maternal context. *Hearing Research*, 252, 21–28. https://doi.org/10.1016/j.heares.2009.04.014

Miranda, J. A., Shepard, K. N., McClintock, S. K., & Liu, R. C. (2014). Adult plasticity in the subcortical auditory pathway of the maternal mouse. *PLoS One*, 9, e101630. https://doi.org/10.1371/journal.pone.0101630

Moreno-Gómez, F. N., León, A., Velásquez, N. A., Penna, M., & Delano, P. H. (2015). Individual and sex distinctiveness in bark calls of domestic chinchillas elicited in a distress context. *The Journal of the Acoustical Society of America*, 138, 1614–1622. https://doi.org/10.1121/1.4929750

Motomura, N., Shimizu, K., Shimizu, M., Aoki-Komori, S., et al. (2002). A comparative study of isolation-induced ultrasonic vocalization in rodent pups. *Experimental Animals*, 51, 187–190. https://doi.org/10.1538/expanim.51.187

Musolf, K., Hoffmann, F., & Penn, D. J. (2010). Ultrasonic courtship vocalizations in wild house mice, *Mus musculus musculus*. *Animal Behaviour*, 79, 757–764. https://doi.org/10.1016/j.anbehav.2009.12.034

Musolf, K., Meindl, S., Larsen, A. L., Kalcounis-Rueppell, M. C., & Penn, D. J. (2015). Ultrasonic vocalizations of male mice differ among species and females show assortative preferences for male calls. *PLoS One*, 10, e0134123. https://doi.org/10.1371/journal.pone.0134123

Neilans, E. G., Holfoth, D. P., Radziwon, K. E., Portfors, C. V., & Dent, M. L. (2014). Discrimination of ultrasonic vocalizations by CBA/CaJ mice (*Mus musculus*) is related to spectrotemporal dissimilarity of vocalizations. *PLoS One*, 9, e85405. https://doi.org/10.1371/journal.pone.0085405

Neunuebel, J. P., Taylor, A. L., Arthur, B. J., & Egnor, S. E. R. (2015). Female mice ultrasonically interact with males during courtship displays. *Elife*, 4. ARTN e06203 https://doi.org/10.7554/eLife.06203

Nevue, A. A., Felix, R. A., 2nd, & Portfors, C. V. (2016). Dopaminergic projections of the subparafascicular thalamic nucleus to the auditory brainstem. *Hearing Research*, 341, 202–209. https://doi.org/10.1016/j.heares.2016.09.001

Newman, S. W. (1999). The medial extended amygdala in male reproductive behavior. A node in the mammalian social behavior network. *Annals of the New York Academy of Sciences*, 877, 242–257.

Nieh, E. H., Kim, S. Y., Namburi, P., & Tye, K. M. (2013). Optogenetic dissection of neural circuits underlying emotional valence and motivated behaviors. *Brain Research*, 1511, 73–92. https://doi.org/10.1016/j.brainres.2012.11.001

Nunez, A. A., & Tan, D. T. (1984). Courtship ultrasonic vocalizations in male Swiss-Webster mice: Effects of hormones and sexual experience. *Physiology and Behavior*, 32, 717–721. https://doi.org/10.1016/0031-9384(84)90184-7

Nyby, J., Wysocki, C. J., Whitney, G., & Dizinno, G. (1977). Pheromonal regulation of male mouse ultrasonic courtship (*Mus musculus*). *Animal Behaviour*, 25, 333–341.

Oberweger, K., & Goller, F. (2001). The metabolic cost of birdsong production. *Journal of Experimental Biology*, 204, 3379–3388.

Okabe, S., Nagasawa, M., Kihara, T., Kato, M., et al. (2013). Pup odor and ultrasonic vocalizations synergistically stimulate maternal attention in mice. *Behavioral Neuroscience*, 127, 432–438. https://doi.org/10.1037/a0032395

Parhar, I. S., Ogawa, S., & Ubuka, T. (2016). Reproductive neuroendocrine pathways of social behavior. *Frontiers in Endocrinology*, 7, 28. https://doi.org/10.3389/fendo.2016.00028

Pasch, B., George, A. S., Hamlin, H. J., Guillette, L. J., & Phelps, S. M. (2011). Androgens modulate song effort and aggression in Neotropical singing mice. *Hormones and Behavior*, 59, 90–97. https://doi.org/10.1016/j.yhbeh.2010.10.011

Pasch, B., Bolker, B. M., & Phelps, S. M. (2013). Interspecific dominance via vocal interactions mediates altitudinal zonation in neotropical singing mice. *American Naturalist*, 182(5), E161-E173. https://doi.org/10.1086/673263

Penn, D., & Potts, W. (1998). MHC-disassortative mating preferences reversed by cross-fostering. *Proceedings of the Royal Society B: Biological Sciences*, 265, 1299–1306.

Petersen, C. L., & Hurley, L. M. (2017). Putting it in context: Linking auditory processing with social behavior circuits in the vertebrate brain. *Integrative and Comparative Biology*, 57, 865–877.

Petric, R., & Kalcounis-Rueppell, M. C. (2013). Female and male adult brush mice (*Peromyscus boylii*) use ultrasonic vocalizations in the wild. *Behaviour*, 150, 1747–1766. https://doi.org/10.1163/1568539X-00003118

Pollard, K. A., & Blumstein, D. T. (2012). Evolving communicative complexity: Insights from rodents and beyond. *Philosophical Transactions of the Royal Society B: Biological Sciences*, 367, 1869–1878. https://doi.org/10.1098/rstb.2011.0221

Pomerantz, S. M., Nunez, A. A., & Bean, N. J. (1983a). Female behavior is affected by male ultrasonic vocalizations in house mice. *Physiology and Behavior*, 31, 91–96. https://doi.org/10.1016/0031-9384(83)90101-4

Pomerantz, S. M., Fox, E., & Clemens, L. G. (1983b). Gonadal hormone activation of male courtship ultrasonic vocalizations and male copulatory behavior in castrated male deer mice (*Peromyscus maniculatus bairdi*). *Behavioral Neuroscience*, 97, 462–469. https://doi.org/10.1037/0735-7044.97.3.462

Poremba, A., Bigelow, J., & Rossi, B. (2013). Processing of communication sounds: Contributions of learning, memory, and experience. *Hearing Research*, 305, 31–44. https://doi.org/10.1016/j.heares.2013.06.005

Pultorak, J. D., Fuxjager, M. J., Kalcounis-Rueppell, M. C., & Marler, C. A. (2015). Male fidelity expressed through rapid testosterone suppression of ultrasonic vocalizations to novel females in the monogamous California mouse. *Hormones and Behavior*, 70, 47–56. https://doi.org/10.1016/j.yhbeh.2015.02.003

Rendall, D., Owren, M. J., & Ryan, M. J. (2009). What do animal signals mean? *Animal Behaviour*, 78, 233–240. https://doi.org/10.1016/j.anbehav.2009.06.007

Rendon, N. M., Keesom, S. M., Amadi, C., Hurley, L. M., & Demas, G. E. (2015). Vocalizations convey sex, seasonal phenotype, and aggression in a seasonal mammal. *Physiology and Behavior*, 152, 143–150. https://doi.org/10.1016/j.physbeh.2015.09.014

Rilling, J. K., & Young, L. J. (2014). The biology of mammalian parenting and its effect on offspring social development. *Science*, 345, 771–776. https://doi.org/10.1126/science.1252723

Riters, L. V. (2012). The role of motivation and reward neural systems in vocal communication. *Frontiers in Neuroendocrinology*, 33, 194–209. https://doi.org/10.1016/j.yfrne.2012.04.002

Robison, W. T., Myers, M. M., Hofer, M. A., Shair, H. N., & Welch, M. G. (2016). Prairie vole pups show potentiated isolation-induced vocalizations following isolation from their mother, but not their father. *Developmental Psychobiology*, 58, 687–699. https://doi.org/10.1002/dev.21408

Roullet, F. I., Wöhr, M., & Crawley, J. N. (2011). Female urine-induced male mice ultrasonic vocalizations, but not scent-marking, is modulated by social experience. *Behavioral Brain Research*, 216, 19–28. https://doi.org/10.1016/j.bbr.2010.06.004

Sakata, J. T., & Vehrencamp, S. L. (2012). Integrating perspectives on vocal performance and consistency. *Journal of Experimental Biology*, 215, 201–209. https://doi.org/10.1242/jeb.056911

Schneiderová, I., Schnitzerová, P., Uhlíková, J., Brandl, P., et al. (2015). Differences in alarm calls of juvenile and adult European ground squirrels (*Spermophilus citellus*): Findings on permanently marked animals from a semi-natural enclosure. *Zoo Biology*, 34, 503–512. https://doi.org/10.1002/zoo.21233

Seyfarth, R. M., Cheney, D. L., Bergman, T., Fischer, J., et al. (2010). The central importance of information in studies of animal communication. *Animal Behaviour*, 80, 3–8. https://doi.org/10.1016/j.anbehav.2010.04.012

Shair, H. N. (2014). Parental potentiation of vocalization as a marker for filial bonds in infant animals. *Developmental Psychobiology*, 56, 1689–1697. https://doi.org/10.1002/dev.21222

Shapiro, L. E., & Insel, T. R. (1990). Infant's response to social separation reflects adult differences in affiliative behavior: A comparative developmental study in prairie and montane voles. *Developmental Psychobiology*, 23, 375–393. https://doi.org/10.1002/dev.420230502

Shepard, K. N., & Liu, R. C. (2011). Experience restores innate female preference for male ultrasonic vocalizations. *Genes, Brain and Behavior*, 10, 28–34. https://doi.org/10.1111/j.1601-183X.2010.00580.x

Shofner, W. P. (2000). Comparison of frequency discrimination thresholds for complex and single tones in chinchillas. *Hearing Research*, 149, 106–114. https://doi.org/10.1007/s10162-010-0237-0

Sisneros, J. A., Forlano, P. M., Deichter, D. L., & Bass, A. H. (2004). Steroid-dependent auditory plasticity leads to adaptive coupling of sender and receiver. *Science*, 305, 404–407. https://doi.org/10.1126/science.1097218

Slobodchikoff, C. N., & Placer, J. (2006). Acoustic structures in the alarm calls of Gunnison's prairie dogs. *The Journal of the Acoustical Society of America*, 119, 3153–3160.

Smith, W. J. (1965). Message, meaning, and context in ethology. *The American Naturalist*, 99, 405–409. https://doi.org/10.1086/282382

Sugimoto, H., Okabe, S., Kato, M., Koshida, N., et al. (2011). A role for strain differences in waveforms of ultrasonic vocalizations during male-female interaction. *PLoS One*, 6, e22093. https://doi.org/10.1371/journal.pone.0022093

Szentgyörgyi, H., Kapusta, J., & Marchlewska-Koj, A. (2008). Ultrasonic calls of bank vole pups isolated and exposed to cold or to nest odor. *Physiology and Behavior*, 93, 296–303. https://doi.org/10.1016/j.physbeh.2007.09.015

Tinbergen, N. (1963). On aims and methods of ethology. *Ethology*, 20, 410–433. https://doi.org/10.1111/j.1439-0310.1963.tb01161.x

Vasudeva, R. K., Lin, R. C., Simpson, K. L., & Waterhouse, B. D. (2011). Functional organization of the dorsal raphe efferent system with special consideration of nitrergic cell groups. *Journal of Chemical Neuroanatomy*, 41, 281–293. https://doi.org/10.1016/j.jchemneu.2011.05.008

von Merten, S., Hoier, S., Pfeifle, C., & Tautz, D. (2014). A role for ultrasonic vocalisation in social communication and divergence of natural populations of the house mouse (*Mus musculus domesticus*). *PLoS One*, 9, e97244. https://doi.org/10.1371/journal.pone.0097244von

Merten, S., Hoier, S., Pfeifle, C., & Tautz, D. (2015). Correction: A role for ultrasonic vocalisation in social communication and divergence of natural populations of the house mouse (*Mus musculus domesticus*). *PLoS One*, 10(1), e0118130. https://doi.org/10.1371/journal.pone.0118130

von Uexküll, J. (1934). *Streifzüge durch die Umwelten von Tieren und Menschen*. Berlin: Verlag von Julius Springer.

von Uexküll, J., von Uexküll, M., & O'Neil, J. D. (2010). *A foray into the worlds of animals and humans, with a theory of meaning*. Translated by J. D. O'Neil from 1934 publication. Minneapolis, MN: University of Minnesota Press.

Voytenko, S. V., & Galazyuk, A. V. (2011). mGluRs modulate neuronal firing in the auditory midbrain. *Neuroscience Letters*, 492, 145–149. https://doi.org/10.1016/j.neulet.2011.01.075

Vyas, A., Harding, C., Borg, L., & Bogdan, D. (2008). Acoustic characteristics, early experience, and endocrine status interact to modulate female zebra finches' behavioral responses to songs. *Hormones and Behavior*, 55, 50–59. https://doi.org/10.1016/j.yhbeh.2008.08.005

Ward, S., Speakman, J. R., & Slater, P. J. B. (2003). The energy cost of song in the canary, *Serinus canaria*. *Animal Behaviour*, 66(5), 893–902. https://doi.org/10.1006/anbe.2003.2250

Wiedenmayer, C. P., Lyo, D., & Barr, G. A. (2003). Rat pups reduce ultrasonic vocalization after exposure to an adult male rat. *Developmental Psychobiology*, 42, 386–391. https://doi.org/10.1002/dev.10112

Willadsen, M., Seffer, D., Schwarting, R. K. W., & Wöhr, M. (2014). Rodent ultrasonic communication: Male prosocial 50 kHz ultrasonic vocalizations elicit social approach behavior in female rats (*Rattus norvegicus*). *Journal of Comparative Psychology*, 128, 56–64. https://doi.org/10.1037/a0034778

Wilson, D. R., & Hare, J. F. (2004). Ground squirrel uses ultrasonic alarms. *Nature*, 430, 523–523. https://doi.org/10.1038/430523a

Wilson, D. R., & Hare, J. F. (2006). The adaptive utility of Richardson's ground squirrel (*Spermophilus richardsonii*) short-range ultrasonic alarm signals. *Canadian Journal of Zoology*, 84, 1322–1330. https://doi.org/10.1139/Z06-120

Woolley, S. C., & Doupe, A. J. (2008). Social context-induced song variation affects female behavior and gene expression. *PLoS Biology*, 6, e62. https://doi.org/10.1371/journal.pbio.0060062

Wright, S. L., & Brown, R. E. (2004). Sex differences in ultrasonic vocalizations and coordinated movement in the California mouse (*Peromyscus californicus*). *Behavioral Processes*, 65, 155–162. https://doi.org/10.1016/j.beproc.2003.09.004

Yoder, K. M., Phan, M. L., Lu, K., & Vicario, D. S. (2015). He hears, she hears: Are there sex differences in auditory processing? *Developmental Neurobiology*, 75, 302–314. https://doi.org/10.1002/dneu.22231

Yu, P., Wang, J., Tai, F., Broders, H., et al. (2011). The effects of repeated early deprivation on ultrasonic vocalizations and ontogenetic development in mandarin vole pups. *Behavioral Processes*, 88, 162–167. https://doi.org/10.1016/j.beproc.2011.09.001

The manufacturer's authorised representative in the EU is Springer Nature Customer Service Centre GmbH, Europaplatz 3, 69115 Heidelberg, Germany. If you have any concerns regarding our products, please contact ProductSafety@springernature.com

Printed and bound by CPI Group (UK) Ltd, Croydon, CR0 4YY
05/10/2026
02228144-0003